DECISION MODELING

IN

POLICY MANAGEMENT

DECISION MODELING

IN

POLICY MANAGEMENT

An Introduction to the Analytic Concepts

Giampiero E.G. Beroggi

Kluwer Academic Publishers
Boston/Dordrecht/London

Distributors for North, Central and South America:
Kluwer Academic Publishers
101 Philip Drive
Assinippi Park
Norwell, Massachusetts 02061 USA
Telephone (781) 871-6600
Fax (781) 871-6528
E-Mail <kluwer@wkap.com>

Distributors for all other countries:
Kluwer Academic Publishers Group
Distribution Centre
Post Office Box 322
3300 AH Dordrecht, THE NETHERLANDS
Telephone 31 78 6392 392
Fax 31 78 6546 474
E-Mail <orderdept@wkap.nl>

 Electronic Services <http://www.wkap.nl>

Library of Congress Cataloging-in-Publication Data

Beroggi, Giampiero E. G., 1960 -
    Decision modeling in policy management : an introduction to the
analytic concepts / Giampiero E. G. Beroggi.
        p.   cm.
    Includes bibliographical references and index.
    ISBN 0-7923-8330-3   (hc. : acid-free paper). -- ISBN 0-7923-8331-1
(pbk. : acid-free paper)
    1. Decision making--Mathematical models.   2. Management science.
I. Title.
T57.95.B47   1998
    658.4'03--dc21   98-42966
                                                                CIP

Printed on acid-free paper.

Printed in the United States of America

In memory of
Emilia Bosi-Pari

TABLE OF CONTENTS

CHAPTER IV: DESCRIPTIVE ASSESSMENT - ALTERNATIVES AND RANKING

CHAPTER V: VALUES AND NORMATIVE CHOICE

CHAPTER VI: CHOICES UNDER UNCERTAINTY

CHAPTER VII: UNCERTAINTY AND NORMATIVE CHOICE

CHAPTER IX: MULTI-ACTOR DECISION MAKING

CHAPTER X: CONSTRAINT-BASED POLICY OPTIMIZATION

Preface

The last decade has experienced major societal challenges at the intersection of technological systems and policy making. Prevalent examples are the liberalization of energy and telecommunications markets, the public aversion towards nuclear power plants, the development of high-speed trains, the debates about global warming and sustainability, the development of intelligent vehicle systems, and the controversies concerning the location of waste depositories, airports, and energy systems.

These challenges, coupled with the call from industry for a systems-engineering oriented approach to policy analysis, motivated Delft University of Technology to launch the first European School of *Systems Engineering, Policy Analysis, and Management* (SEPA). The purpose was to educate engineering oriented policy analysts in bridging the gap between engineering systems and policy decision making processes, both for the public and private sector.

Up to now, more than 500 first-year students and 30 Ph.D. students have enrolled in the program. In 1993, I set up a class called *Quantitative Methods for Problem Solving* which had to address the most relevant issues in decision making for policy management, such as linear and non-linear optimization, multiattribute utility theory, multicriteria decision making, concepts from game theory, outranking relations, and probabilistic influence diagrams.

It became quickly clear that no classical textbook covers or approaches what became *Decision Modeling in Policy Management*. The objective was to have a text which would help solve problems rather than discuss a set of theories. I decided to call the underlying approach *Visual Interactive Decision Modeling* (VIDEMO). The VIDEMO approach is based on three premises.

(1) <u>Policy Problems</u>: the increased complexity of technological systems with societal impact calls for analytic support for decision and policy making.

(2) <u>Analytic Tools</u>: the large arsenal of analytic tools in decision and policy analysis can be presented from a problem-solving perspective.

(3) <u>Information Systems</u>: advanced information systems provide the technological basis for making analytic tools accessible for managers and decision makers.

An illustrative example for the first premise is the decree defining the Dutch safety goals for technological systems in terms of stochastic dominance; the concept is regularly debated in parliament and altered to account for needs by industry and public interests. The second premise calls for a unification of the terminology across the different schools. A consistent terminology is used throughout the book which supports the communication among analysts, and also between analyst and decision maker. The third premise is based on estimates that more than 80 percent of managers develop their own software systems which incorporate some sort of analytic concepts.

Approaching only two of these three aspects at the same time leads to the three classical modeler for problem solving: *management modeler* at the intersection of policy problems and information systems; *engineering modeler* at the intersection of information systems and analytic tools; and *economics modeler* at the intersection of policy problems and analytic tools.

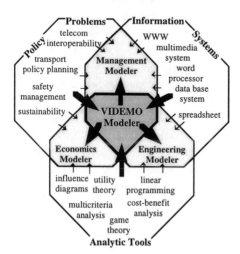

The purpose of the VIDEMO approach, however, is to provide a communication vehicle at the intersection of all three aspects. This makes the large arsenal of analytic techniques comprehensible to and accessible for both the analysts as well as the less analytically skilled decision makers. Moreover, starting at the intersection of the three aspects makes this book valuable for managers, engineers, and policy analysts.

I have now been using this text with accompanying case studies in class since 1994. The approximately 100 students taking this class every year are third year university students with quite some quantitative background. However, the quantitative and analytic aspects used in this volume require merely basic knowledge of linear algebra and calculus. The detailed numerical examples and the 100 worked out problems make the book viable as reader, reference, and textbook.

Along with the writing of this text I have developed a multimedia software system which I used to perform all the computations in the text. The reason to choose a multimedia authoring environment was that the system can be kept open, accessible, and easily adaptable. The software can be downloaded from my web page. More information on the VIDEMO approach, the multimedia software, and the syllabus can be found at: www.sepa.tudelft/webstaf/beroggi/VIDEMO.htm.

School of Systems Engineering,
Policy Analysis, and Management
Delft University of Technology

Giampiero E.G. Beroggi
Haarlem, The Netherlands

Acknowledgments

This book has emerged from two contemporary developments: the upcoming of the visual-interactive modeling methodology and the commitment of the School of Systems Engineering, Policy Analysis, and Management to make this methodology its central modeling paradigm for problem analysis.

For the theoretical treatise I have relied on several classic works, including Raiffa's *Decision Analysis*, Luce and Raiffa's *Games and Decisions*, Keeney and Raiffa's *Decisions with Multiple Objectives*, Vincke's *Multicriteria Decision Aid*, French's *Decision Theory*, Nicholson's *Microeconomic Theory*, Saaty's *Analytic Hierarchy Process*, Kendall and Gibbons' *Rank Correlation Methods*, Schachter's *Probabilistic Influence Diagrams*, Ecker and Kupferschmid's and Hillier and Lieberman's introductions to *Operations Research*, and Kirchgraber's *Linear Algebra*.

The inspiration for applying decision modeling in policy management comes from my former teacher and current colleague William A. Wallace who introduced me to and later on let me teach decision analysis and influence diagrams.

My special thanks for fruitful discussions and helpful comments go to Warren E. Walker for reminding me what policy analysis is all about, and to Pitu B. Mirchandani for assuring me that analytic tools have a serious stand in policy problem solving.

I am especially grateful to Laurie Waisel and Frans Van Daalen for their remarks and suggestions. My utmost thank goes to Penny N. Spring for her encouragement and motivation.

Special thanks go to my colleagues at the School of Systems Engineering, Policy Analysis, and Management for their collegial collaboration during these intense years of building up this new educational program. Finally, I want to thank my students for their patience while the material was growing to its current form and for their remarks that should make it viable for students, teachers, and practitioners.

Introduction

Decision modeling in policy management is a broad field that can be addressed from many different disciplines. Each discipline provides its own terminology and set of tools. The purpose of this book is to present the most relevant concepts in decision modeling for policy management in a unified and discipline-spanning approach - *Visual Interactive Decision Modeling* (VIDEMO). VIDEMO consists of a three-step decomposition of the analytic modeling process (see Figure 1).

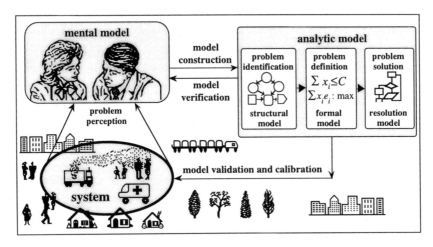

Figure 1: Three-step analytic modeling process.

In a first step, the problem is **identified** by translating an observer's (analyst or decision maker) mental model of a real-world decision problem into a **visual-structural model**. The structural model illustrates the elements of the decision problem and their relations. In a second step, the identified problem is **defined** by deriving an **analytic-formal model**. The formal model describes the relations which were identified in the structural model. In a third step, the defined problem is **solved** by specifying an **algorithmic resolution model**.

The purpose of this three-step decomposition of the analytic modeling process is to let the analyst, together with the decision maker, focus first on the problem identification, then on the problem definition, and only then on the problem resolution. Instead of seeing the world through specific analytic glasses, the analytic tools should be matched to the problem at hand. The complexity of engineering management processes and the increasing public awareness for critical large-scale technological systems call for such a VIDEMO approach. VIDEMO's objective is to bridge the gap between the powerful set of analytic decision making concepts and the discursive process of policy making. The separation of problem identification, formalization, and resolution opens the path for analytic tools to serve in the policy process, rather than vice versa.

• Identifying the Decision Problem: Structural Modeling

For reasons of simplicity, transparency, and consistency, the most reasonable compromise is made for the terminology (Chapter I). The six **elements of decision modeling** used to visualize a decision problem are:

- **Alternatives** (decision options, actions, tactics, strategies, policies).
- **Criteria** (attributes) used to evaluate the alternatives.
- **Scenarios** partitions of the uncertainty space.
- **Content goals** (objectives) stating how alternatives should perform with respect to the criteria.
- **Structural goals** (objectives) stating how the structure of the alternatives should look.
- **Decision makers** (actors) expressing preferences and making decisions.

These elements of decision modeling are used to construct the structural model. Each class of decision elements has its own icon. Figure 2 shows an example of a structural model. The identified elements are: three decision makers, four alternative locations for processing wastes, three criteria to evaluate the four alternatives, four scenarios, three content goals, and two structural goals. The content goal element tells that safety and costs must meet certain constraints, and that the aspiration is to maximize benefits. The structural goal says that two out of the four alternatives (locations) must be chosen and that location *A* can only be chosen in combination with location *D*.

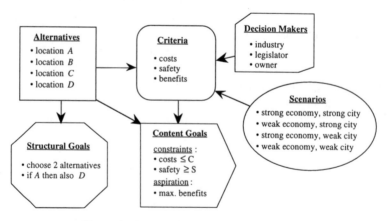

Figure 2: An example of a structural model.

An arrow from one element (predecessor) to another (successor) means that the successor element is formalized (defined) in terms of the predecessor element. Consequently, we read from the structural model in Figure 2 that three decision makers evaluate four alternatives with respect to three criteria and four scenarios. The solution consists of the two locations which maximize the benefits and which satisfy the safety and cost constraints. Moreover, if A is one of the locations then D must be the other one. Obviously, the decision makers agree on the goals and the scenarios because there are no arrows from the decision makers to those elements.

The structural model in Figure 2 is quite simple and the analyst, together with the decision maker, might want to derive a more detailed structure. For example, the criteria might be decomposed into more specific aspects, such as investment costs, operational costs, safety to humans, and safety to the environment. The same could be done with the other elements; relations among actors could be portrayed as an actor model, scenarios as influence diagrams, and goals as a goal hierarchy.

When the structural model is completed, we say that the problem is *identified*. However, a structural model does not tell the whole story yet; rather, it provides the basis to *define* the problem in terms of a formal model.

• Defining the Identified Decision Problem: Formal Modeling

Formalizing the problem means specifying the elements and the relationships as identified in the structural model. There is not a unique way to define a problem, just like there is not a unique way to structure a problem.

The major characteristic of a decision problem in policy management is that it contains at least one alternative element. Each alternative a_j (j-th alternative) has a *decision variable* x_j associated with it. The first step in the formalization of the problem is to specify the type of the decision variables.

If the decision is to chose some out of the four locations, we are dealing with binary decision variables $x_j \in \{0,1\}$, j=1,2,3,4. If location a_j is selected then we have $x_j=1$, and if it is not selected then we have $x_j=0$. However, the decision variables could also refer to the amount of waste processed per year (e.g., in tons), where $x_1=1$ means that 1 ton of waste is processed at location a_1, and $x_2=3.5$ that 3.5 tons are processed at location a_2. The decision variables can now take on any positive real value. Some assumptions about the criteria have now to be made. For example, if processing 1 ton costs \$10, does processing 3.5 tons cost \$35? Moreover, if $x_1=1$ and $x_2=3.5$, are the total costs \$45?

The specification of such relationships is the central part of the formal model. The formalization goes then on, for example, by defining the decision maker element. We could assign a weight w_l to each decision maker d_l, l=1,2,3. This weight could be used to determine the weighted average of the decision makers' assessments. Another formal model would be to define that the legislator specifies the assessment procedure for the alternatives with respect to the criteria, industry does the evaluation, and the owner the subjective interpretation. The scenario, criteria, and goal elements can also be formalized in different ways.

Also part of the formal model is to define the principle for aggregating preferences across scenarios (e.g., expected value), across decision makers (e.g., voting), across criteria (e.g., multiattribute utility theory), and across alternatives (e.g., linear model).

• Solving the Defined Decision Problem: Resolution Modeling

Only at this point of the modeling process do we start worrying about how to resolve the formalized problem. That is, we do not try to frame the problem during the problem identification (structuring) and definition (formalization) phase to match one of the known algorithms. This freedom might lead to problem definitions for which to resolve no traditional resolution model exists yet. The analyst must then make some responsible judgment if the formulation can be changed so that a known resolution approach can be employed. For example, a cost model might be non-linear and we might wonder whether it could be approximated by a linear model. If this were the case, we would have to define (formalize) the details (e.g., coefficients) of the linear model.

Another crucial aspect of resolving a problem is the process of eliciting subjective preferences. A decision maker might be risk averse if the decision options involve major financial consequences. The analyst must then devise a process to elicit the decision maker's risk attitude, and to aggregate them across multiple alternatives, criteria, scenarios, and decision makers.

Also part of the resolution model is to search through the space of potential solutions for feasible solutions. For example, the structural goal in Figure 2 says to choose two out of the four alternatives, and if A is chosen then the second alternative must be D. This means that there are four combinations of two locations which satisfy these two constraints: (1) A and D, (2) B and C, (3) B and D, and (4) C and D. Had we had 20 or even more locations and some additional constraints, the number of feasible solutions could become very large. In such cases, we would like to devise an algorithmic procedure which either constructs all feasible solutions or which finds the best combinations that also satisfy all content goals. For the example in Figure 2, we would want to find the most beneficial solution which, in addition to the structural goals, also complies with the safety and cost constraints.

• Overview of the Chapters

The first two chapters give an overview to the VIDEMO modeling paradigm. The other eight chapters address different aspects of decision modeling in policy management. They all comply with the three-step decomposition of the analytic modeling process into structural, formal, and resolution modeling. An overview of the concepts addressed in the ten Chapters is given in Figure 3.

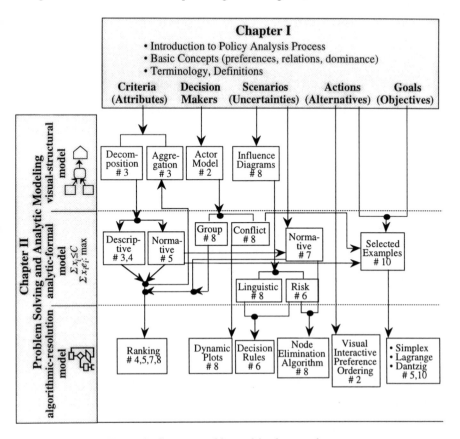

Figure 3: Concepts addressed in the ten chapters.

Chapters II-X address principles that refer to:

(1) preference elicitation,

(2) aggregation of preferences across decision makers, criteria, scenarios, and alternatives; and

(3) search for a solution, especially when the feasible alternatives are defined implicitly.

Two major schools are known for preference elicitation and aggregation: the **descriptive** (European) and the **normative** (American) schools. Chapters III and IV discuss several descriptive preference elicitation and aggregation methods, where the aggregation is done across criteria (attributes).

Chapter V introduces a normative preference elicitation and aggregation concept for multiple criteria (attributes). Before this concept is extended for aggregation across scenarios in Chapter VII, the concepts of uncertainty are introduced in Chapter VI. Chapter VIII addresses the manipulation of probabilistic influence diagrams using the concepts introduced in Chapters VI and VII.

Chapter IX addresses the aggregation of the assessments of multiple decision makers and aspects of conflict analysis. Chapter X, finally, discusses the aggregation of values within criteria and across actions, and the solution of problems with implicitly represented alternatives.

The VIDEMO approach provides a consistent treatment of quite diverse topics in decision analysis and operations research, such as multicriteria methods, linear, non-linear, and integer programming, value and utility theory, dynamic decision problems, group decision problems, conflict analysis, etc. Notations and definitions are consistent and kept to a minimum, and the different topics are addressed in the appropriate context. For a more in-depth study of the specific analytic methods and tools, appropriate references to classic works are made throughout the text.

The figure below shows a screen view of the main menu of the VIDEMO software which can be downloaded from the web. It consists of 12 modules and relations to the book are given. All numerical examples in this book were computed with this software. The software can also be used to perform computations for different problems.

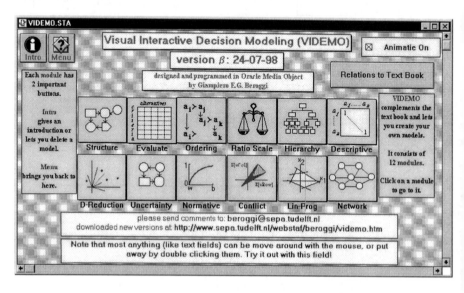

http://www.sepa.tudelft.nl/webstaf/beroggi/VIDEMO.htm

CHAPTER I

THE PROBLEM SOLVING PROCESS

1. The Context of Problem Solving

1.1. What is a Problem?

Problems are situations that everyone, including managers and decision makers, face many times each day; however, there is no generally agreed-upon definition of what a problem is. What may be a problem for one person is not necessarily a problem for another. Problems appear at some moment in time, remain present for a certain period, and eventually disappear. Problems get resolved by 'natural' means or through human intervention. Sometimes problems remain undetected, while other situations are assessed as problems when in fact they are not.

A problem can be defined as the need to investigate changes to a system, where a system is a part of the real world, defined by elements and their relations. The motivation for wanting to investigate a change to a system can be that a deterioration of the system has been perceived or that an improvement of the system's behavior is desired. The emphasis of this definition of a problem is on changes and needs to evaluate decision options in a way which leads to actions that improve the system in the desired way. In this book we will focus on problems that call for decisions concerning the implementation of actions. In other words, we will deal primarily with **decision problems.**

A decision problem can be resolved by the selection (choice) of appropriate actions. However, the effects of the actions may not last forever and the system under investigation may continue to change, calling for new decisions. The fact that the effects of actions can cease is very crucial from a practical point of view and has to be kept in mind by the decision analysts and decision makers. Moreover, an action might resolve one problem while creating others, requiring a continuous monitoring of the system.

To be able to identify potential actions and to select the best one, the analyst must first gain an in-depth understanding of the system, its elements, and their relations. The analysis of the system is done by collecting data and information and processing them to a meaningful 'picture' of the elements and their relations in the system. Systems analysis can be based on (1) statistical data analysis, (2) simulation analysis, or (3) decision analysis. Problem solving can also be addressed from these three perspectives. The **statistical approach** to problem solving focuses on how to collect and analyze data, how to derive conclusions from these data, and how to communicate the results to the decision makers [Chatfield, 1995]. Hypotheses about the causes and consequences of relationships in the system are tested for their significance, from which actions can be derived. The **simulation approach** to problem solving focuses on the dynamic behavior of the system and the consequences of changes in the system [Senge, 1990]. Usually, a small number of different system layouts are simulated, and the best performing layout is proposed as a solution to the problem. The **decision analysis** approach to problem solving focuses on the interactions among decision makers, the formulation of objectives, the identification of alternatives, and the role of uncertainties. Once the problem has been formulated and evaluated in terms of **decision variables**, the challenge lies in the search of the best suited decision options.

Although all three approaches interact with one another, it is quite natural to focus on one of the three, without excluding aspects of the remaining two approaches. In this book we will address problem solving in policy management from a decision analysis perspective. We will focus particularly on visual modeling, decomposed into structural, formal, and resolution modeling. In this chapter we will introduce the basic concepts of decision modeling, and in Chapter II we will discuss their role in the analytic modeling process.

Most of the systems that we will address refer to public and industrial large-scale systems where the decision makers are administrators, managers, or politicians. Examples of systems and decision problems are:

- Selection of best suited rail construction project
- Location of service providers, chemical plants, waste disposal sites, hazardous material storage places, and waste landfills
- Designation of routes for shipping hazardous materials
- Determination of sequential transportation policies with uncertain states of the system
- Conflict resolution in political debates
- Assessment of safety goals in environmental risk management
- Decision making from an economic perspective
- Prioritization of evaluation criteria

Special emphasis is placed on the modeling aspects of decision analysis (as will be defined in Chapter II), including the visualization of the elements and their relations, the assessment and aggregation of preferences, the decomposition and aggregation of evaluation criteria, the handling of uncertainties, the resolution of conflicts among multiple decision makers, and the reduction of complexity in problem solving. Unlike most policy analysis books, however, we will stress visual modeling and decision analysis aspects over economic and financial issues. In addition, we will focus on established theories from a practical point of view, although successful problem solving and decision making is to a large extent a matter of experience. While experience cannot be captured by any theory, the decision maker can improve the quality of his/her decision making capabilities by combining personal experience and analytic concepts.

1.2 The Problem Solving Process in Policy Management

Problem solving is a process that involves three different domains (see Figure I.1): (1) the decision makers' domain in the center; (2) the policy analysts' domain surrounding the decision makers, where the policy analysts interact with the decision makers and the environment; and (3) the environment or the system under investigation (for example a transportation system), surrounding the problem solving process. Ultimately, problems occur and need to be solved in the physical environment or the real-world that we define as the system under investigation. Although problems are analyzed in an abstract and analytic manner, their identification and the consequences of the decisions belong to the real-world. The decision makers do not necessarily perform themselves all the analyses that have to be done as part of the problem solving process. They often are supported by specially trained decision or policy analysts. The policy analysts' responsibility is to find solutions not only that are technically feasible, but also that can last in the environment considering other aspects, such as acceptance by all actors.

Problem solving is a process which begins with the perception of a problem. It is initiated by politicians, managers, or other decision makers of public institutions or private organizations. The purpose of **policy analysis** is to provide a basis for better decision making. This is done using systematic analytic methods to better understand the real problem, to identify more easily the objectives and conflicts, to analyze more quickly viable actions, and to reach faster consensus on complex decisions.

Policy analysis involves at least an analyst and a decision maker. In general, the two are not identical. The literature distinguishes three types of policy analysts [Patton and Sawicki, 1993]: (1) the research oriented technician with excellent analytic skills but few political skills; (2) the analyst turned bureaucrat (e.g., politician); and (3) the highly skilled entrepreneur with both analytic and political skills.

Decision makers are situated at all levels of government and private organizations. They work in finance, transport, energy, environment, social, and other departments. Decision makers must deal with many different stakeholders with conflicting interests. The complexities in the decision making process in policy management stem both from technological issues and from the dissent among the parties involved.

Policy management is not merely a scientific-analytic task [Heineman et al., 1997]. In general, all stages of policy management are influenced by the decision makers. This adds a political, subjective, and sometimes even irrational dimension to the scientific-analytic problem solving process. For example, a decision maker might require a priori that certain alternatives must be included in the analysis, or that some constraints must be considered with highest priority, while others are more 'flexible.' Policy analysis is therefore a process of learning and understanding, of problem evaluating and redefining, and of goal formulating and strategy changing.

The problem solving process in policy planning is a cycle with iterations and feed-backs within the five steps and between analysts, decision makers, and the environment (see Figure I.1). Problem identification, problem definition, and problem solution make up the decision making process, which is a sub-sequence of the problem solving process. In this book we will focus on these three steps of decision making.

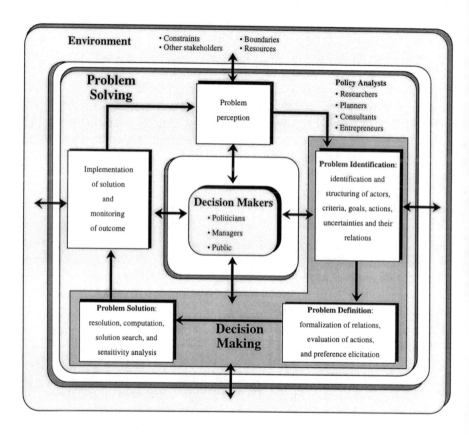

Figure I.1: The problem solving process in policy management and decision making.

In Chapter II (Section 1) we will put this decision making sequence of problem solving in the context of analytic modeling. There we will discuss how to visualize the elements of decision making and how to structure their relationships. This is what is referred to in Figure I.1 as problem identification. Problem definition refers to the formalization of the problem. This means that we have to evaluate the alternatives, elicit subjective preferences, and aggregate the preferences to arrive at an overall ranking of the decision options. Finally, to solve the decision problem, we have to search for a solution and perform sensitivity analysis. This three-step decomposition of the decision making process (identification, definition, solution) will be applied in all chapters.

The purpose of this book is to discuss different analytic concepts that can be employed at each stage of the problem solving process. Depending on the problem at hand and the given 'environmental' constraints, some concepts are more appropriate than others. Classical considerations for choosing the appropriate concept are the available means (e.g., human resources, budget, knowledge, time), the purpose of the study (e.g., preliminary analysis, detailed study, relative comparisons), and the constraints defined by the decision makers and other stakeholders. These can refer to alternatives and criteria that must be considered for the study, the necessary depth of the study, the form of presenting the results, and the ways of communicating with third parties.

To find solutions to a problem that satisfy the decision makers and that last in the environment, the decision makers (e.g., politicians, public representatives, managers) and other stakeholders should be actively involved in all stages of the problem solving process. Experience shows that many analytic studies have failed only because the gap between decision makers and environment has not been bridged successfully. Moreover, it should be kept in mind that the analytic approach chosen to address a specific decision problem should not only depend on the technical suitability of the approach but also on the way results can be communicated to the decision makers and the form in which the decision makers would like the results. Less analytically skilled decision makers usually prefer more intuitive approaches, while analytically skilled decision makers tend to rely on theoretically more elaborated concepts.

Another important aspect to be considered in problem solving is the level of decision making. Decisions are made for **strategic**, **tactical**, or **operational** purposes. Operational decisions can be part of a tactical measure, and tactical measures can be part of a broader strategy. For example, the preventive reduction of the speed limit in heavy traffic is an operational decision which is part of a traffic flow management tactic. This traffic management tactic could be part of a pollution reduction strategy which in turn is part of a national traffic policy. There are many more practical considerations that play a crucial role in successful problem solving. However, our focus is on the modeling aspects and less on the successful implementation of decisions. It is the analyst's responsibility to interpret the model and the results in the context of practical, real-world problem solving.

Before we address the basic concepts of the three decision making steps shown in Figure I.1, we will briefly address the problem-perception and solution-implementation phases. The **perception** of problems can be done by **humans** or by **sensors**. A problem is perceived if some deviations from normal or planned conditions of the system under investigation have been noticed. The analyst is familiar with those deviations which have already occurred in the past, or with events that have not yet occurred but which can be expected due to the design of the system (e.g., some accidents in energy generation facilities). For such cases, remedial actions can be devised prior to a potential event. The **pre-planning** of actions can be done for different kinds of problems. For example, traffic in some urban areas is restricted when the air pollution level exceeds certain values, and the number of fire fighting units that are dispatched depends on the intensity and type of fire.

Automatic detection systems, whether they are based on sensors or on human observers, are employed mostly for operational activities, such as traffic management and emergency response. It is often not economical to pre-plan decisions and policies which occur infrequently, with environmental constraints changing very rapidly. Our major focus lies on the **one-at-the-time decisions**, although the discussed paradigms and methods can also be applied to repetitive decision situations.

The **implementation** of a solution can involve physical construction, changes of practices and logistics, or communications with third parties (e.g., public). Implementing a new policy calls for special attention to the **transition** phase. Poorly planned transition phases may lead to opposition by third parties or to the failure of the whole project. In order to minimize the chances of failure in the transition phase, all parties should be involved in all the steps of problem solving. A successful implementation of a solution requires that it can be implemented under the assumed conditions. If for any reason these conditions change (e.g., increase of labor costs), the policy should be reinvestigated and the changes should be presented to the decision makers.

Monitoring the policy outcomes means to check whether the implemented actions have resolved the problem. A policy is considered unsuccessful if the sought outcomes have not been achieved, even though the conditions for which the policy has been designed might still hold. A failed policy can mean that something went wrong with the analysis. A policy can also result in unexpected outcomes if the assumed conditions change. However, a comprehensive policy should be able to deal with varying state conditions. This means that the results, especially when computed quantitatively, must be interpreted very carefully.

2. Problem Identification: The Elements of Decision Making

2.1 Actors and Decision Makers

The general term for people who participate in the decision making process is **actors**. Actors can be people who analyze the decision problem, people who have the power to make decisions, people who implement the decisions, or anyone else who plays an active or passive role in the decision process. For example, special interest groups, such as environmentalists, political parties, and professional associations, play a very important role in policy decision processes.

Actors who actively participate in the decision making process by assessing the decision options and having a say in the choice process are the **decision makers**. Actors who are not decision makers are referred to as **stakeholders**. Chapter IX is dedicated to decision situations with multiple decision makers. If the decision makers are willing to accept formal processes to reach a consensus decision, we have what we call **group decision making**. This means that the decision makers accept formal voting and preference aggregation procedures. For example, if members of an organization gather in a group decision support room to analyze different business strategies, all of the actors have the common interest of profit maximization. However, despite this common goal, there might be different opinions on how to achieve this goal. The aggregation of the actors' assessments is therefore the major challenge in group decision making. Group decision making settings are typical for organizations, collaborating parties, and groups of decision makers where only one solution can be implemented. The construction of large infrastructure projects is a typical example of a group decision making setting; even if the different parties do not agree, only one infrastructure project can eventually be realized.

Similar decision situations are encountered for decision problems where society is trying to reach a decision in a democratic manner. Here too, the means to achieve a goal are often conflicting, although the goals might be identical. This kind of decision situation is referred to as **social choice**. In Chapter IX (Section 2) we will discuss several paradoxes that occur in social choice situations, including Arrow's famous **impossibility theorem** [Arrow, 1951], which states that there cannot be a welfare function (preference aggregation function over all actors) which complies with a set of social choice axioms. This means that the solution (as a result of the preference aggregation) could be imposed, meaning that all decision makers favor one alternative but the welfare function favors another one, or dictatorial, meaning that the welfare function favors the choice of only one decision maker.

When each of the decision makers can choose from his/her own set of alternatives, we have a **conflict situation**, often called a **game setting**. The preference for an alternative depends also on the strategies of other decision makers. Conflict situations will be discussed in Chapter IX (Section 3), and we will make a distinction between **strictly competitive** and **cooperative** conflict settings. A two-person decision situation where the win of one decision maker equals the loss of the other is called a strictly competitive or **zero-sum** game. An example of a two-person game is a risk analyst who must decide on the best risk reduction policy. The two actors in this conflict situation are the risk analyst and nature. Each of them will eventually choose a strategy. Their benefits depend on the analyst's choice for a policy and the outcome of nature. Game theory is a major field in economic decision analysis and we will be able to address only its basic concepts.

2.2 Attribute, Criterion, Objective, and Goal

Decision alternatives can be described by a set of **attributes**. For example, transportation routes can be characterized by the attributes: name of route, length, width, maximum speed, number of lanes, average population density along the route, maintenance costs, surface condition, traffic safety, and construction year. To decide which route to choose, not all attributes will be relevant. One might consider only maintenance costs, traffic safety, length of route, and condition of route (which can be derived from the two attributes construction year and surface condition).

Obviously, a decision maker would like to choose a route which has the lowest maintenance costs, highest traffic safety, shortest length, and best condition. These aspects are called the **criteria** of a decision problem. Technically speaking, criteria are functions, indicating a direction of increasing or decreasing preference. For example, if $c(a_j)$ is the function that assigns maintenance costs and $s(a_j)$ is the function that assigns safety values to the j-th alternative, a_j, then possible criteria are: $c(a_j)$, $s(a_j)$, and $k_c \times c(a_j) - k_s \times s(a_j)$ (where k_c and k_s are scaling constants), all to be minimized.

It is of fundamental importance to identify carefully the issues of a decision problem and to select the appropriate criteria. The decision options must then be evaluated with respect to these criteria. The evaluations (measured in certain units, such as $, miles, pounds, etc.) are represented in an **evaluation table**, also called **evaluation matrix**, or **score card**. The evaluation values must then be transformed into subjective preferences with which decision options can be compared and prioritized.

Also important when identifying criteria is the definition of the units of measurement and the scales (Section 3.2). This, however, depends on the degree of uncertainty with which the decision options can be measured. For example, costs can often be expressed with rather high certainty in monetary units, while risks can often be expressed only in orders of magnitude (with rather high uncertainty). Or, as another example, risks to humans could be evaluated numerically, but risks of ground water contamination may be evaluated only with linguistic values (e.g., high, medium, and low).

If a small number of criteria breed additional criteria, it is advisable to perform a **decomposition** of criteria into crisp and measurable criteria. For example, the criterion risk can be decomposed into the sub-criteria risk to human, risk to the environment, and risk to material values. These sub-criteria can then be decomposed into more specific aspects. For example, risks to human could be decomposed into risks to life and injuries, long-term and short-term risks, reversible and irreversible risks, etc. Figure I.2 shows a possible decomposition of the criterion risk. In Chapter III we will discuss the hierarchical decomposition of criteria as proposed by the analytic hierarchy process [Saaty, 1980]. The criteria at the bottom of each branch are the **root** or **evaluation criteria**.

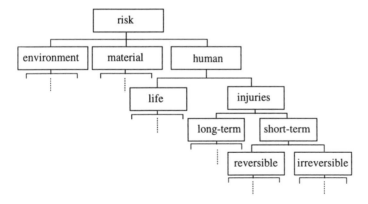

Figure I.2: Decomposition of criteria.

Conversely, it is often possible (or even necessary) to do an **aggregation** of criteria because they reflect (and thus over-emphasize) similar aspects. For example, if ten criteria have been chosen to analyze a decision problem, where one criterion is risk and the other nine refer to cost aspects, costs could be over-emphasized. This can lead to a biased interpretation of the results if the costs are not of overwhelming consideration. In Chapter III we will discuss how to reduce the number of criteria, and how to visualize the preference-order of decision options if the original set of criteria can be reduced to only two relevant dimensions.

The discussion so far makes clear that for the terminology used in this book we refer to criteria as important characteristics of alternatives. However, some research communities use the term criteria in a broader sense. For example, a criterion could be to find as many routes as possible which are not longer than a given value. This criterion does not refer to a characteristic of the individual routes but to the solution of the decision problem. Such a broader definition of the term criterion is also referred to as an objective. In fact, objectives and criteria are used in the decision analysis literature interchangeably [Keeney and Raiffa, 1993]. In the following chapters, however, we will address concepts from many different research communities. Therefore, we will use the term criterion in the narrower sense, referring to important attributes.

The decision analysis literature uses the term goal to state what should be achieved. A goal could be to select a landfill which does not cost more than a certain amount. The difference between a goal and an objective is that a goal can be achieved or not (e.g., costs below a certain amount), while the objective indicates a direction of increasing preference (e.g., to minimize costs). A goal is a much stronger statement than an objective. For example, assume we have a certain budget to invest in a year. If we spend it exactly we will receive the same amount next year, if we spend more or less, then the difference will be deducted from next year's budget. Thus, our *goal* is to spend the exact budget (or to have zero deviations), while our *objective* is to minimize the positive and negative deviations.

It should be noted, however, that what decision analysts refer to as goals (something that should be achieved) is in everyday language often referred to as criteria. For example, one of the criteria for a country to enter the European monetary union is that the national budget deficit must be smaller than 3% of the gross domestic product. On the other hand, what decision analysts call attributes is often referred to as criteria; for example, the criteria to judge the quality of different policies might be acceptability, feasibility, and sustainability.

The policy analysis community uses also different definitions: *goals* are broadly worded statements about what we desire to achieve in the long run; *objectives* are more focused and concretely worded statements about end states; *criteria* are specific statements about the dimensions of the objectives that will be used to evaluate alternatives; and *measures* are tangible, if not quantitative, operational definitions of criteria [Patton and Sawicki, 1993].

Unfortunately, each scientific community has its own terminology, drawing fine (often fuzzy) lines between the definitions of attributes, criteria, objectives, and goals. Because we address concepts from many different research communities, we will use the following definitions:

- **Criterion**: an important characteristic of an alternative, expressed in terms of some attributes, which *implies* increasing or decreasing preferences (e.g., risk, cost, negative impact, opportunity). Decision options are *evaluated* with respect to different criteria.

- **Goal**: an *explicit* description of preferences for finding a solution, expressed in terms of the criteria (e.g., minimize costs, costs must be lower than $20), or referring to the alternatives (e.g., minimize the number of transportation modes during a journey, or choose at least two transportation modes). The term **objective** will be used less frequently.

- **Evaluation measure**: a value expressed on some *scale* which reflects preferences for alternatives (for example, costs of $20 which are less preferred than costs of $15; gains of $20 which are more preferred than gains of $15).

Goals can be defined in terms of multiple criteria. If one tries to copy a successful project that has been accomplished in the past, s/he would identify the characteristics of that project (in terms of the criteria) as the goals for the new project. For example, s/he would try again to minimize costs, maximize safety, invest the same amount of money, etc. However, the success of the previous project might have been caused primarily by a combination of the criteria and not by their individual contribution. Thus, the effective balancing of criteria is often more relevant than their individual consideration. We will discuss in Chapters IV, V, and VII how to solve problems with multiple criteria, and in Chapter X how to combine multiple goals into one overall objective.

A special class of goals are constraints and aspirations. **Constraints** define the minimum or necessary requirement that a solution must satisfy. For example, we might require that costs must be below a certain level (<), or that the budget must all be spent (=), or that a journey can involve at most two transportation modes. **Aspirations** (also called aspiration levels) indicate directions of increasing preference. They refer to maximization (with aspiration level ∞), minimization (with aspiration level -∞), or values that should be achieved (e.g., spending a certain budget).

The definition of the overall goal is done in terms of the chosen criteria. For example, a goal could be to find the safest routing strategy, neglecting the costs involved, or treating costs as second priority to break ties. Another goal could be to find the best suited location for a landfill considering the criteria costs, safety, and economic benefits. But goals can also refer to the form (structure) of the solution. For example, a routing strategy could be defined as feasible if it consists of highway road segments only. Goals can thus refer to the criteria and to the structure of the decision options.

Goals which refer to criteria (e.g., minimize risks) will be called **content goals** (in the decision analysis literature referred to as criteria), goals which refer to the structure or form of the decision options (e.g., to consider only highway segments, or to take at least two independent actions) will be called **structural goals** (in the decision analysis literature referred to as constraints). A feasible route from one point on a road network to another must consist of connected road segments between two intersections. Thus, a structural goal for a feasible route is the connectivity of the road segments. The requirement to rank all decision options is also expressed as a structural goal.

Structural goals can also refer to content goals. For example, a decision maker might decide that the overall goal is to maximize profit. With this goal in mind, s/he would add one of the two constraints to minimize costs or to maximize safety. Whichever of the two constraints gives better results will be considered as part of the decision problem.

A decision option which complies with all structural goals is called a **potential solution**. For example, assume that a decision maker wants to find the best infrastructure project. Then, each individual project is a potential solution. On the other hand, if s/he wants to find the two best projects, then any combination of two projects is a potential solution.

A potential solution which also complies with all constraints is called a **feasible solution**. For example, if the problem is to get as fast as possible from city A to city B

by train, and the only constraint is defined by a certain amount of money (e.g., $100) which cannot be exceeded, then any train connection between city A and B (potential solution) which costs at most $100 is a feasible solution. The best decision option, called the **optimal solution**, is the one among the feasible solutions which performs best in terms of the aspiration levels. In our example, the optimal solution could be the fastest (aspiration level is to minimize travel time) rail connection between A and B which costs at most $100.

Goals may conflict with each other. For example, benefits and safety aspects in transportation refer usually to conflicting goals. A fast and therefore cheap route can involve higher risks than a slower and therefore more expensive route. The performance of the decision options measured with respect to each criterion could be transformed into one single measure of performance. The goal could then be to find the decision option which maximizes (or minimizes) this one-dimensional measure of performance.

Goals can be stated as hard or soft, meaning that **hard goals** must be satisfied, while **soft goals** can be relaxed. For example, if the soft goal is to keep the total costs of a solution below a given threshold, solutions with slightly higher costs may also be acceptable. Or, if the soft goal is to minimize risks, a solution with reasonably low risks might be considered also.

The Nobel laureate Herbert A. Simon [1972] changed the focus of decision research by questioning the **optimal decision** concept. He did this by arguing that the human decision process is characterized by **bounded rationality**. This means that decision makers adopt a **sub-optimal behavior** due to the limited cognitive resources of humans. Consequently, human decision makers do not naturally comply with 'rational' decision rules. Simon proposed to replace the concept of optimal solutions with the concept of **'satisficing' solutions**, meaning that one should be willing to compromise on soft goals in order to attain hard goals.

Using the definition of hard and soft goals we can say that constraints are hard goals, while aspiration levels are soft goals. Thus, a way to circumvent hard goals is to define all goals as aspiration levels. Then, the decision maker can 'tighten' the aspiration levels until only a small set of feasible solutions remains. Note that using a soft approach does not mean compromising on analytic tools. The challenge is to use the most appropriate analytic model, to make proper interpretations of the numeric results, and to communicate the results to the decision makers in a meaningful way. In terms of the introduced terminology, goals can be categorized as follows:

- **content goals** (referring to the criteria):
  - **aspirations** (e.g., maximize benefits or minimize risks, costs, etc.)
  - **constraints** (e.g., risks, costs, etc. must be smaller than, larger than, or equal to given values)

- **structural goals** (expressed in terms of aspirations and constraints):
  - referring to decision options (e.g., choose two (or as many as possible) actions)

- referring to content goals (e.g., consider only one out of *m* content goals)

Goals are often defined in hierarchies, where the overall goal stands at the top of the hierarchy. Lower levels in the hierarchy reflect more concrete goals and associated criteria with specified units of measurements. On the lowest level, the goals are stated in terms of measures of performance of the decision options. Therefore, the goals determine the criteria with respect to which the decision options must be evaluated.

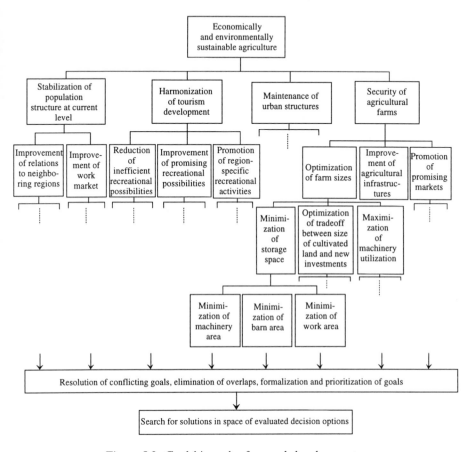

Figure I.3: Goal hierarchy for rural development.

Figure I.3 shows an example of a goal hierarchy for a rural development project which has as the overall goal an 'economically and environmentally sustainable agriculture.' The second goal level addresses the population structure, the tourism development, the urban structures, and the agricultural farms. At the third goal level, the goals become more concrete, referring to criteria, such as recreational possibilities and farm sizes. At the fourth level, aspirations and constraints are identified. At the fifth level, finally, specific units of measurement are introduced.

At the bottom of the goal hierarchy, conflicts and redundancies of the goals are identified. Conflicts must be resolved, redundancies eliminated, and the remaining goals formalized and prioritized. With this goal system, the search for the most preferred solution in the solution space (where the feasible alternatives have been evaluated with respect to the identified criteria) can begin.

2.3 Actions and Decision Variables

It will be taken as a premise that in order to solve a decision problem, something must be *done*, even if it is to wait until certain circumstances change or disappear. Thus, solving a decision problem means to identify **actions**. In other words, the solution to a decision problem is described in terms of which actions must be taken. To each action a_j we assign a **decision variable** x_j where the value of the decision variable stands for the intensity of the action:

$$a_j \longrightarrow x_j$$

For example, $x_j=1$ means that action a_j is chosen, while $x_j=0$ means that action a_j is not chosen. Thus, $x_j=2$ means that action a_j is chosen twice and $x_j=0.2$ means that one fifth of action a_j is chosen. We distinguish three types of decision variables:

- **Binary** decision variables, $x_j \in \mathbf{B}=\{0,1\}$; i.e., take or leave action a_j, such as the selection of the project a_j.

- **Integer** decision variables, $x_j \in \mathbf{Z}$; i.e., take action a_j multiple times, such as the number of people assigned to project a_j.

- **Real** decision variables, $x_j \in \mathbf{R}$; i.e., take fractions of action a_j, such as the time spent on project a_j.

For example, a policy maker must choose from two projects. Each of the two projects (actions) may be represented as a binary decision variable, $x_j \in \mathbf{B}$. One possible solution to the decision problem is $x_1=1$ and $x_2=1$, meaning that both projects are chosen. If the decision maker is now asked to assign people to the two projects, then the number of people assigned to each of the two projects is defined by the decision variables, $x_j \in \mathbf{Z}^{\geq 0}$. A possible solution to the decision problem is $x_1=24$ and $x_2=12$, meaning that 24 people work on the first project and 12 on the second. If the decision maker is asked to allocate time to the two projects, then the days assigned to each of the two projects are the two decision variables, $x_j \in \mathbf{R}^{\geq 0}$. A possible solution to the decision problem is $x_1 = 500/3$ and $x_2 = 200/6$, meaning that 166 days and 16 hours should be spent on the first project and 33 days and 8 hours on the second project.

The potential solutions to a decision problem are called **alternatives**. Depending on the type of solution, alternatives are referred to as **policies, tactics,** or **strategies**. A strategy is usually composed of multiple tactics, while a policy is a more generic term for strategy. There are different ways to identify alternatives. Patton and Sawicki [1993] propose several approaches for identifying alternatives: analysis and experimentation; no-action (status quo) analysis; quick surveys, literature review; comparison of real-world experiences; passive collection and classification; development of typologies; analogy, metaphor, and synectics; brainstorming; and comparison with an ideal. For an in-depth discussion concerning the generation and screening of alternatives see [Walker, 1988].

An important characteristic of alternatives is whether they are defined explicitly or implicitly. A decision maker faced with the problem of locating a landfill in a community can describe the decision problem in different ways. One way is to list potential locations and to evaluate each location with respect to the following criteria: (1) the total distance to the waste collection centers, (2) the emission impact, (3) the esthetic value, and (iv) the risk to the population and the environment. Another way to define the same problem is to state constraints, such as: (1) the location must be in the industrial area, (2) the sum of the distances to the waste collection centers should be less than 100 kilometers, (3) the emission impact cannot exceed a predefined threshold value, (4) the location must fit esthetically into the environment, and (5) the risks to population and environment must be kept below given threshold values.

The difference between these two descriptions of the same decision problem is that in the first description, the alternatives are given **explicitly**, while in the second description, the alternatives are defined **implicitly** in terms of the constraints. Constraint-optimization problems will be discussed in Chapters V and X. The challenge is to define a solution method which constructs, assesses, and compares implicitly represented alternatives.

Another example of implicitly represented alternatives is a road map. If we want to go from one city to another, we have to find several explicit routes and compare the different routes to one another. Often in constraint-optimization problems, the decision variables take on real and integer values, where the constraints and goals are defined analytically.

Sequential actions reflect also implicit alternatives. For example, a manager plans a visit to different countries. In each country s/he wants to visit one out of a small number of cities. The problem is to find a route that goes through each country, visiting one city in each country, so that the total travel costs are minimized. The decision variables of this problem are the cities, and the potential solutions to this problem consist of a sequence of cities, one for each country. Approaches to solve these types of decision problems in the presence of uncertainty will be discussed in Chapter VIII.

2.4 Uncertainties, States, and Scenarios

Uncertainty in problem solving reflects two aspects. The first is the uncertainty about the (future) state of the system, that is, the ambiguity about the assumptions concerning the environment of the decision problem. For example, one might assume a certain demand or economic growth for the next ten years. All evaluations and decisions are then based on this assumption. If the assumption changes, the results could become meaningless.

The second type of uncertainty refers to the evaluation of the alternatives. Given that the future development of the environment is known, there still is the possibility of an incorrect evaluation of the decision options.

Our focus is on the first aspect of uncertainty. In Chapters VI and VIII we discuss the concepts of probability theory to deal with uncertainty; the concept of fuzzy logic is used if the uncertainty should be expressed in terms of linguistic variables (Chapter VI).

Uncertainty is described in terms of **partitions** of the total uncertainty space. For example, assume a city government must decide where to locate a hospital. The preferences for the potential locations (alternatives) depend on the uncertain development of the city. This uncertainty is captured by the partitions *economic growth* (low or high) and *inhabitants* (many or few). Thus, each of the two partitions defines two **states** of the uncertainty space. Any combination of states is called a **scenario**, and the states themselves are called **marginal scenarios**.

With the four marginal scenarios, four **joint scenarios** can be defined: s_1: low economic growth and many inhabitants, s_2: high economic growth and many inhabitants, s_3: low economic growth and few inhabitants, and s_4: high economic growth and few inhabitants. For any two out of the four scenarios we have $s_i \cap s_j = \varnothing$, so we say that the scenarios are **mutually exclusive** which means that they do not overlap. Because we also have $s_1 \cup s_2 \cup s_3 \cup s_4 = total\ uncertainty$, we say that the scenarios are **collectively exhaustive**, which means that they completely describe the entire uncertainty space. Scenarios which must be considered for the evaluation of the alternatives are called **evaluation scenarios**.

A scenario is therefore an uncertain condition of the system which cannot be influenced by the decision maker. Alternatives must be evaluated with respect to all scenarios. Three levels of problem solving under uncertainty are distinguished in the decision analysis literature:

- **Certainty**: The alternatives can be evaluated without consideration of any scenarios (Chapters IV, V, and X).

- **Informed Uncertainty (Risk)**: The alternatives must be evaluated with considerations of some scenarios; the chance of occurrence of each scenario can be quantified with probability or possibility values (Chapters VI, VII, VIII, and IX).

- **Complete Uncertainty**: The alternatives must be evaluated with considerations of some scenarios but the chance of occurrence of each scenario cannot be quantified (Chapter VI).

It should be noted that both problem solving under risk and under complete uncertainty are often referred to as problem solving under uncertainty.

3. Problem Definition

3.1 Evaluation Measures

To solve a decision problem, the potential actions must be evaluated. This is done with an **evaluation measure** which describes the performance of action a_j ($j=1,...,n$) with respect to criterion c_i ($i=1,...,m$), scenario s_k ($k=1,...,v$), and decision maker d_l ($l=1,...,w$). The evaluation measure is a function

$$e_i^{kl}(a_j) := e_{ij}^{kl}$$

that assigns values (numeric or symbolic) to action a_j. The evaluation of an action thus depends on three elements: the criteria, the scenarios, and the decision makers.

It is quite obvious that there must be at least one representative of each element, that is, one criterion, one scenario, and one decision maker. Multiple representatives of each element can be seen as dimensions or points of view. For example, if there are three criteria, we have three dimensions or points of view. Thus, each of the three elements spans a different space. For example, if we have three criteria, four scenarios, and two decision makers, then we have a 3-dimensional criterion space, a 4-dimensional uncertainty space, and a 2-dimensional decision maker space. Figure I.4 shows that the evaluation of actions is done with respect to different criteria, scenarios, and decision makers, each providing different points of view.

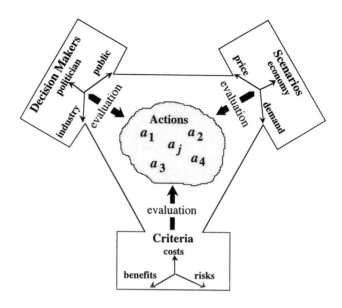

Figure I.4: Dimensions of evaluating actions.

An example of an $m \times n$ **evaluation matrix** E (score card), written as $E^{m \times n}$, for m criteria (rows) and n actions (columns) is given in Table I.1.

| | a_1 .... a_j .... a_n | | | | | | | a_1 | a_2 | a_3 | a_4 | a_5 |
	x_1 .... x_j .... x_n							x_1	x_2	x_3	x_4	x_5
c_1			.				risk [-]	0.3	0.9	0.6	0.2	0.7
.			.				zone [class]	II	I	III	I	III
c_i	. . . . . . e_{ij}						cost [10^6 \$]	15	17	20	30	10
.							landuse [km^2]	30	42	18	24	47
c_m							profit [value]	low	low	high	med.	high

Table I.1: Evaluation table under certainty (left) with example (right).

Let's assume that the five alternatives given in Table I.1 refer to five infrastructure projects from which we want to choose the best one. In this case, the decision variables are binary, $x_j \in \mathbf{B} = \{0,1\}$; that is, take or leave infrastructure project a_j. We further assume that the entry e_{ij} in the table shows the impact for $x_j = 1$ with respect to c_i, while no impact would result if we renounce to an alternative. Thus, the resulting impact is $x_j \times e_{ij}$. We can further assume that the total effect of implementing multiple alternatives with respect to criterion c_i is $\Sigma_j x_j e_{ij}$. Thus, we can formulate the decision problem as follows: min $e_i = \Sigma_j x_j e_{ij}$, for $i=1,...,5$, and $\Sigma_j x_j = 1$ (saying that exactly one out of five projects must be chosen).

If the alternatives are evaluated with respect to m criteria, the overall evaluation of each alternative, $e(a_j)$, is expressed as an m-dimensional vector: $e(a_j)=[e_{1j},...,e_{mj}]^T$. If uncertainty is considered in terms of multiple scenarios, s_k $(k=1,...,v)$, the evaluation of each alternative is expressed as an $m \times v$ matrix, with the corresponding evaluation matrix, $E^{m \times v \times n}$. Finally, if the evaluation is based on multiple criteria, scenarios, and decision makers, we get a 4-dimensional evaluation matrix, $E^{m \times v \times w \times n}$, as illustrated in Table I.2.

		d_1 — $a_1 . a_j . a_n$ / $x_1 . x_j . x_n$	.		d_l — $a_1 . a_j . a_n$ / $x_1 . x_j . x_n$	.		d_w — $a_1 . a_j . a_n$ / $x_1 . x_j . x_n$
s_1	c_1 / c_i / c_m	$\ldots . e_{ij}^{11}$	.	s_1	c_1 / c_i / c_m	$\ldots . e_{ij}^{1l}$	.	s_1	c_1 / c_i / c_m : $\ldots . e_{ij}^{1w}$
.		.	. .	.		.	. .	.	.
s_k	c_1 / c_i / c_m	$\ldots . e_{ij}^{k1}$	.	s_k	c_1 / c_i / c_m	$\ldots . e_{ij}^{kl}$	.	s_k	c_1 / c_i / c_m : $\ldots . e_{ij}^{kw}$
.		.	. .	.		.	. .	.	.
s_v	c_1 / c_i / c_m	$\ldots . e_{ij}^{v1}$	.	s_v	c_1 / c_i / c_m	$\ldots . e_{ij}^{vl}$	.	s_v	c_1 / c_i / c_m : $\ldots . e_{ij}^{vw}$

Table I.2: Evaluation table for n alternatives with respect to m criteria, v scenarios, and w decision makers.

There is a lot to say about the practical process of evaluating alternatives, such as cost, effort, accuracy, reliability, uncertainty, consistency, and significance of the evaluation and measurement process. An in-depth discussion of these issues would go beyond the scope of this book. The evaluation matrix is the starting point of the analytic treatment of a decision problem. No matter how sophisticated the analytic treatment is, the quality of the proposed solutions will always remain highly dependent on the quality of the evaluation matrix. In fact, the effort spent on the analytic part is fruitless if the evaluation matrix has not been assessed properly. The same caution holds for the selection of the criteria and the identification of potential actions, scenarios, and decision makers, and the definition of the goals. If a crucial criterion has been forgotten, an important scenario omitted, a relevant decision maker neglected, or a potential action disregarded (quite often the zero-action which means to do nothing), the analytic treatment will never be able to make up for this deficiency.

3.2 Measurement Scales

The evaluation of the alternatives with respect to each criterion is done using a **scale** and an appropriate measurement unit, such as dollar ($) for monetary values, kilometer (km) for distance measurements, and pound (lb) for masses. A scale is a function that assigns numbers or symbolic expressions to the alternatives. For example, the temperature can be measured with a thermometer in degrees Celsius (°C) or Fahrenheit (°F), or estimated in terms of warm, tepid, or cold. A scale can be transformed into another one. For example, we could transform temperatures from °C to °F, distances from km to miles, or currencies from dollar to yen.

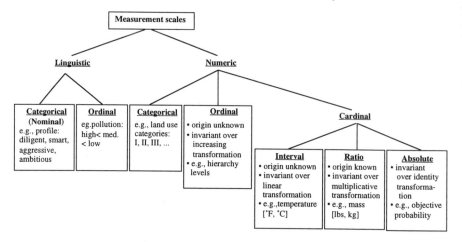

Figure I.5: Measurement scales.

A **linguistic scale** is either categorical or ordinal. A **categorical** linguistic scale assigns **nominal** values to the alternatives, such as the description of the profile of employees in terms of diligent, smart, aggressive, and ambitious. An **ordinal** linguistic scale implies a preference order. For example, medium air pollution is preferred to high air pollution.

A **numeric scale** is categorical, ordinal, or cardinal. A **categorical** numeric scale groups the alternatives into numeric categories. For example, the land of a community may be divided into types I, II, and III. Often, a categorical linguistic scale is transformed into a categorical numeric scale. For example, 'high' becomes 3, 'medium' 2, and 'low' 1. The characteristics of scales are expressed in terms of **admissible transformations**, $\phi(e)$, of the original evaluation values, e.

An **ordinal scale** describes the preference relation between two entities. Thus, any **strictly increasing transformation**, $\phi(e)$, is admissible for an ordinal scale. For example, if $e_1=80$ and $e_2=100$, then $\phi(e_1)=50$ and $\phi(e_2)=200$ is an admissible transformation because it preserves the order ($e_1<e_2$, and $\phi(e_1)<\phi(e_2)$). An example of a strictly increasing (monotonic) transformation is the logarithm. Thus, $\log(e)$ is an admissible transformation for an ordinal scale.

An **interval scale** preserves the proportions between differences. Thus, two interval scales can be transformed into each other by a **positive linear transformation**. This means that the positive linear (affine) transformation is admissible: $\phi(e)=a\times e+b$, with $a>0$. For example, if three evaluations have the values (3,8,12) and a transformation leads to the values (9,19,27), then the transformation is positive linear which means that we are dealing with an interval scale (see Figure I.6, left, in this case $a=2$ and $b=3$).

The temperature scales degrees Celsius (°C) and Fahrenheit (°F) are interval scales: $F=9/5\times C+32$. It should be noted that the zero points of two interval scales are not identical (e.g., 0 °C$\neq$0 °F). It also does not make sense to say it is twice as cold if the temperature drops from 20°C to 10°C, because if we transform this to degrees Fahrenheit, the drop factor is: 68/50=0.74$\neq$0.5. However, it makes sense to compare temperature changes on both scales. For example, a temperature drop of 10°C (from 20°C to 10°C) is twice as much as a drop of 5°C (from 15°C to 10°C). This factor is the same in degrees Fahrenheit: (68°F-50°F)/(59°F-50°F)=2. We have in fact the following relation between the two temperature scales: $\Delta 1°F=\Delta 5/9°C$.

A **ratio scale** preserves proportions between outcomes. Thus, if one evaluation is twice as good as another, it also scores twice as good on another ratio scale. For example, weights are evaluated on ratio scales. If one object is twice as heavy as another object in kilograms, it is also twice as heavy in pounds. The zero point of a transformed ratio scale is still zero. For example, the length of objects can be measured in meters, centimeters, yards, miles, or kilometers, all with the same zero point. The ratio scale is a special case of the interval scale ($b=0$): $\phi(e)=a\times e, a>0$. Thus, the ratio scale is invariant under positive **multiplicative transformations** (similarity transformations). Figure I.6 shows the difference between interval and ratio scales.

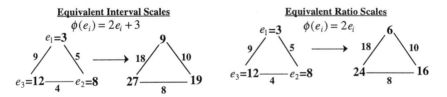

Figure I.6: Equivalent interval and ratio scales.

An **absolute scale** is invariant under the **identity transformation**; in other words, it is unique. An example of an identity scale is the subjective probability, where the probability of the sample space is equal to one. Thus, $\phi(e)=e$, meaning that the absolute scale is a special case of the ratio scale, where $a=1$. Figure I.7 shows an overview of the invariant transformations for the ordinal scale (strictly increasing transformation), the interval scale (linear transformation), the ratio scale (multiplicative transformation), and the absolute scale (identity transformation).

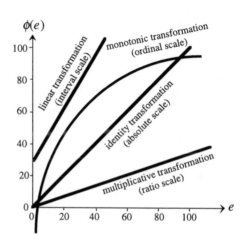

Figure I.7: Examples of invariant transformations for different numeric scales.

3.3 Preference Elicitation

In order to aggregate multidimensional evaluations to an overall evaluation for each alternative, subjective interpretations of the evaluations must be done. For example, a tripling of the gains might for one decision maker mean also a tripling of the preference for this alternative, while for another it might just mean a doubling, and for yet another a quadrupling. Consequently, the evaluation matrix must be transformed into a subjective preference matrix to reflect the subjective preferences of each decision maker.

Two schools prevail for the elicitation of subjective preferences and their aggregation across criteria and decision makers: descriptive and normative. **Descriptive preference elicitation** means to describe the advantages (dominance) and disadvantages (inferiority) of each alternative over the others. To do this, **paired comparisons** of all the alternatives must be done with respect to all criteria, scenarios, and decision makers. This means that we compare each alternative a_j with all other alternatives, a_1, ..., a_{j-1}, a_{j+1}, ..., a_n, and state its advantages and disadvantages over the others. An advantage could mean that the preference value of the alternative is 1 and a disadvantage that the preference value is -1. The aggregation across all criteria gives the overall preference value for alternative a_j for a specific scenario and decision maker. In Chapter III we will discuss the descriptive approach to elicit and aggregate preferences for criteria (weights), and in Chapter IV we will discuss the descriptive approach to elicit and aggregate preferences for alternatives. The descriptive approach of preference elicitation and aggregation is referred to as the *European school* of decision analysis.

Normative preference elicitation means to assess a decision maker's subjective **preference function**. With this function, all evaluation values can be transformed into subjective preference values. The normative approach is based on a set of **axioms**. If a decision maker complies with these axioms we say that s/he is **rational**; for a rational decision maker, a personal preference function can be elicited which prescribes his/her subjective preferences for different alternatives. In Chapter V we will discuss the normative approach of preference elicitation under certainty, called value theory. In Chapter VII we will discuss the normative approach of preference elicitation under uncertainty, called utility theory. The normative approach of preference elicitation is referred to as the *American school* of decision analysis. It is important to note that both quantitative and qualitative approaches can be used both for normative and descriptive models. Table I.3 gives examples for these circumstances.

	quantitative	qualitative
normative	preference functions (Chapters V and VII)	laws and regulations
descriptive	paired comparisons (Chapters III and IV)	professional codes

Table I.3: Normative/descriptive vs. qualitative/quantitative approaches.

Our focus is on quantitative approaches. The preference values for both the descriptive and normative approaches usually range from 0 to 1, where 1 is assigned to the most preferred and 0 to the least preferred outcome. However, any other range, such as from 0 (worst) to 100 (best) or from -1 (worst) to 0 (best), can also be used.

Both normative and descriptive approaches have their merits and their shortcomings. Normative approaches are criticized because they are based on the assumption of rational decision makers. However, many reports about real-world decision problems show that a normative or rational approach is useful even for very complex technological and societal problems. Descriptive approaches are criticized because they lack a sound decision-theoretical basis. To minimize the limiting factors of any preference elicitation methods, it is important to involve the decision maker, keep the methods transparent, and perform sensitivity analysis.

3.4 Binary Preference Relations

The fundamental concept for comparing and ordering alternatives according to their preference is the **binary relation**. A binary relation $\Re$ on a set of alternatives $A = \{a_1,...,a_n\}$, is defined as $(\Re,A) := a_i \Re a_j$, $(a_i,a_j) \in A$. The following types of preference relations are frequently used:

- **strong preference**: $a_i \succ a_j$ (a_i is more preferred than a_j).
- **indifference**: $a_i \sim a_j$ (a_i and a_j are equally preferred).
- **weak preference**: $a_i \succsim a_j$ ($a_i \succ a_j$ or $a_i \sim a_j$: a_i is at least as preferred as a_j).
- **incomparability**: $a_i <> a_j$ (a_i and a_j cannot be compared).

The following relations are frequently used to compare numeric evaluation values:

- **greater than**: $e_i > e_j$ (e_i is greater than e_j).
- **equal**: $e_i = e_j$ (e_i and e_j are equal).
- **at least as great as**: $e_i \geq e_j$ (e_i is at least as great as e_j).
- **incomparability**: $e_i <> e_j$ (e_i and e_j cannot be compared).

Preference functions, $\pi(e_i)$, as used in normative preference theories, are generally based on an ordinal scale. Consequently, the preference values are invariant under strictly increasing transformations. For gains we would prefer \$100 over \$50, thus: \$100 $\succ$ \$50; but for costs, we would prefer \$50 over \$100; thus: \$50 $\succ$ \$100. This means that we have:

$$\text{for gains:} \quad a_i \succsim a_j \leftrightarrow \pi(e_i) \geq \pi(e_j) \leftrightarrow e_i \geq e_j$$
$$\text{for losses:} \quad a_i \succsim a_j \leftrightarrow \pi(e_i) \geq \pi(e_j) \leftrightarrow e_j \geq e_i$$

For the general case, where the evaluation measure is not specified (but monotonic) we use the following preference relation between evaluation values:

$$a_i \succsim a_j \leftrightarrow e_i \succsim e_j$$

The strong preference ($\succ$), weak preference ($\succsim$), and the indifference ($\sim$) relation defined on a set A, $(\Re,A)$, can have different characteristics, which depend on the set of elements and on the decision maker, where $\{a_i, a_j, a_k\} \in A$:

- **Transitivity**: $a_i \succ a_j$ and $a_j \succ a_k \to a_i \succ a_k$.
- **Weak Transitivity**: $a_i \succ a_j$ and $a_j \succ a_k \to \underline{\text{not}} \{a_k \succ a_i, a_k \sim a_i\}$; i.e., they are either transitive $(a_i \succ a_k)$ or incomparable $(a_j <> a_k)$.
- **Transitivity**: $a_i \sim a_j$ and $a_j \sim a_k \to a_i \sim a_k$.
- **Transitivity**: $a_i \succsim a_j$ and $a_j \succsim a_k \to a_i \succsim a_k$.
- **Symmetry**: $a_i \sim a_j \to a_j \sim a_i$.
- **Asymmetry**: $a_i \succ a_j \to \underline{\text{not}} \; a_j \succ a_i$.
- **Reflexivity**: $a_i \sim a_i$.
- **Irreflexivity**: $\underline{\text{not}} \; a_i \succ a_i$.
- **Completeness** ($\succsim$): either $a_i \succ a_j$ or $a_j \succ a_i$ or $a_i \sim a_j$.
- **Completeness** ($\succ$): either $a_i \succ a_j$ or $a_j \succ a_i$.
- **Comparability**: $a_i \succsim a_j$, or $a_j \succsim a_k$, or both.
- **Substitution**: $a_i \succ a_j$ and $a_j \sim a_k \to a_i \succ a_k$; $a_i \sim a_j$ and $a_j \succ a_k \to a_i \succ a_k$.
- **Indifference**: $a_i \succsim a_k$ and $a_k \succsim a_i \leftrightarrow a_k \sim a_i$.
- **Consistency**: $a_i \succ a_j \to \underline{\text{not}} \; a_j \succsim a_i$.

If a strong (irreflexive) and asymmetric preference relation ($\succ$), defined on a set of elements A, is transitive and complete, we have a **complete strong preference order**. This means that we can rank all elements from most preferred to least preferred without ties. If there are elements which are incomparable, but transitivity still holds, we have a **partial strong preference order**.

If a consistent weak preference relation ($\succsim$), defined on a set of elements A, is transitive and complete (indifference holds), we have a **complete weak preference**. This means that some elements are compared with the strong, asymmetric, and transitive preference relation ($\succ$), and some with the transitive, symmetric, and reflexive indifference relation ($\sim$). Moreover, substitution also holds. In other words, all elements can be ordered from most preferred to least preferred with possible ties. If a consistent weak preference relation is transitive but not complete (indifference and substitution hold), we have a **partial weak preference order**.

Normative decision theories assume that all alternatives can be ordered from most preferred to least preferred, with possible ties. This means that transitivity must hold and that all elements are comparable.

This normative view of a human decision maker has been challenged by behavioral studies which show that human decision makers often cannot compare elements and violate transitivity. Especially the indifference relation has been found to be intransitive, in that $a_i \sim a_j$ and $a_j \sim a_k \to a_i \succ a_k$. Consequently, descriptive decision models have been proposed that allow a decision maker to assess elements to be incomparable. A relation on a set where the indifference relation is not transitive and the strong preference relation is weak transitive is called a (complete or partial) **pseudo order**.

An example of a pseudo order (intransitive indifference relation) is to define indifference ($\sim$) not by a point threshold but by an interval threshold, for example: $a_i \sim a_j \leftrightarrow |e_i-e_j| \leq \Delta$. If, for example, $e_i-e_j=\Delta/2$, and $e_j-e_k=3\Delta/5$, then we have: $a_i \sim a_j$ and $a_j \sim a_k$, but $a_i \succ a_k$ instead of $a_i \sim a_k$. It makes sense to relax the stringent point threshold for indifferences because it is very difficult to be consistent with subjectively assessed preferential indifferences between almost identical elements. The consequence, however, is that the indifference relation leads no longer to a proper order in terms of the normative assumptions of rational decision making.

4. Problem Solution

4.1 Preference Aggregation

In order to solve a problem, the identified alternatives must be compared to one another with respect to all criteria, scenarios, and decision makers. These comparisons eventually will lead to the choice of one or several actions for implementation; that is, a solution to the decision problem. Ideally, the most preferred alternative outperforms all other alternatives for all criteria, scenarios, and decision makers. This, however, rarely happens. The compromise objective is to generate an overall preference value for each alternative, such that the alternative with the highest value is the best and the one with the lowest value the worst. Both the descriptive and the normative approaches of preference assessment compute a single overall preference value for each alternative.

Some alternatives consist of multiple actions. For example, a road is made up of multiple road segments, each of which is assessed independently as an individual action. Then, the overall preference of the road must be determined from the individual preferences of the road segments which are part of this road. Thus, we sometimes have to aggregate preferences over multiple actions.

For a decision problem with multiple criteria (m), actions (n), decision makers (w), and scenarios (v), we want to address the following issues:

- given e_{ij}^{kl}, <u>for all i:</u> aggregation of preferences <u>across criteria</u> c_i ($i=1,...,m$); e.g., with linear weight function, linear scaling model, or multiplicative model (Chapters IV, V, VII).

- given e_{ij}^{kl}, <u>for all j:</u> aggregation of preferences <u>across actions</u> a_j ($j=1,...,n$); e.g., non-linear and linear model (Chapters V and X).

- given e_{ij}^{kl}, <u>for all k</u>:   aggregation of preferences <u>across scenarios</u> s_k ($k=1,...,v$); e.g., with expected value (Chapters VI and VII).

- given e_{ij}^{kl}, <u>for all l</u>:   aggregation of preferences <u>across decision makers</u> d_l ($l=1,...,w$); e.g., through voting (Chapter IX).

The different methods for preference aggregation are (besides the process of assessing the subjective preferences and the search for a solution) a central subject which will be addressed in almost all chapters. The simplest models for preference aggregation are **linear models**. Thus, different theories and approaches will be discussed that lead to a linear model. The overall preference is then the weighted sum (descriptive models, Chapters III and IV) or the scaled sum (normative models, Chapters V and VII). Another example of a linear model is the expected value (Chapters VI, VII, and VIII).

4.2 Dominant and Efficient Alternatives

Before we aggregate the preferences over all criteria, scenarios, and decision makers, it is worthwhile to check if one alternative outranks all other alternatives with respect to all criteria, scenarios, and decision makers. An alternative a_j which for all criteria c_i's is weakly preferred to another alternative a_k and strongly preferred with respect to at least one criterion is called **criteria dominant**. Thus, alternative a_j dominates alternative a_k if:

- $e_{ij} \succcurlyeq e_{ik}$ for all i, and
- $e_{ij} \succ e_{ik}$ for at least one i.

Alternatives which are not dominated are called **efficient**. The set of efficient alternatives is called the **Pareto optimal set** (Vilfredo Pareto, 1848-1923, Italian economist).

The alternative a_3 in Figure I.8 (left) dominates all other alternatives. The alternatives a_3 and a_4 Figure I.8 (right) are Pareto optimal, while the alternatives a_1 and a_2 are dominated by a_4 and a_3, respectively. For the efficient alternatives, the preferences must be aggregated to an overall preference value for each alternative.

If one is interested in only the most preferred alternative (and not in the preference order of all alternatives) then the dominated alternatives can be discarded from further analysis. However, it should be noted that a dominated alternative can be more preferred than a Pareto optimal alternative. The reader should verify this with the example in Figure I.8 on the right (assume the weight of 'time' is orders of magnitude larger than the weights of all other criteria, then the dominated a_1 will outperform the Pareto optimal a_3), and look at the numeric example in Chapter II, Section 3.2.

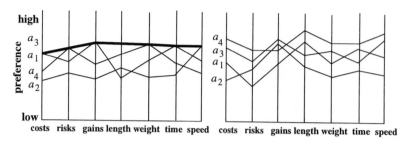

Figure I.8: Dominant and efficient alternatives.

The analogous definition of dominance and efficiency can be applied to the scenarios and decision makers. A dominant alternative in terms of the uncertainty is called **first order stochastically dominant** (Chapter VI). In Chapter IX we will use the principle of dominance and inferiority to resolve strictly competitive conflict situations between two actors.

4.3 Preference Graphs

The preferences between elements, such as alternatives, criteria, and decision makers, can be visualized with an **oriented preference graph**. A graph consists of two sets; the set of all elements (the nodes or vertices), and the set of all preferences between these elements (links or edges). When the elements are the alternatives, we have as the set of nodes, A, the set of all alternatives: $A=\{a_j| j=1,...,n\}$. The set of oriented links, L, is the set of all preference relations: $L=\{l_{jk} \mid a_j \succ a_k$ (strong) or $a_j \succeq a_k$ (weak)$\}$.

A preference graph can be represented by an adjacency matrix, G, in which $l_{jk}=1$ and $l_{kj}=0$ if $a_j \succ a_k$, and l_{jk}="-" and l_{kj}="-" if the relation between the two is not defined. Figure I.9 shows the adjacency matrix and the corresponding strong preference graph for the following strong preference relations on the set of the six alternatives: $L=\{a_1 \succ a_5, a_1 \succ a_6, a_2 \succ a_5, a_3 \succ a_4, a_3 \succ a_5, a_4 \succ a_2, a_6 \succ a_3\}$.

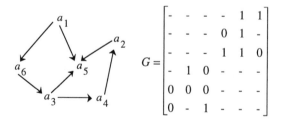

Figure I.9: Strong preference graph and corresponding adjacency matrix.

If transitivity could be assumed for the preference graph in Figure I.9 which relations must be added to have a complete strong preference order (see Problem 8)?

Examples of graphs for the preference orders introduced in Section 3.4 are given in Table I.4. The corresponding rankings are also shown for the preference orders. The ranks of the pseudo orders, however, are not clear.

	$\succ$: strong preference order	$\succsim$: weak preference order	pseudo order
complete	rank: $A \succ B \succ C \succ D$	rank: $(A \sim D) \succ B \succ C$	rank: unclear
partial	ranks: $D \succ A$, and $D \succ C \succ B$	ranks: $(D \sim C) \succ B$, and $D \succ A$	rank: unclear

Table I.4: Different types of preference orders with examples.

The adjacency matrix of a complete strong preference order has the characteristic that $l_{jk}+l_{kj}=1$ or l_{jk}="-" for $i \neq k$. This means that (given the relation is defined) one of the two symmetric entries has the value one, and the other has the value zero. If the preference relation between all pairs of alternatives is known, the sum of entries in the row for the best alternative is n-1, for the second best n-2, down to zero for the worst alternative. In terms of the preference graph, a strong preference order has no cycles. Thus, there is no path starting in one element and leading back to the same element.

Preferences between elements are sometimes expressed by an **intensity measure** on a ratio scale. For example, the decision maker might be able to express not only that a_j is preferred to a_k ($a_j \succ a_k$), but also the intensity of this preference; for example, a_j is n times more preferred than a_k: $\pi(a_j)=n\pi(a_k)$. We can also represent intensities of relations in matrix and graph forms. In this case, the graph is called **network** or **weighted preference graph**, and the corresponding matrix is the weighted adjacency matrix, G_w. Figure I.10 shows the weighted adjacency matrix and the corresponding preference graph for the following preference relations among six alternatives: $L=\{\pi(a_1)=6\pi(a_3), \pi(a_1)=24\pi(a_5), \pi(a_1)=2\pi(a_6), \pi(a_2)=2\pi(a_5), \pi(a_3)=1\pi(a_4), \pi(a_3)=4\pi(a_5), \pi(a_4)=2\pi(a_2), \pi(a_4)=4\pi(a_5), \pi(a_6)=3\pi(a_3), \pi(a_6)=12\pi(a_5)\}$.

If two alternatives are indifferent, then they have the relative weight 1, and if they are incomparable (or their relation is not defined), then they have no relative weight at all (-). It should be noted that the reflexive preferences, $\pi(a_j)=1\pi(a_j)$, which stand for the 1's in the diagonal of the matrix, should also be shown in the graph, but they have been omitted to avoid overload of the figure.

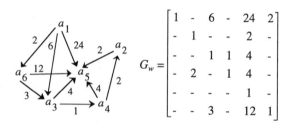

Figure I.10: Weighted preference graph and corresponding adjacency matrix.

In a weighted preference graph, there are two types of transitivities. The first corresponds to the transitivity of non-weighted preferences, called **ordinal transitivity**. This means that if $\pi(a_i)/\pi(a_j) \geq \pi(a_i)/\pi(a_k)$, then $\pi(a_k)/\pi(a_j) \geq 1$. The second refers to the intensities, called **intensity transitivity**, saying that if $\pi(a_i)=n\pi(a_j)$ and $\pi(a_j)=m\pi(a_k)$ then it must follow $\pi(a_i)=nm\pi(a_k)$. For example, if A is twice as heavy as B, and B is three times as heavy as C, then we would expect that A is six times as heavy as C. For subjective assessments, it is probably easier for a decision maker to be ordinal transitive than intensity transitive. In Chapters III and IV we will discuss how to assess preferences of criteria based on paired assessments, where the preferences are expressed on a ratio scale. Moreover, we will discuss how to resolve inconsistencies concerning intensity (multiplicative) transitivity.

Weighted graphs may have to comply with other characteristics, such as symmetry. If the weights refer to preference intensities, we might require that if $\pi(a_j)=m\pi(a_k)$ then it must follow that $(1/m)\pi(a_j)=\pi(a_k)$, which is called **reciprocal symmetry**. Symmetry can reduce the number of paired assessments significantly, as will be discussed for reciprocal symmetry in Chapters III and IV, and for ordinal transitivity in Chapter II, Section 5.

4.4 Search for Solutions

The search for a solution depends on the types of decision variables (binary, integer, or real) and the implicit or explicit definition of feasible alternatives. Finding a solution involves a search process and an appropriate **search strategy**. Clearly, the problem identification and definition puts requirements on both the search procedure as well as the search strategy.

If we have alternatives which are defined implicitly, we would like to have a search strategy that simultaneously generates and evaluates alternatives. That is, we would not want to generate alternatives first and then compare them to one another. For example, imagine a traveler who tries to find on a road map the best route to the planned destination. Writing down all feasible roads and then evaluating them would be too work intensive. In Chapter X we will discuss a systematic procedure that simultaneously constructs and evaluates roads, starting at the origin until the destination is reached.

Any search strategy is tailored to the structure and form of a decision problem. There are several important types of problems for which special search strategies have been devised:

- **Ranking a finite set of explicit alternatives**: rank all elements (alternatives, criteria, or decision makers) from most to least preferred; e.g., ranking of construction projects, service locations, policies, decision makers, etc. (Chapters IV, V, VII, IX).

- **Constraint-optimization with implicit alternatives**: find only the most preferred alternative; e.g., search for fastest route on a network, search for cheapest combination of investment policies, search for optimal work-shift schedule, etc. (Chapters V and X).

- **Clustering of alternatives**: find all efficient alternatives; e.g., search for non-dominated alternatives which can be presented to the decision maker for further analysis (Chapter IV).

- **Optimization of finite set of sequential actions**: find a policy consisting of a chronologically ordered sequence of actions; e.g., search for the best travel route, sequential investment policies, sequential risk abating strategies, etc. (Chapter VIII).

- **Resolution of conflicts among decision makers**: find a solution which satisfies all decision makers; e.g., resolution of conflicts in a conflict situation; search for a solution in a risky situation (e.g., decision maker against nature), search for a solution in a strictly competitive setting (Chapter IX).

In the following chapters we will discuss different types of solution search strategies. The critical aspects of a solution search strategy are whether it can find the best (or most preferred, or good enough) solution, and how much effort it will take to do so.

4.5 Sensitivity Analysis

Every solution to a decision problem has a certain **stability**. The stability of a solution is determined by how sensitive the solution is to changes in the assumptions, parameters, evaluation values, preference elicitations, etc. Whenever we discuss solution approaches and present results we will address the issue of **sensitivity analysis**, or at least talk about the limitations of the different methods. Descriptive assessments are based on paired comparisons. Consequently, if a new element (alternative, criterion, decision maker, scenario) is added, the original ranking of the alternatives might change. This **rank-reversal** can result from the structure of the problem, which can be affected by adding near replicas of alternatives. Another possible cause of rank-reversal is the dependency between elements. These types of phenomena will be discussed in Chapter IV. Rank-reversal can also occur in voting procedures. In Chapter IX we will discuss the limitations of the oldest voting procedure, the Borda count, which suffers from potential rank-reversal.

An important type of sensitivity refers to the parameters of a decision model. In Chapter V and X we will address parameter sensitivity in constraint-optimization problems, such as the search for the best investment policy. The stability of a solution can be expressed in terms of **allowable changes** (allowable increases and decreases) of the parameters so that the solution will not change. If the goal is to maximize benefits, the sensitivity of the optimal policy is expressed in terms of **shadow prices**. These shadow-prices reflect how much can be gained if some constraints are relaxed.

5. Summary

The problem solving process in policy planning is an iterative process that involves the analyst, the decision makers, and the environment (including other actors). It typically starts with the perception of a problem, goes then through three decision analytic steps, and ends with the implementation of a solution. A distinctive part of the policy planning process refers to decision making, which consists of problem identification, problem definition, and problem solution.

Making a decision means to take some actions. Actions are evaluated with respect to criteria, for different scenarios, and from the point of view of different decision makers.

Criteria can be decomposed into sub-criteria or aggregated into clusters. Goals can be defined in a hierarchy; content goals refer to the criteria in terms of aspiration levels (max, min) and constraints (>, ≥, =); structural goals are used to define the structure of potential solutions (alternatives) and to describe how content goals must be prioritized. Moreover, goals can be hard, saying what must be achieved, or soft, saying what should be achieved. Constraints are often defined as hard goals and aspiration levels as soft goals.

Decision options are made up of single or multiple actions. They are called alternatives, policies, tactics, or strategies, and can be represented explicitly or implicitly. Alternatives are evaluated for different possible states of the system under investigation. The occurrence of these states can be certain, risky, or uncertain.

The evaluation of the alternatives is represented in an evaluation matrix or score card, and the evaluations are based on measurement scales. Scales are either linguistic (categorical, ordinal) or numeric (categorical, ordinal, or cardinal). Cardinal scales are interval (invariant under positive linear transformations), ratio (invariant under positive multiplicative transformations), or absolute (invariant under identity transformation).

The major challenge in decision modeling is the transformation of the evaluations into subjective preferences and their aggregation across actions, criteria, scenarios, and decision makers. There are two fundamentally different approaches to decision making: the descriptive approach is based on paired comparisons, and the normative approach is based on subjective preference functions for a rational decision maker.

The objective of any choice process is to come up with an incomplete or complete ranking of all feasible alternatives, using strong, weak, and indifference relations. Preference relations can be represented as preferences or intensities on a graph, called a network. Preference orders over a set of alternatives are either complete or incomplete. Moreover, they are either strong preference orders, weak preference orders, or pseudo orders.

If the evaluations for the different criteria are not aggregated, the goal is to find dominant and efficient alternatives. The set of alternatives which are not dominated by any other alternative is called the Pareto optimal set; its elements are the efficient alternatives.

Any solution to a decision problem has a certain stability or sensitivity. Factors affecting the sensitivity of a solution are the evaluation of the alternatives, possible dependencies across the elements, and the presence of near replicas of alternatives. In Chapter IV we will discuss these types of instabilities for when the preference intensities of alternatives and criteria are assessed with paired comparisons.

6. Problems

1. Discuss a real-world example of a goal hierarchy with at least four levels.

2. Give examples of decision problems in policy planning with explicit alternatives. Identify possible criteria (multiple) and arrange the goals in a hierarchy.

3. Give an example of a decision problem in policy planning with implicit alternatives, where the decision options are binary, real, and integer. Identify the criteria (multiple) and goals (hierarchy).

4. Give a real-world example of an interval scale (such as the temperature scale). Discuss a realistic admissible transformation.

5. A community is investigating potential sites to locate a landfill. In a first screening, the community has identified eight sites. The sites have been evaluated with six criteria, given in the table below.

		a_1	a_2	a_3	a_4	a_5	a_6	a_7	a_8
c_1	operational costs [$$10^6$]	17	20	22	27	32	15	40	10
c_2	human risk [10^{-6}]	2.2	1.6	2.8	1.9	3.2	1.2	3.0	1.5
c_3	expected profit [$$10^5$]	72	68	65	50	54	75	45	80
c_4	loss of land [km^2]	2.0	1.7	1.1	2.8	1.5	0.8	2.5	1.0
c_5	new jobs [-]	89	135	150	115	65	178	46	200
c_6	land value [class]	low	med.	med.	very good	worst	best	very low	good

a) Determine the feasible alternatives considering the following constraints: $e_{2j} \leq 3.0$, $e_{5j} \geq 50$.

b) Determine the Pareto optimal alternatives (i.e., the non-dominated alternatives) out of the set of feasible alternatives.

c) If all criteria are equally important, which of the Pareto optimal alternatives would you judge as the best one, and why?

6. For the evaluation table of Problem 5:

a) Define a quantitative normative model which transforms, for each criterion c_i, the numeric evaluation values, e_{ij}, into normalized preference values, n_{ij}, where the best evaluation value has the value 1 and the worst the value 0, with a linear relation in-between: $e_{ij} \rightarrow n_{ij} \in [0,1]$. Write down the mathematical model. What do you do with the linguistic scale for land value?

b) Assume that the aggregated preference over the criteria is additive; that is, $n_j = \Sigma_i n_{ij}$. Rank the alternatives with this normative choice model.

c) Discuss the shortcomings of this normative choice model. How would you incorporate the importances (weights) of the criteria into your mathematical preference model?

7. Visualize the normalized evaluations from Problem 5 and 6 in a multidimensional graph as illustrated below, where lower preferences are closer to the center and higher preferences farther away. How would you identify if an alternative is 'better' than another?

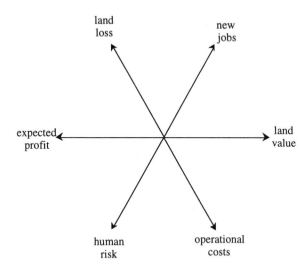

8. Assuming that transitivity holds for the strong preference relation ($\succ$) in Figure I.9, which relations must be added to have a complete strong preference structure and what is the complete strong preference order?

9. Discuss real-world examples for the six preference orders of Table I.4.

10. Give numeric examples (3×3 matrices) for the following relations:

• multiplicative-transitive preference relation,
• ordinal-transitive but not multiplicative transitive preference relation.

CHAPTER II

THE ANALYTIC MODELING PROCESS

1. From Problem to Model

1.1 The Model as Abstraction of Reality

Problems occur and must be solved in the real world environment. This environment can be perceived and described as a collection of elements and their relations. The elements are persons, organizations, cities, etc., all interacting with one another. The collection of these elements and their relations is called a **system**. To describe and solve a problem, a **model** of the system is developed. A model is an abstraction of a system under investigation, made by an observer of the system. This observer can be an analyst, or any other person who studies the system to derive some conclusions about its behavior. The construction of the model and the identification of the criteria, goals, scenarios, decision makers, and actions depend strongly on the perception of the model builder. Human perception of the system under investigation may be influenced by factors unrelated to the problem, such as the analyst's mood, recent experiences, or personal attitudes toward the problem. Although the analytic approach to modeling is based on rules, axioms, and formalisms, modeling will always remain a critical craft skill [Willemain, 1995].

Figure II.1 shows the relations between real world, system, mental model, and analytic model. The perceived problem in the system under investigation is represented as a **mental model**, which then gets translated into an **analytic model**, and from which a decision will be derived to solve the problem.

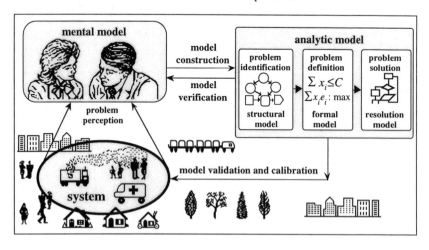

Figure II.1: The decision modeling process.

Modeling is an iterative process. First, the mental model is translated into an analytic model which consists of a **structural model**, a **formal model**, and a **resolution model**. The analytic model is then verified. **Verification** means to check if the mental model has been correctly translated into the analytic model. In other words, one must check whether the structure represents the system, whether the formal definition describes the structure and whether the resolution approach can lead to the envisioned type of solution. If this is the case, the analytic model is validated by comparing the results it produces (its behavior) to the system under investigation, or, if this is not possible, by simulating the system's behavior. If the **validation** does not satisfy the expectations, either the mental model should be revised (restarting the analytic modeling process again) or the analytic model should be calibrated.

Calibration means to make changes to the analytic model such that it produces the expected results. A consequence of the changes may be that the analytic model no longer represents the mental model appropriately. Calibration is solely based on the performance of the model, in contrast to validation which requires that any mismatch between mental model, analytic model, and system be eliminated. Consequently, the changes performed on the mental model as part of validation are explainable alterations focusing on the constructs of the model, rather than adjustments related only to the behavior of the model (as done for calibration).

1.2 The Analytic Modeling Process

The process of building the analytic model can be decomposed into three steps (Figure II.2): **structuring** (structural model, depicting the elements of the system and their relations), **formalization** (formal model, defining the relations identified in the structural model), and **resolution** (resolution model, searching for a solution based on the formalization of the problem).

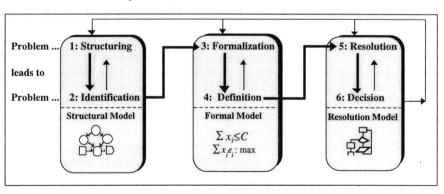

Figure II.2: The analytic modeling process.

The thick arrows in Figure II.2 show the principal process of building the three-component analytic model. From Figure II.2 it can be seen that problem structuring leads to the **identification** of the problem in form of the structural model; problem formalization leads to the **definition** of the problem in form of a formal model; and model resolution leads to the **decision** in form of the resolution model.

In the following chapters, some of the most important quantitative methods used to build the three sub-models will be discussed. However, this three-step decomposition of the analytic modeling process also holds for non-quantitative approaches. For example, if someone is about to organize a meeting, the structure of a problem consists of the people who will be invited and their relationships. The formal model consists of a set of rules concerning the principles of discussion and negotiation for the meeting. Finally, one resolution model consists of the agenda of the meeting which defines the procedural aspects of the meeting. The results of the meeting could then be implemented in the real world. If the results do not meet the expectations, another meeting could be arranged, starting a first iteration of the modeling and problem solving process.

This three-step decomposition of the analytic model building process is in accordance with the problem solving process introduced in Chapter I (Figure I.1). There we discussed the three steps of decision making as part of problem solving. These three steps correspond to the three steps of the analytic model building process. Problem identification is done with the structural model, problem definition with the formal model, and problem solution with the resolution model. The first step we mentioned in Figure I.1 was problem perception, which is addressed in the mental model. The final step was implementation of the solution and monitoring of the outcomes, which refers to model validation and calibration.

2. Structural Models

2.1 Definitions and Elements

The analytic modeling process starts with the identification of the problem in form of a structural model. This step is also called the **conceptualization phase** of the problem solving process. The purpose is to identify the relevant elements and their relations, such as the criteria, actions, scenarios, goals, and decision makers.

Some of the well-known diagrammatic methods that analysts use for the conceptualization phase are the soft-systems-approach [Checkland, 1988], [Avison et al., 1992], probabilistic influence diagrams [Schachter, 1986], knowledge maps [Howard, 1989], cognitive maps [Eden, 1988], causal models [Bagozzi, 1980], and systems thinking [Senge, 1990]. These approaches are based on specific modeling paradigms, some of them focusing more on data analysis, others more on systems analysis, and again others more on decision analysis. The decision analysis modeling approaches often address specific types of decision problems, such as decision making under uncertainty or group decision making. However, because we address a wide range of decision problems, we will introduce a novel notation for structural modeling as part of the analytic modeling process.

The purpose of a structural model is to illustrate or visualize (diagrammatically) the **elements** of a decision problem and their **relations**. In Chapter I we introduced the elements that we will consider in the problem solving process. These are the **decision makers** (persons involved in the decision making process), the **actions** (which, when chosen, can resolve the problem), the **criteria** (which are necessary to evaluate the actions), the **scenarios** (partitions of the uncertainty space), the **content goals** (which state in terms of the criteria what a good action should achieve), and the **structural goals** (which state how actions have to be combined to form tactics, strategies, or policies). Table II.1 gives an overview of the six basic elements for structural modeling, their name, notation, and icon with symbol.

Elements:	decision makers	actions and decision variables	criteria	scenarios	content goals	structural goals
Notation:	$D=\{d_l\}$ $l=1,...,w$	$A=\{a_j\}$ $X=\{x_j\}$ $j=1,...,n$	$C=\{c_i\}$ $i=1,...,m$	$S=\{s_k\}$ $k=1,...,v$	$G_C=\{g_{Cp}\}$ $p=1,...,r$	$G_S=\{g_{Sq}\}$ $q=1,...,s$
Icon with symbol:						

Table II.1: The elements of decision modeling.

It should be noted that the shapes of the different elements are unique. The decision makers are represented as polygons, the actions as rectangles, the criteria as rounded rectangles, the scenarios as circles (or ovals), the content goals as a combination of rectangle and triangle, and the structural goals as octagons.

A structural model consists of an oriented **graph** (see Chapter I, 4.3) where the elements are the nodes and the influences are the arrows (arcs). A node from which an arrow originates is called a **predecessor** node of the node where the arrow points to, called the **successor** node. Each node is identified by its name and symbol and evaluated; the evaluations might then be transformed to reflect subjective preferences. The **identification** is done in terms of the name and symbol, the **evaluation** is done in terms of all predecessor nodes, and the **transformation** is done based on the formal model. Multiple elements of one element class (e.g., actions) can be represented by one node if they affect the same successor nodes. This leads to a **compact** structural model (see Figure II.3).

Figure II.3 shows that there are two types of nodes: **marginal** nodes (with no incoming arrows) and **conditional** nodes (with at least one incoming arrow).

Conditional nodes must be defined in terms of the predecessor nodes. In Figure II.3, the actions, decision makers, and scenarios are marginal nodes, while the criteria, content goal, and structural goals are conditional nodes.

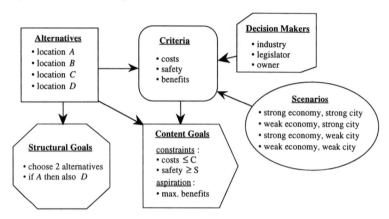

Figure II.3: Example of structural model.

The structural model in Figure II.3 shows that three decision makers (grouped into one node) must choose two out of four locations (e.g., for a hospital in a city). The decision will depend on the uncertain economic growth of the region (strong or weak) and also on the future development of the city (strong or weak). These two partitions of the uncertainty space make up a total of four scenarios. Three criteria are used to evaluate the four locations, which means that each decision maker must determine for each of the four scenarios twelve evaluations. Overall, the four alternatives will be evaluated 36 times (3 criteria, 3 decision makers, and 4 scenarios).

Structural, formal, and resolution aspects can never be completely separated. Often, it is advisable to write information about the formal or resolution model right into the structural model. Although most of the issues addressed in the following chapters are divided into structural, formal, and resolution model, it is neither possible nor desirable to draw clear lines between these three models. For example, we will introduce the basic concepts of eliciting subjective preferences as part of the formal model; however, the process of doing the preference elicitation with a decision maker will be addressed as part of the resolution model.

Just as a problem can be defined in different ways, the models used to solve the problem also are not unique. The purpose of using this three-step decomposed modeling approach is to improve communication by making the modeling process more transparent. In the following sections, we will give an overview of the topics addressed in Chapters III to X in the context of their structural models. There are no rules for how 'detailed' the structural model should be. Detailed means that, at least theoretically, we could define one node for every single alternative, criterion, scenario, decision maker, and goal. We can also insert the formalized relations into the structural model. Such a structural model (for the evaluation matrix in Table I.1, Chapter I) is given in Figure II.4.

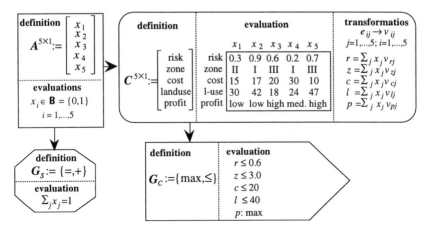

Figure II.4: Example of structural model with parts of formal model.

The five actions a_j (e.g., landfill locations) are represented (evaluated) by binary decision variables, $x_j \in \{0,1\}$. These actions are then evaluated with respect to five criteria (risk, zone, cost, landuse, and profit). The evaluations e_{ij} (alternative a_j, criterion c_i) are transformed into v_{ij}. For example, the zone values (I, II, III) could be transferred into (1, 2, 3), and the profit values (low, medium, high) into (1, 2, 3). Total values for all criteria are then computed. The total risk, for example, is defined as: risk=$x_1 \times v_{r1} + x_2 \times v_{r2} + x_3 \times v_{r3} + x_4 \times v_{r4} + x_5 \times v_{r5}$, which means that if we would choose all five landfills (i.e., $x_j = 1$, $j=1,...5$) the total risk would be the sum of all landfills' risks. For each criterion a content goal is defined. The structural and content goals say to choose the landfill location which maximizes the profit and satisfies the constraints.

2.2 Actions

The actions, $A=\{a_j\}$, are represented as rectangles. For each action, a_j, we introduce a **decision variable** x_j. These decision variables may be binary (**B**=$\{0,1\}$), integer (**Z**=$\{..., -2,-1,0,1,2,...\}$), or real (**R**=$\{...,-0.2,...0,...,0.3,...\}$). Binary variables indicate that the action can either be part of the solution or not. For example, if the problem is to identify potential (or optimal) landfill locations, and the locations are the actions, then each action is represented by a binary decision variable (i.e., if a_3 is chosen, then $x_3=1$, otherwise $x_3=0$). Integer decision variables indicate that actions can be included multiple times in the solution. For example, the number of people (x_j) working on project a_j is a non-negative integer variable, $x_j \in \mathbf{Z}^{\geq 0}=\{0, 1, 2, 3, ...\}$.

Real-valued decision variables ($x_j \in \mathbf{R}$) indicate that also fractions of actions can be chosen. For example, if the problem is to find the optimal land sizes in a land zoning project for industrial zone (a_1), recreational zone (a_2), and inhabited zone (a_3), the solution could be to use $x_1 = 3.5$ acres industrial zone, $x_2 = 2.9$ acres recreational zone, and $x_3 = 4.3$ acres inhabited zone. The value x_j reflects thus the **intensity** of action a_j, in the sense that 1 acre is the unit action, and $x_1 = 3.5$ means that action a_1 is taken 3.5 times. An issue to be addressed is whether the consequence (e.g., air pollution) of $x_1 = 3.5$ also is 3.5 times the consequence of $x_1 = 1$ (proportionality of intensities within actions). Another issue is whether the consequence of $x_1 = 3.5$ *and* $x_2 = 2.9$ is equal to their sum (additivity of consequences across actions). If proportionality and additivity hold, we speak of a **linear model**.

A special class of decision problems, called **dynamic decision problems**, requires that a chronologically ordered sequence of actions be implemented. At different points in time, the decision maker must choose an action from different finite sets of actions. A straightforward way to solve dynamic decision problems is to determine all combinations of actions that make up feasible alternatives and choose the best one. However, this can be very time consuming. In Chapter VIII we will discuss a resolution model to solve dynamic decision problems more efficiently.

Figure II.5 shows a structural model of a dynamic problem under certainty with three sets of actions. For example, a manager must visit one out of many cities in each of three countries (the first country has n_1 cities, the second n_2 cities, and the third n_3 cities). The structural goal tells that the traveler must visit exactly one city in each country (first in country one, then in country two, and end in country three). To illustrate the order in which the decisions must be made, dashed arrows are drawn between the action nodes. They are called **informational arcs** because they indicate that the state of the preceding node is known at the time the decision must be made.

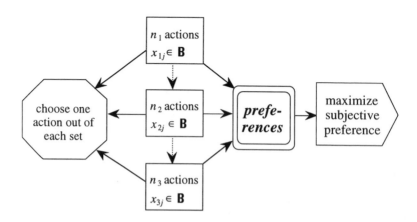

Figure II.5: Structural model of dynamic decision problem under certainty.

2.3 Criteria

The criteria, $C=\{c_j\}$, are represented as rounded rectangles. The criteria are used to evaluate the alternatives. The chosen measurement scales can be indicated in the structural model: ordinal (**O**), interval (**I**), ratio (**R**), categorical (**C**), or absolute (**A**).

The criteria node can also show how the evaluation values are transformed into subjective preference values, how these preferences are aggregated, and whether the criteria have been decomposed or aggregated. To emphasize that a node contains an own small structural model or that a transformation is performed, the node is drawn with double lines. Figure II.6 shows such an example, which refers to Figure II.3.

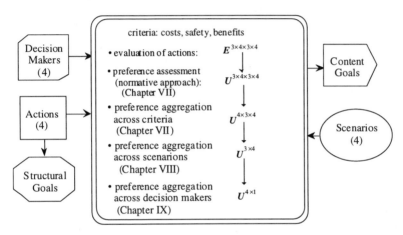

Figure II.6: Example of compact evaluation node.

Often, the decomposition of criteria is visualized in the structural model. Figure II.7 shows an example of criteria decomposition into sub-criteria. The criteria c_1 and c_2 are the originally identified criteria. These meta criteria are decomposed into two levels.

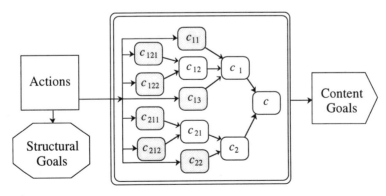

Figure II.7: Decomposition of meta criteria into evaluation criteria.

The criteria at the bottom of the hierarchy are the **root criteria**, also called the **evaluation criteria**. In Figure II.7, these are (shaded) c_{11}, c_{13}, c_{22}, c_{121}, c_{122}, c_{211}, and c_{212}. The actions are evaluated with respect to these evaluation criteria. For the example in Figure II.7, it is simply assumed that there is only one decision maker and one scenario (i.e., no uncertainty). The structural goal might be to choose the best action and the content goals to optimize the performance of the actions for a combination of criteria.

When a large number of criteria is used to evaluate actions in a decision problem, the decision analyst might question whether all those criteria are really necessary. The purposes of criteria are: (1) to reflect important aspects, and (2) to differentiate the alternatives (e.g., even if cost is an important criterion but all alternatives have the same costs, the criterion does not provide any information). Ideally, the information provided by the multicriteria evaluation can be reduced to m' ($< m$) criteria in such a way that as little information as possible is lost. The structural model highlighting a reduction of the number of criteria looks just like the one for decomposition, as shown in Figure II.7. If $m'=1$, a complete ordering of all actions is obtained. Formal models to decompose and aggregate criteria will be discussed in Chapter III.

2.4 Scenarios

The **scenarios**, $S=\{s_k\}$, represented as circles, can be derived from several **partitions** of the uncertainty space, each describing multiple **states**. For example, let's assume that a chemical company must choose one out of three insurance policies (see Figure II.8).

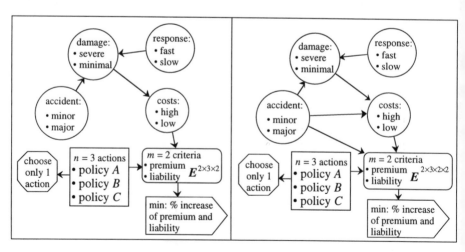

Figure II.8: Elements of uncertainty and their relations.

The objective is to minimize the relative increase of the insurance premium and the liability. Premium and liability depend on the costs and possibly on the type of accident. The damage resulting from an accident depends on the speed of the response of the company's fire fighters. The costs depend on the damage and possibly on the accident type.

There are four elements contributing to the overall uncertainty of the decision problem, each of them defining a partition of the uncertainty space (represented as a circle or oval). Two slightly different versions of the insurance problem are shown in Figure II.8.

Both structural models in Figure II.8 indicate that the accident type and the response speed affect (influence) the degree of damage, and that the degree of damage affects the accident costs. The model on the right has two additional arrows, one from accident to costs and one from accident to the criteria node. The interpretation of these additional arrows is that the costs depend not only on the damage but also on the accident type, and premium and liability depend not only on the costs but also on the accident type.

With two damage degrees (severe and minimal), we must do two assessments for each cost class in the left model. For the right model, costs must be assessed for four combinations of damage and accident (2×2). The three policies in the left model are evaluated for only two scenarios (high costs, low costs) and two criteria (premium, liability) which gives 2×3×2=12 assessments, while in the right model they are evaluated for four scenarios (minor accident with high costs, major accident with high costs, minor accident with low costs, and major accident with low costs) and two criteria which gives 2×3×4=24 assessments. It is clear that the structure of the uncertainty space is a major issue in and of itself which will be addressed in more depth in Chapters VI and VIII.

2.5 Decision Makers

The decision makers, $D=\{d_l\}$, are represented as polygons. The symbol for a decision maker is a face. An arrow from a decision maker element to any another element indicates that the successor elements must be evaluated from the point of view of that specific decision maker. If only one decision maker is involved, we simply omit the icon of the decision maker. If multiple decision makers are involved in the decision making process, a distinction is made between group decision making and conflict resolution. In a **group decision making** setting, all decision makers agree at some point on the actions, criteria, scenarios, and goals; they eventually come up with a single group solution (Figure II.9, left). In **conflict decision making**, on the other hand, each decision maker defines his/her own problem (Figure II.9, right). Each of the conflicting decision makers can choose his/her own solution, but the benefits are determined by what they all decide to do. In Chapter IX we will discuss the problems related to voting principles in a group setting and approaches for conflict resolution.

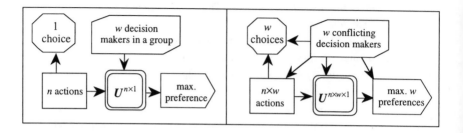

Figure II.9: Group decision making (left) and conflict decision making (right).

3. Formal Models

3.1 General Aspects

A major part of a formal model is to formalize the evaluation of the alternatives, the elicitation of subjective preferences, and the aggregation of preferences across multiple criteria, decision makers, and scenarios. A problem is defined when it is formally described (see Figure II.2). Describing a problem formally means therefore specifying the relations as indicated in the structural model in terms of functions. Formalization lays the basis for the resolution of the problem; defining a problem includes also specifying the basic approaches for its solution. For example, if the uncertainties are defined in terms of probabilities, the resolution approach must be based on probabilistic reasoning. Consequently, structural, formal, and resolution models are strongly related to one another, and, as mentioned earlier, it is neither possible nor necessary to separate them fully.

Figure II.10 shows the issues that will be addressed in the remaining chapters; these issues include both problem definition (plain text), and resolution modeling (*italic* text). The full meaning of Figure II.10 becomes clear only after studying all the chapters. Figure II.10 can then be used to comprehend the relations between the different approaches and serves as an overview of the most frequently used approaches in decision modeling in policy management. We will now briefly address some of the major issues depicted in Figure II.10.

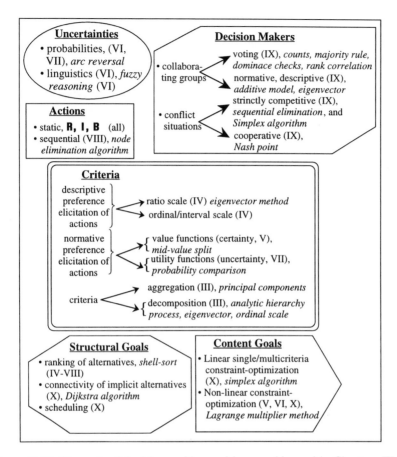

Figure II.10: Elements of decision making and issues addressed in Chapters III-X.

3.2 Descriptive vs. Normative Preference Elicitation

It was mentioned in Chapter I that the elicitation of subjective preferences can be done with either a normative or a descriptive approach. With a descriptive approach, the advantages and disadvantages of each action over the others are elicited using paired comparisons with respect to all criteria. The results of these paired comparisons are represented in a table or matrix. If the preferences are expressed numerically, different scales may be used. If an ordinal scale is used, we might assign a value of 1 to an alternative if it is better than (or at least as good as) another one, and 0 otherwise. If a ratio scale is used, we might say that an alternative is two or three times more preferred than another one. For an interval scale, we could assign a value between 0 and 1 to an alternative, based on the difference to another alternative.

Let's assume that we compare four policies based on three criteria: risk reduction (RR, %, note that higher values are preferred to lower ones), long term costs

(CO, $\$10^6$), and implementation time (IT, months). Figure II.11 shows the structural and formal model and the numeric values, where x_j is the decision variable of alternative a_j ($x_j=1$ if a_j is part of the solution, $x_j=0$ otherwise), e_{ij} is the evaluation value of a_j for criterion c_i, k_{jkli} and n_{ij} are normalized subjective preference values, w_i is the weight of criterion c_i, and λ_i is the scaling constant of criterion c_i. It should be noted that a_4 is dominated by a_3; thus, the Pareto optimal set is $\{a_1,a_2,a_3\}$. The reader should verify this.

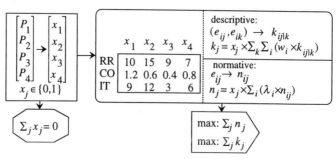

Figure II.11: Structural and formal model.

A) Descriptive Preference Elicitation

With a **descriptive** approach all policies must be compared **pairwise** for all criteria. If one policy, P_j, performs better than another one, P_k, with respect to criterion c_i, we would assign to it as its **preference** or **dominance** measure the normalized distance measure:

$$k_{jkli} = \begin{cases} \dfrac{e_{ij} - e_{ik}}{e_{i,best} - e_{i,worst}}, & \text{if it is } > 0 \\ 0, & \text{otherwise} \end{cases},$$

where e_{ij} is the evaluation value of policy P_j with respect to criterion c_i, and $e_{i,best}$ and $e_{i,worst}$ are the best and worst outcomes of the four policies for criterion c_i.

risk reduction ($k_{jkl\text{RR}}$):

	P_1	P_2	P_3	P_4
P_1	0	0	.125	.375
P_2	.625	0	.750	1.0
P_3	0	0	0	.250
P_4	0	0	0	0

costs ($k_{jkl\text{CO}}$):

	P_1	P_2	P_3	P_4
P_1	0	0	0	0
P_2	.750	0	0	.250
P_3	1.0	.250	0	.500
P_4	.500	0	0	0

implementation time ($k_{jkl\text{IT}}$):

	P_1	P_2	P_3	P_4
P_1	0	.333	0	0
P_2	0	0	0	0
P_3	.667	1.0	0	.333
P_4	.667	.667	.333	0

The descriptive approach uses **weights** for the criteria, to express their mutual importance. Let's assume that we have determined the weights of the criteria as (methods to determine weights of criteria will be discussed in Chapter III): $w_1=0.3$,

$w_2=0.5$, and $w_3=0.2$. Then we multiply the entries of the matrices $(k_{jk|i})$ with the weight (w_i) of the corresponding criterion (c_i) to get the weighted preferences:

risk reduction, $w_1=0.3$

	P_1	P_2	P_3	P_4
P_1	0	0	.0375	.1125
P_2	.1875	0	.225	.300
P_3	0	0	0	.075
P_4	0	0	0	0

costs: $w_2=0.5$

	P_1	P_2	P_3	P_4
P_1	0	0	0	0
P_2	.375	0	0	.125
P_3	.500	.125	0	.250
P_4	.250	0	0	0

implementation time: $w_3=0.2$

	P_1	P_2	P_3	P_4
P_1	0	.667	0	0
P_2	0	0	0	0
P_3	.133	.200	0	.667
P_4	.133	.133	0	0

These weighted preferences must be aggregated to an overall preference for each policy. There are two ways to do this. One way is to add up all the values in a row for each policy and criterion. This is called **preference aggregation within criteria**. Another way is to put the three matrices on top of each other and to add up the three entries for each cell. This is called **preference aggregation across criteria**. Below are the two resulting matrices (these two approaches will be discussed in Chapter IV):

preference aggregation within criteria:

	risk	costs	time	Σ
P_1	.150	0	.6667	0.8167
P_2	.7125	.500	0	1.2125
P_3	.075	.875	1.0	1.9500
P_4	0	.250	.2666	0.5166

preference aggregation across criteria:

	P 1	P 2	P 3	P4	Σ
P_1	0	.6667	.0375	.1125	0.8167
P_2	.5625	0	.225	.425	1.2125
P_3	.633	.325	0	.9917	1.9500
P_4	.3833	.1333	0	0	0.5166

Using these matrices, we can now compute the overall subjective preference value for each of the four policies by adding up the entries for each row (Σ). Note that the resulting preference for each policy is the same whether we aggregate across or within criteria. Based on the aggregated preferences we see that P_3 is the best policy, followed by P_2, P_1, and P_4 as the least preferred policy.

Inferiorities of alternatives can be determined analogously to the dominances of alternatives. More on the principle of descriptive preference elicitation and the advantages of preference aggregation across and within criteria will be discussed in Chapter IV. There we will discuss not only interval measures, as in this example, but also ratio and ordinal measures. Several approaches for assessing the weights of criteria will be discussed in Chapter III.

B) Normative Preference Elicitation

The **normative** approach to preference elicitation is not based on a paired comparison principle. Instead, it is based on a set of **axioms** about the decision maker's preference structure. If the decision maker complies with these axioms, s/he is called **rational**, which implies that a **preference function** exists which reflects his/her preferences over a range of potential alternatives. Such a preference function is determined for each criterion, where the preference of the best outcome is assigned a value of 1 and the one of the worst outcome a value of 0 (any other interval is also possible, e.g., 100 to 0). As we will discuss in Chapters V and VII, a neutral decision maker has the following preference function:

$$n_{ij} = \frac{e_{ij} - e_{i,worst}}{e_{i,best} - e_{i,worst}}$$

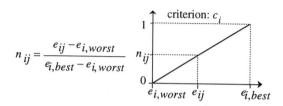

The value e_{ij} is the evaluation value of policy P_j with respect to criterion c_i. For the example introduced above, we get the following preference values (in parentheses):

actions (policies)	Risk Reduction 0.3	Costs 0.5	Implemen- tation Time 0.2	Σ
P_1	10 (.375)	1.2 (0)	9 (.333)	.1791
P_2	15 (1.0)	0.6 (.750)	12 (0)	.6750
P_3	9 (.250)	0.4 (1.0)	3 (1.0)	.7750
P_4	7 (0)	0.8 (.500)	6 (.667)	.3834

The next step in the preference aggregation procedure is to derive from these three single-criterion preference functions a **multicriteria preference function**. For the normative approach, in contrast to the descriptive approach, **scaling constants** (λ_i) rather than weights (w_i) are determined for each criterion. Under some specific conditions which will be discussed in Chapter V, the multicriteria preference model is additive. In the right-most column in the table above, the result of the additive model is given, where the scaling constants are assumed to have the same numeric values as the weights used for the descriptive example: $\lambda_1=0.3$, $\lambda_2=0.5$, and $\lambda_3=0.2$. The results in the table above show that the preference order for the four policies is different, with P_3 being the most preferred policy followed by P_2, P_4, and P_1.

The normative approach is often criticized because human decision makers do not always behave rationally. However, when they do act rationally, the normative approach is often considered superior to the descriptive approach because it is based on formal theories of rational decision making. In Chapter V we will discuss the normative approach when there is no uncertainty about the evaluation of the alternatives, and in Chapter VII we will discuss the normative approach in the presence of uncertainty. The difference between the two is that under complete certainty, the decision maker is asked to express preferences for known outcomes, while under uncertainty, s/he is asked to express preferences for uncertain outcomes, called **lotteries**.

Chapter II: The Analytic Modeling Process

3.3 Single versus Multiaction Decision Problems

So far we have tacitly assumed that the decision variables, x_j, take on binary values; this means that an alternative (e.g., policy) is either chosen or not. In this case, the structural goal would be to choose only one policy (or rank all policies), and the content goal would be to maximize the total preference. The formalizations of these two goals are shown in Table II.2, where $x_j \in \mathbf{B}$, and n_j is the preference of policy P_j (see also Figure II.4):

goal	formal model	structure
content	maximize: $\sum_{j=1}^{4} x_j n_j$	maximize total preference
structural	$\sum_{j=1}^{4} x_j = 1$	choose 1 out of 4

Table II.2: Formalization of content and structural goals.

The result for the normative model (where P_3 has the highest preference value) is then $x_1=0$, $x_2=0$, $x_3=1$, $x_4=0$, and $\Sigma_j x_j \pi_j = 0.7084$.

Figure II.12 shows a slightly different definition of the problem given in Figure II.11, where e_{rj} stands for the risk reduction, e_{cj} for the long term costs, and e_{tj} for the implementation time when choosing a_j.

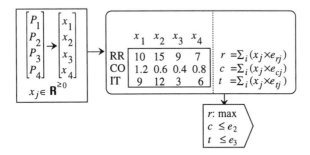

Figure II.12: Structural and formal model for multiaction decision problem.

Instead of choosing one policy out of the four, we ask how much, or how intensely, each policy should be employed. That is, the x_j values are positive real numbers. For example, imagine that the four policies refer to four different programs and we want to know how much of the available resources (e.g., money) should be allocated to each of the four programs. For this definition of the problem, we must assess how the resource allocation will affect risk reduction, costs, and implementation time. Here too, we assume for the time being a **linear model,** saying that: (i) the resulting impact with respect to the three criteria is proportional (**proportionality**) to the amount invested in the projects, and (ii) the resulting impact of all four projects is the sum of all the individual impacts (**additivity**). This means that the total costs are simply the sum of the costs resulting from each policy, where the costs of each policy are the product of amount invested and unit costs.

The three policies are assumed to have the following unit-impact or **unit-effectiveness** for the three criteria, when a unit-investment ($x_j=1$) is done (note that the values are the same as the ones in Figure II.11):

- c_1: A unit-investment (e.g., \$100.000) on P_1 reduces the risk by 10 units, a unit-investment on P_2 reduces the risk by 15 units, a unit-investment on P_3 reduces the risk by 9 units, and a unit-investment on P_4 reduces the risk by 7 units.

- c_2: A unit-investment on P_1 requires additional costs of \$1.2×10⁶, a unit-investment on P_2 requires additional costs of \$0.6×10⁶, a unit-investment on P_3 requires additional costs of \$0.4×10⁶, and a unit-investment on P_4 results in costs of \$0.8×10⁶.

- c_3: A unit-investment on P_1 requires 9 weeks of work, a unit-investment on P_2 requires 12 weeks of work, a unit-investment on P_3 requires 3 weeks of work, and a unit-investment on P_4 requires 6 weeks of work.

We introduce three content goals, which include two constraints: (1) a limited amount of funds for costs (c_2), e_2; (2) a limited amount of time to be spent on the implementation of the policies (c_3), e_3; and the aspiration level: maximization of risk reduction (c_1). Again, the objective here is not to choose the best out of the four policies but to find an optimal mix among the four policies.

The three **content goals,** where x_j is the decision variable of policy P_j, are:

aspiration level	• c_1 (risk reduction):	max: $10x_1 + 15x_2 + 9x_3 + 7x_4$
constraint	• c_2 (costs):	$e_2 \geq 1.2x_1 + 0.6x_2 + 0.4x_3 + 0.8x_4$
constraint	• c_3 (time):	$e_3 \geq 9x_1 + 12x_2 + 3x_3 + 6x_4$

The decision problem has four **decision variables** (amounts to be invested in each of the four policies) $x_1, x_2, x_3,$ and x_4 ($x_j \in \mathbf{R}^{\geq 0}$, j=1,2,3,4) which are evaluated with three criteria and where a content goal (as aspiration or as constraint) is defined for each criterion.

It should be noted that there are no structural goals. The solution to this problem consists of a combination of $x_1, x_2, x_3,$ and x_4. Therefore, the solution will be a generic policy which tells how much to invest in each of the four policies. If we choose e_2=\$1.2×10⁶, and e_3=20 hours, then the optimal solution is to invest \$4/3×10⁵ units in the second policy (x_2=1.333), \$2×10^5 units in the third policy (x_3=2), and nothing in the first and fourth policies (x_1=x_4=0). By doing so, the resulting risk reduction is 35.5 units. **Linear constraint-optimization** problems of this nature will be discussed in Chapter X. Non-linear constraint-optimization problems will be addressed in Chapters V and X.

4. Resolution Models

4.1 General Resolution Approach

The resolution model is a description of how to solve the formalized (defined) problem. A resolution model can be specified at different levels of detail. At the most general level, the resolution model provides an approach to the whole problem solving process. This is often accomplished in form of a **resolution map** (e.g., Figure II.6) which shows what must be done when, and how to know when a good enough choice has been made with the potential to solve the problem.

Figure II.6 contains parts of a resolution model. To begin, the alternatives are evaluated with the m criteria. Then, the evaluations are transferred into subjective preferences. The next step is to aggregate these subjective preferences, first across criteria, then across scenarios, and finally across decision makers. The most preferred alternative is the one with highest preference value.

Resolution models also refer to specific search strategies for finding the most preferred alternatives in the space of potential alternatives. One such case is the constraint-based formulation of the optimization problem discussed in the previous section. The solution was to invest in the second and third policies. However, we have not discussed how we did find this solution. In Chapter X we will discuss systematic computational approaches (i.e., resolution models) to solve constraint-optimization problems.

An iterative computational procedure is called an **algorithm**. Algorithms consist of three basic steps: **initialization, iteration**, and **stop condition**. The initialization in the example of the constraint-optimization problem is to set all decision variables equal to zero (x_j=0, assuming this is a feasible solution). The iteration step consists of a rule for modifying the x_j values (decision variables) in a consistent way. The stop condition tells when the computation can stop. Most search algorithms improve at each step the preference value (optimization algorithms) such that the stop condition corresponds to an optimality test.

An example of a high-level resolution model for the problem of finding a meeting date is shown in Figure II.13. The structure of the meeting (structural model) shows which people will participate, where they live, what their relationships are, where the location of the meeting is, and so on. The formal aspects (formal model) refer, for example, to the fact that weekends must remain free, that the meeting goes from 8 AM to 6 PM, that lunch will be provided, and that participants must be contacted by mail. The iteration step in an optimization algorithm is intended to move towards the optimal solution, such that after some iterations, the best date for the meeting can be found.

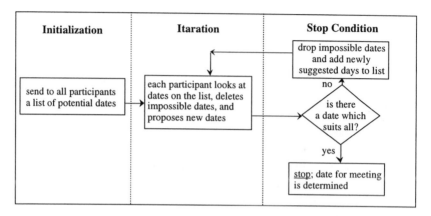

Figure II.13: Algorithm for finding a date for a meeting.

As another example, imagine you are placed in the middle of the night somewhere in a park of a foreign city which is completely closed by straight-line rows of buildings forming a hexagon. You are told to meet someone at a bar which lies the furthest north in the park. Assuming there is no row of buildings that is precisely east-west oriented, you know that the bar lies at an intersection of two rows of buildings (corner point).

One way to get to that bar is to start walking from wherever you are in direction north until you hit a row of buildings. Then, continue along that row of buildings in the direction which leads you north (since there are no east-west rows of buildings, the other direction leads you south) until you reach a corner point with another row of buildings. If moving along this next row of building brings you still further north then you continue, otherwise (if it leads you south) you know you have reached the bar.

The same approach can be used for optimization algorithms if the space of potential solutions meets certain conditions. One condition for the example above was that there are no east-west rows of buildings. Another condition is that the hexagon has no "indentations," meaning that you will never have to go south for a while before going north again to reach the bar.

Optimization algorithms are formulated in such a way that at each iteration step (a corner point) the computed value (the temporary coordinates) moves closer to the optimum (the bar). The algorithm stops when a new iteration does not improve the situation (the next row of buildings leads south).

4.2 Resolution Complexity

The resolution of a defined decision problem leads to the 'optimal' alternatives. The solution may consist of a list of priorities, a ranking of all the sufficiently good policies, or a recommendation of actions to be taken. A major challenge for policy analysis is to identify conditions which indicate that a problem can be solved (i.e., that a solution exists). If this is the case, an algorithm can be designed to resolve the formalized decision problem. If a solution exists, the algorithm should be designed to find it as quickly as possible.

Algorithms are categorized by how much effort they require to find a solution. We could, for example, speed up the process of finding a date for a meeting by sending out a very large list of possible dates and by adding additional dates at each iteration. This would be a change in our **search strategy**. Often, however, the problem is so complex that even a very efficient algorithm is still quite slow. In terms of the three-step analytic modeling process (Figure II.2), we say that a problem is complex if it is hard to structure, formalize, or resolve. A problem which is hard to structure is called **ill-structured**. One that is hard to formalize (and thus to define) is called **ill-defined**. Finally, a problem which is hard to solve is called **computationally complex**.

Resolution models are either **strategic** (informed or non-informed) or non-strategic (random or subjective). **Informed** resolution models use a measure of performance in the iteration step. Examples of informed resolution models are the simplex algorithm (Chapter X) and the shortest path algorithm in a network (Chapter X). **Non-informed** resolution models search not for *the* best solution, but for a solution that fulfills the stated requirements. The search principle of strategic non-informed resolution models consists of a systematic search through the solution space until a solution is found.

Non-strategic resolution approaches can often be employed informally during the initial phase of problem solving. For example, brainstorming can be seen as a non-strategic search procedure. Based on the initial results, a strategy can be devised that helps to find a '**satisficing**' solution in a systematic manner.

The reason to employ resolution models is to reduce the search effort without compromising the quality of the decisions. A common way to measure a problem's complexity with regard to the resolution process is to estimate the number of potential solutions. The efficiency of a resolution model is characterized by the number of iterations it takes to solve the problem. In the end, one might have to trade off complexity, efficiency, and quality of solutions when building a resolution model. The most important thing when not necessarily looking for the 'best' possible solution, is to know how 'far off' the chosen solution is, and how much improvement additional search effort would provide.

5. Interactive Complete Strong Preference Ordering

The problem solving methods that we will discuss in the following chapters show their practical merits when they are used interactively. To support real-time interactivity, the methods are implemented in computer systems. Special care must be taken with the human-machine interface to provide efficient and effective decision support. A computer system with some sort of built-in 'intelligent' decision support is called **decision support system** (DSS). The challenge in designing a DSS lies in identifying the appropriate task-sharing between the human decision maker and the computer system. Ideally, the computer does the computationally intensive tasks and the human the cognitive subjective assessments.

The following example illustrates how a DSS can be designed that interactively checks the consistency of a decision maker's preference assessments. Suppose a decision maker must sort five items with the strong preference relation ($\succ$), for example, set priorities for five infrastructure projects, rank five content goals, or prioritize five decision support methods for a specific problem. We assume that a complete strong preference order can be established, if the decision maker is given enough time to perform the necessary assessments. We allow the decision maker to contradict transitive inferences, but require that these contradictions be resolved before the ranking is completed.

5.1 Structural Model

A simple way to work interactively towards the ranking of five items is to pick repeatedly two items, perform paired comparisons, and draw transitive conclusions until the ranking is completed (Figure II.14). The decision maker may pick only pairs which s/he has not yet assessed. If s/he picks a pair which has been inferred by transitivity, s/he is asked if s/he agrees with this preference. For example, assume that the decision maker assesses $a_1 \succ a_2$ and $a_2 \succ a_3$, then the system would conclude $a_1 \succ a_3$. If the decision maker now wants to compare a_1 and a_3, the system would ask if s/he agrees with $a_1 \succ a_3$. If not, the system deletes all direct and inferred assessments which are related (directly or indirectly) to a_1 and a_3, and uses $a_3 \succ a_1$ as a new assessment. The preference relations that are not affected by a contradiction remain unchanged (e.g., $a_5 \succ a_4$ would not have been affected by the preference reversal).

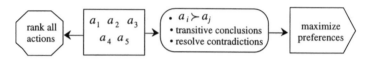

Figure II.14: Structural model of interactive complete strong preference ordering.

The structural model tells that the alternatives are compared pairwise with the strong preference relation, $a_i \succ a_j$. The goal is to rank all alternatives (structural goal) according to their preference (content goal).

The preference assessments are captured in a five-by-five matrix, $L^{5 \times 5}$, where $a_i \succ a_j$ gives $l_{ij}=1$ and $l_{ji}=0$. The ranking is completed when the sum of all entries l_{ij} is 10. The reader should explain why this is a feasible stop condition.

5.2 Formal Model

We assume that the decision maker can determine the strong preference between any two elements of a set of elements $(a_i, a_j) \in A$, where either $a_i \succ a_j$ or $a_j \succ a_i$, and also that transitivity in strong preference holds; that is, $a_i \succ a_j$ and $a_j \succ a_k \rightarrow a_i \succ a_k$. The formal process consists of picking any pair (never repeating an assessment), determining the strong preference relation between the two items, and making all transitive inferences, until all elements are completely ranked from most to least preferred. The result is a complete strong preference order.

The set S contains all strong preferences between two elements $(a_i, a_j) \in A$ that the decision maker has assessed directly. The set I contains all transitive inferences that have been made after each new assessment by the decision maker. Figure II.15 shows a screen view of an interactive DSS for complete strong preference ordering.

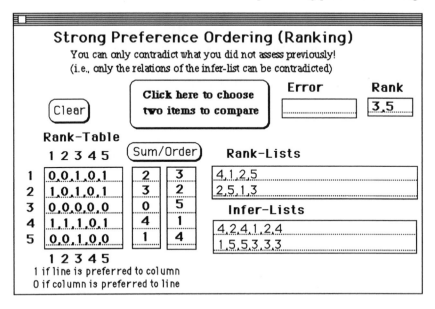

Figure II.15: Screen view of interactive complete strong preference ordering.

5.3 Resolution Model

A resolution model to solve the interactive complete strong preference assessment is based on an algorithm. It works with the two sets, S (for direct strong preference assessment, called *Rank-Lists* in Figure II.15) and I (for inferred preferences, called *Infer-Lists* in Figure II.15), and a matrix L (for final rankings, called *Rank-Table* in Figure II.15), where the elements of S and I are ordered pairs.

The sets and the matrix are initialized as follows: l_{ij}='-' (meaning that the pair $(a_i, a_j) \in A$ is not assessed yet), $S=\emptyset$, and $I=\emptyset$. The decision maker then assesses any not yet assessed pair, $(a_i, a_j) \in A$, $i \neq j$, with the strong preference relation. At each stage of the assessment process, the newly ordered pair $a_i \succ a_j$ is put into the set S. Then, all inferences with pairs from S and I are determined and put into I. There are three types of transitive inferences:

1: from the new pair $a_i \succ a_j$ with all pairs which have been directly assessed by the decision maker,
2: from the new pair $a_i \succ a_j$ with all pairs which have been transitively inferred in previous assessments, and
3: from any newly inferred preference from 1 or 2 with all directly assessed preferences.

To illustrate the principle of transitive inference let's look at the example in Figure II.16. The first two assessments are $a_4 \succ a_2$ and $a_1 \succ a_5$. These two assessments have no common elements and are thus called **disjoint preferences**. They can be represented as two graphs with no common elements (Figure II.16, left). The next preference assessment by the decision maker is $a_2 \succ a_1$. Now, three preferences can be inferred transitively. The fourth assessment is $a_5 \succ a_3$ which completes the preference order.

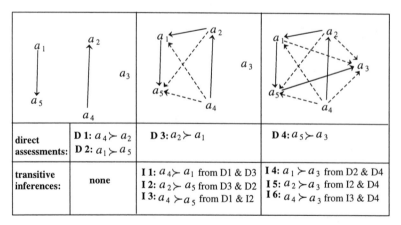

direct assessments:	**D 1:** $a_4 \succ a_2$ **D 2:** $a_1 \succ a_5$	**D 3:** $a_2 \succ a_1$	**D 4:** $a_5 \succ a_3$
transitive inferences:	none	**I 1:** $a_4 \succ a_1$ from D1 & D3 **I 2:** $a_2 \succ a_5$ from D3 & D2 **I 3:** $a_4 \succ a_5$ from D1 & I2	**I 4:** $a_1 \succ a_3$ from D2 & D4 **I 5:** $a_2 \succ a_3$ from I2 & D4 **I 6:** $a_4 \succ a_3$ from I3 & D4

Figure II.16: Interactive strong preference order with transitive inferences.

It should be noted that the example in Figure II.16 is the same as the one in Figure II.15. As we see, the sum of direct and inferred preferences for $n=5$ items always equals 10. The minimum number of assessments to rank n elements with this approach is $n-1$. In general, the number of interactive assessments will be larger. If no contradictions of the transitively inferred preferences are made, the maximum number of assessments is $(n^2-n)/2$. If contradictions are made, the number of assessments can even be larger.

If the decision maker picks a pair whose preference has already been transitively inferred, s/he can either accept the preference relation between the two elements or reject it. If it is accepted, the assessment is confirmed and a new, not yet assessed, pair can be chosen for assessment. If the inferred preference order $a_i \succ a_j$ is rejected, then all pairs of S and I with at least one element being related (directly or indirectly) to a_i or a_j are deleted. The new pair $a_j \succ a_i$ is put into S. The related elements are found by going through the sets S and building up iteratively a list, called *Error* in Figure II.15. Eventually, only disjoint preferences remain after the resolution of a contradiction. For example, if a decision maker assesses $a_2 \succ a_5$, $a_1 \succ a_3$, $a_3 \succ a_4$, the model concludes $a_1 \succ a_4$. If, in the next direct assessment, the decision maker decides $a_4 \succ a_1$, which is a contradiction, the model retains $a_4 \succ a_1$ and $a_2 \succ a_5$. The resolution model in algorithmic form is given in Figure II.17.

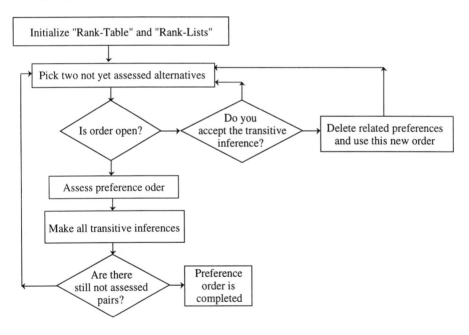

Figure II.17: Resolution model for interactive complete strong preference ordering.

It should be noted that the resolution model is based on the assumption that previously made direct assessments cannot be revised, while transitive inferences can be rejected at any time. Moreover, the decision maker is not supported in a strategy that minimizes the number of steps required to complete the ranking of all five items. For more on this approach of preference ordering see [Beroggi, 1997 a].

6. Summary

Problems are analyzed with a model which is an abstraction of the real world (system) under investigation. The analytic modeling process in problem solving is an iterative process, consisting of three steps: structuring, formalizing, and resolving. For each of these steps a model is constructed: structural model, formal model, and resolution model.

The structural or conceptual model illustrates the relations between the elements: decision makers, actions, criteria, scenarios, content goals, and structural goals. Each element has a unique icon and symbol. An arrow from one element to another element means that the evaluation of the successor element must be done in terms of the predecessor element.

The formal model states how to evaluate the alternatives, elicit and aggregate preferences; it also shows the structure of the solutions and the aspirations. Different modeling paradigms are used to formalize a problem. Aspects from the formal and resolution model are often visualized as part of the structural model to facilitate communications.

The resolution model defines the procedure to resolve the problem based on the formalization. It states algorithmically how to search for the 'best' solution. An algorithm consists of an initialization step, an iteration step, and a stop condition. Algorithms may be informed or heuristic. An optimization algorithm has the characteristic feature that at each iteration step it improves the temporary solution. Algorithms should provide an efficient search strategy. A resolution map tells how to elicit preferences and how to aggregate them across criteria, decision makers, scenarios, and actions.

Interactive methods for problem solving are implemented in decision support systems. They guide the decision maker through the problem solving process and make consistency checks. A decision support system helps the decision maker reduce complexity by relieving him/her from tedious tasks. The decision maker's task is to focus on the assessment process, while the system takes care of the computational aspects.

7. Problems

1. Assume you are in charge of organizing a meeting about policy planning for sustainable development. Ten experts representing the range of decision makers involved in this issue must be selected.

a) Discuss several structural, formal, and resolution models of such a meeting.

b) How would you verify, calibrate, and validate these models?

c) Discuss why laws and regulations can be seen as formal models, and law enforcement as a resolution model.

2. Given is the road network below and assume you want to travel from city A to H.

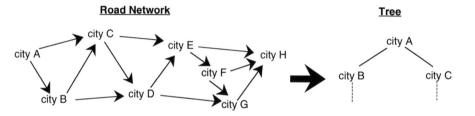

a) Discuss the difference between implicit and explicit representation of alternatives for this problem.

b) Illustrate all explicit roads from city A to city H in a tree that starts in city A.

c) Define an algorithmic resolution model to find the shortest road from city A to city H, assuming that the length of each segment between any two cities is known.

3. The transportation ministry (TM) and an environmental group (EG) must decide on a new highway between two industrial areas. EG uses environmental impacts (E) and safety (S) as the most relevant meta criteria, while TM considers financial implications (F) and safety (S) as the relevant meta criteria. Discuss a hierarchical decomposition of these meta criteria from the points of view of EG and TM, and identify possible evaluation criteria. Discuss the critical aspects for deriving such hierarchical decompositions.

4. What are the differences in the structural models a) and b) in Models I and II?

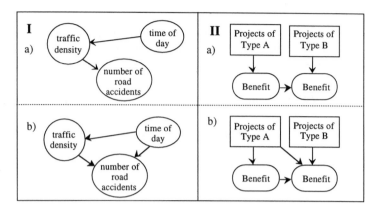

5. Discuss the example in Section 3.2. Change the dominance and preference functions such that the descriptive and normative approaches produce the same preference order.

6. For the example in Section 3.3 select some values for the variables x_j and compute the resulting risk reduction. Make sure the solution is feasible. Add additional constraints for the criteria, such as energy use, benefit, etc. Find a feasible solution by inspection.

7-10. Draw and discuss the structural models for the following decision problems:

- Multi-Criteria Decision Making (MCDM): The costs of five infrastructure projects will be assessed. The project with the lowest costs will be chosen.

- Group Decision Making (GDM): Four friends rank six restaurants according to their preferences for lunch and dinner. The two restaurants with the highest preferences will be chosen.

- Decision Making Under Uncertainty (DMU): The costs of eight cars will be assessed. The car with the lowest costs will be chosen. The costs depend on the demand and the production costs. Both demand and production costs are uncertain and depend on the economic growth which is also uncertain.

- Constraint-Optimization (CO): Fifty square meters fabric are available to make pants and shirts. The question is how many pants and shirts should be made. The constraining factors are the total work time, the total amount of fabric, and the goal is to maximize the total gain.

• Dynamic Programming (DP): A new employee must be chosen from three candidates. Depending on who gets the job, one out of two projects will be accepted. The chosen candidate and the project make up the success for the company; the objective is to maximize the success of the company.

• Conflict Analysis (CA): Sue and Joe must choose today one out of two classes (A and B), but unfortunately they cannot communicate to each other to discuss their choice. Their satisfaction depends on which class they choose and whether they end up in the same class. Both want to maximize their satisfaction.

CHAPTER III

DESCRIPTIVE ASSESSMENT - CRITERIA AND WEIGHTS

1. Relative Intensities and Weights

In Chapter II we introduced the descriptive approach to preference elicitation which is based on paired comparisons of the items to be ordered from most to least preferred. These items may be alternatives, criteria, decision makers, goals, or scenarios (to assess their likelihood). In the first part of this chapter we will address descriptive approaches for determining the preferences for criteria, called **weights**. These descriptive preference elicitation concepts will also be used in subsequent chapters to determine the preferences or importances of alternatives (Chapter IV) and of decision makers (Chapter IX).

1.1 Consistent Assessment

A weight, w_i, expresses the relative importance of criterion c_i over all other criteria, where for m criteria:

$$\sum_{i=1}^{m} w_i = 1.$$

There are different ways to determine weights, depending on how we can express the relative importances of the criteria. The most fundamental way to express importances of criteria is to determine the preference order of the criteria. For example, we might be able to say that risks to humans (c_1) are more important than costs (c_2), followed by energy use (c_3) and time (c_4). Thus, we have $c_1 \succ c_2 \succ c_3 \succ c_4$ and therefore $w_1 > w_2 > w_3 > w_4$, where $\Sigma_i w_i = 1$. One way to arrive at this preference order is through the interactive strong preference ordering model discussed in Chapter II, Section 5. There we saw that we must do at least $m-1$ direct assessments to get the complete preference order for m items. Thus, the maximum number of paired assessments that we have to do with m criteria is the sum of direct and transitively inferred assessments:

$$\sum_{i=1}^{m-1} i = \binom{m}{2} = \frac{m^2 - m}{2}.$$

We would like now to derive numeric **preference intensities**, k_i, and **weights**, w_i, for each criterion c_i, from the given the preference order of the criteria. Rietveld and Ouwersloot [1992] propose the following formula:

$$w_i = \frac{k_i}{\sum\limits_{j=1}^{m} k_j}, \text{ where } k_i = \sum_{r=i}^{m} \frac{1}{r}.$$

The preference intensity of the most important criterion (risks to human, c_1) is $k_1 = 1 + 1/2 + 1/3 + 1/4 = 25/12$, the preference intensity of the second most preferred criterion is $k_2 = 1/2 + 1/3 + 1/4 = 13/12$, and so on.

The idea behind this formula is the following. Assume a person starts a one-person company. After each year, a new person is hired. Seniority or importance of the persons depends on the number of years they are with the company. At the end of each year each person must bring a cake to the anniversary and share it with the more senior persons. Thus, at the first anniversary, the founder eats one cake alone. At the second anniversary, the founder gets the own cake plus half of the cake of the new person, while the new person gets half of the own cake. At the third anniversary, the founder gets the own cake, plus half of the second person's cake, plus one third of the third person's cake; the second person gets half of its cake plus one third of the third person's cake, and the third person gets one third of its own cake. The above formula stands for the m-th anniversary, where m people are working at the company; their importance (i.e., k_i) is determined in terms of how much cake they get to eat. Table III.1 gives the preference intensities (k_i) and weights (w_i) for up to six criteria.

number of criteria: m	k_1 w_1	k_2 w_2	k_3 w_3	k_4 w_4	k_5 w_5	k_6 w_6	Σk_i Σw_i
2	3/2 .75	1/2 .25	- -	- -	- -	- -	4/2 1.00
3	11/6 .61	5/6 .28	2/6 .11	- -	- -	- -	18/6 1.00
4	25/12 .52	13/12 .27	7/12 .15	3/12 .06	- -	- -	48/12 1.00
5	137/60 .45	77/60 .26	47/60 .16	27/60 .09	12/60 .04	- -	300/60 1.00
6	147/60 .41	87/60 .24	57/60 .16	37/60 .10	22/60 .06	10/60 .03	360/60 1.00

Table III.1: Preference intensities and expected weights for uniformly distributed preference orders (values are rounded so that $\Sigma_i w_i = 1$).

This formula to compute the weights can be modified if ties are present. Let's assume that the second and third criteria are equally preferred. Then, we have the following preference order: $c_1 \succ (c_2 \sim c_3) \succ c_4$. In this case, we would multiply the preference intensity by the number of tied criteria, which for this example is a factor of two.

Thus, we have the following preference intensities: $k_1=11/6$, $k_2=k_3=5/6$, and $k_4=2/6$. We would then compute the resulting weights as: $w_1=0.48$, $w_2=w_3=0.22$, and $w_4=0.08$, with $\Sigma_i w_i = 1$.

If the preference decrease between two consecutive criteria is not constant, artificial criteria can be introduced. For example, assume that the preference decrease between the first and second criterion is three times as much as the preference decrease between the second and the third, and the third and the fourth criterion. In such a case, we would simply introduce two artificial criteria, z_1 and z_2. Thus, we get the following preference order: $c_1 \succ z_1 \succ z_2 \succ c_2 \succ c_3 \succ c_4$. Computing the preference intensities as before gives: $k_1=147/60$, $k_2=37/60$ $k_3=22/60$, and $k_4=10/60$. We would then compute the weights as: $w_1=0.68$, $w_2=0.17$, $w_3=0.10$, and $w_4=0.05$, with $\Sigma_i w_i = 1$.

A complete paired assessment of m elements can be quite work intensive for large values of m. In practical situations, however, more straightforward procedures are used to assess the weights of criteria. The simplest approach to determine the weights is to use a scale from 0 to 100 and to assign to each criterion a preference intensity $k_i \in [0,100]$, where a high value corresponds to a high importance for criterion c_i. These preference intensities are then normalized by dividing them by the sum of all preference intensities; this yields the weights, w_i:

$$
w_i = \frac{k_i}{\sum\limits_{j=1}^{m} k_j},
$$

with $\Sigma_i w_i = 1$. The problem with this approach is that the decision maker does not really compare all of the criteria to one another. Because the results are not based on paired comparisons, this approach is not really a descriptive approach.

Another simplified approach to determine weights for criteria is to distribute a fixed value, for example 100, among all the criteria. Then, we have $\Sigma_j k_j = 100$, and the weight of criterion c_i is $w_i = k_i/100$. This approach is closer to a descriptive approach because it forces the decision maker to consider all criteria simultaneously when assigning a preference intensity to one criterion. Assigning a preference intensity value k_i to criterion c_i is done by considering the importance of this criterion and, at the same time, by considering how much is left ($100-k_i$) to distribute among the remaining criteria. The decision maker would probably order the criteria first and then assign preference values to them.

Because the process of determining a preference order (as discussed in Chapter II) does not allow inconsistencies in the assessments (e.g., offending intransitivity), these types of approaches to assess weights are called **consistent**. In the following two sections we will discuss a descriptive approach based on a ratio scale where inconsistencies concerning both ordinal and intensity transitivity can occur.

1.2 Relative Importance and Ratio Scale

The idea of using a **ratio scale** to assess the relative importance between two criteria stems from the **analytic hierarchy process** (AHP), as introduced by Saaty [1980, 1996]. Assessing the relative importance of m criteria pairwise involves assessing the ratios of the (unknown) weights of the criteria c_i and c_j: $k_{ij} = w_i/w_j$. These assessments are then inserted into a quadratic matrix $K^{m \times m}$, where m is the number of criteria.

The ratio scale approach is based on the assumption that the decision maker can assess how many more times one criterion is preferred to (or more important than) another one. Although the practical application of this approach is subject to criticism, Harker and Vargas [1987] emphasize that the literature in psychology accepts the use of a ratio scale to measure the relative intensity of stimuli.

The scale to assess the relative importance of elements, as proposed by Saaty, ranges from 1/9 to 9. The interpretation of this scale is described in Table III.2 (intermediate values are used appropriately).

Relative Importance $k_{ij} = \dfrac{w_i}{w_j}$	Definition (w_i is the weight of c_i, and w_j the weight of c_j)
9	extreme importance of c_i over c_j
7	strong importance of c_i over c_j
5	essential importance of c_i over c_j
3	moderate importance of c_i over c_j
1	equal importance of c_i and c_j
1/3	moderate inferiority of c_i over c_j
1/5	essential inferiority of c_i over c_j
1/7	strong inferiority of c_i over c_j
1/9	extreme inferiority of c_i over c_j

Table III.2: Ratio scale for descriptive assessment as proposed by the AHP.

There are several reasons for using a scale from 1 to 9. Experiments conducted by Miller [1956] showed that an individual can not compare simultaneously more than seven (plus or minus two) elements without being confused. This result motivated Saaty to introduce 9 (=7+2) levels of importance for the comparison of elements. Although this scale seems to do a good job of capturing the preferences of human decision makers, Saaty [1980] acknowledges that it could be altered to suit specific needs. For example, Lootsma [1992] proposes a geometric scale (1/16, 1/4, 1, 4, 16) to express the importance of one element over another.

The assessment of the preference intensities through paired comparisons with a ratio scale is based on two **axioms**:

- **Axiom 1**: The preference intensities are **finite**: $k_{ij} = \frac{w_i}{w_j} < \infty$.

- **Axiom 2**: The preference intensities are **reciprocal-symmetric**: $k_{ij} = \frac{1}{k_{ji}}$.

The first axiom states that any two criteria are comparable, meaning that one of the two is preferred. Incomparability, as introduced in Chapter I, is assumed not to exist. From the second axiom it follows that if we assess the relative importances of m criteria in a matrix $\boldsymbol{K}$, where the entry $k_{ij}=w_i/w_j$ stands for the relative importance of criterion c_i over criterion c_j, then $\boldsymbol{K}$ is a positive reciprocal matrix.

As introduced in Chapter I, the matrix $\boldsymbol{K}$ might not be ordinal or intensity transitive. The following characteristics of the matrix $\boldsymbol{K}$ are important:

- The matrix $\boldsymbol{K}$ is **ordinal transitive** if for $k_{ij} \geq k_{ik}$ it follows that $k_{kj} \geq 1$.

- The matrix $\boldsymbol{K}^{m \times m}$ is **intensity transitive** if $k_{ij} \times k_{jl} = k_{il}$.

- An intensity transitive reciprocal-symmetric matrix $\boldsymbol{K}^{m \times m}$ is **consistent**.

Note that the AHP requires the matrix of relative intensities, $\boldsymbol{K}^{m \times m}$, to be reciprocal-symmetric but neither ordinal transitive nor intensity transitive. Below are examples of matrices $\boldsymbol{K}^{m \times m}$: $\boldsymbol{A}$ is reciprocal-symmetric and intensity transitive, and therefore consistent; $\boldsymbol{B}$ is reciprocal-symmetric, inconsistent, and ordinal (but not intensity) transitive; $\boldsymbol{C}$ is reciprocal-symmetric but neither intensity nor ordinal transitive; and $\boldsymbol{D}$ is not reciprocal-symmetric and therefore not acceptable.

consistent:

A	c_1	c_2	c_3
c_1	1	3	6
c_2	1/3	1	2
c_3	1/6	1/2	1

inconsistent but ordinal transitive:

B	c_1	c_2	c_3
c_1	1	2	7
c_2	1/2	1	3
c_3	1/7	1/3	1

inconsistent and not ordinal transitive:

C	c_1	c_2	c_3
c_1	1	5	2
c_2	1/5	1	3
c_3	1/2	1/3	1

not acceptable (not reciprocal-symm.):

D	c_1	c_2	c_3
c_1	1	5	2
c_2	1/7	1	3
c_3	1/3	1/2	1

Because we require descriptive assessments based on a ratio scale to be reciprocal-symmetric, a decision maker with m criteria needs to assess only the preference intensities k_{ij}, $i=1,...,m-1$, $j=i+1,...,m$, because $k_{ii}=1$, and $k_{ij}=1/k_{ji}$. The total number of assessments required by the decision maker with m criteria is then $(m^2-m)/2$. If the assessments of the relative intensities between criteria are **consistent** (i.e., intensity transitive) then the weights of the criteria, w_i, are computed as:

$$w_i = \frac{\sum\limits_{j=1}^{m} k_{ij}}{\sum\limits_{j=1}^{m}\sum\limits_{k=1}^{m} k_{kj}}, \quad \text{or} \quad \boldsymbol{w} = \begin{bmatrix} w_1 \\ : \\ w_m \end{bmatrix} = \frac{\boldsymbol{Ke}}{\boldsymbol{e}^T \boldsymbol{Ke}}, \quad \boldsymbol{e}^T = [1,\ldots,1],$$

where $\sum_{i=1}^{m} w_i = 1$. In words, the weight w_i of criterion c_i is the sum of the entries in row i of matrix $\boldsymbol{K}$, divided by the sum of all entries in matrix $\boldsymbol{K}$. However, if multiplicative (intensity) transitivity does not hold ($k_{ij} \times k_{jl} \neq k_{il}$), the assessments are inconsistent and an approach to resolve the inconsistencies must be employed.

1.3 Resolution of Inconsistencies

Inconsistencies may be resolved in one of several ways. One approach, fairly intuitive, is to introduce a distance measure (e.g., Euclidean metric) and to minimize the sum of the squared deviations between the assessed k_{ij}, and the true ratio of the weights, w_i/w_j (least squares method):

$$\text{min:} \quad \sum_{j=1}^{m}\sum_{i=1}^{m} \left(k_{ij} - \frac{w_i}{w_j} \right)^2.$$

The approach to resolve inconsistencies proposed in the AHP is based on the analogy of paths and their intensities. The relative intensity measures, k_{ij}, are seen as intensities on a path between the criteria c_i and c_j. Thus, the intensity matrix $\boldsymbol{K}$ corresponds to the intensity matrix $\boldsymbol{G_w}$ introduced in Chapter I. The matrix $\boldsymbol{K}$ is reciprocal-symmetric and complete but not necessarily ordinal or intensity transitive. It should be noted that the decision maker is allowed to be inconsistent in his/her assessment; that is, s/he can offend both multiplicative and ordinal transitivity.

A numerical example of a preference intensity matrix $\boldsymbol{K}^{3 \times 3}$ with the corresponding preference intensity graph is shown in Figure III.1. The value k_{ij} stands for the preference intensity for going from c_i to c_j. For example, $k_{12}=5$ means that it is "essentially nice" (see Table III.2) to go from c_1 to c_2, or, in terms of the criteria, that c_1 is "essentially more important" than c_2. The preference intensity for going from c_j to c_i is $k_{ji}=1/k_{ij}$.

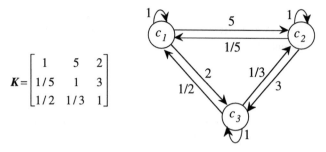

$$K = \begin{bmatrix} 1 & 5 & 2 \\ 1/5 & 1 & 3 \\ 1/2 & 1/3 & 1 \end{bmatrix}$$

Figure III.1: Intensity matrix with corresponding intensity graph.

A **path** on a graph is defined as a connected sequence of oriented links (arcs). For example, the links (c_1,c_2), (c_2,c_3), and (c_3,c_1) build a path that starts and ends in c_1. An **r-walk** is defined as a path consisting of r links. The **intensity** of an r-walk is defined as the product of the links' intensities. Thus, the intensity of the 3-walk c_1-c_2-c_3-c_1 is: $k_{12} \times k_{23} \times k_{31} = 5 \times 3 \times 0.5 = 7.5$. The 3-walk starting and ending in c_1 but going around in the other direction is: $k_{13} \times k_{32} \times k_{21} = 2 \times 1/3 \times 1/5 = 1/7.5$. The **total intensity of all r-walks** from node c_i to node c_j is the sum of the intensities of all r-walks. It is well known from graph theory that the entry $k_{ij}^{(r)}$ of the matrix $K^r = K \times ... \times K$ (r times) is the total intensity of all r-walks from node c_i to node c_j. To see this, let's compute the total intensity of all 3-walks starting in c_1 and ending in c_2 (which is the entry in row 1 and column 2 of the matrix K^3):

- c_1-c_1-c_1-c_2: $1 \times 1 \times 5$ = 5
- c_1-c_1-c_2-c_2: $1 \times 5 \times 1$ = 5
- c_1-c_2-c_2-c_2: $5 \times 1 \times 1$ = 5
- c_1-c_2-c_1-c_2: $5 \times 1/5 \times 5$ = 5
- c_1-c_2-c_3-c_2: $5 \times 3 \times 1/3$ = 5
- c_1-c_3-c_1-c_2: $2 \times 1/2 \times 5$ = 5
- c_1-c_1-c_3-c_2: $1 \times 2 \times 1/3$ = 2/3
- c_1-c_3-c_3-c_2: $2 \times 1 \times 1/3$ = 2/3
- c_1-c_3-c_2-c_2: $2 \times 1/3 \times 1$ = 2/3

$$\Sigma = 32$$

For the example of Figure III.1 we get the following matrices, K^r, $r = 1,2,3$:

$$K^1 = \begin{bmatrix} 1 & 5 & 2 \\ 1/5 & 1 & 3 \\ 0.5 & 1/3 & 1 \end{bmatrix}, \quad K^2 = \begin{bmatrix} 3 & 10\frac{2}{3} & 19 \\ 1.9 & 3 & 6.4 \\ 1\frac{1}{15} & 3\frac{1}{6} & 3 \end{bmatrix}, \quad K^3 = \begin{bmatrix} 14\frac{19}{30} & 32 & 57 \\ 5.7 & 14\frac{19}{30} & 19.2 \\ 3.2 & 9.5 & 14\frac{19}{30} \end{bmatrix}.$$

From the matrix K^3 we see that the sum of all 3-walks from c_1 to c_2 is 32 (entry in row 1 and column 2) as determined above. The total intensity of all 3-walks from c_2 to c_3 is 19.2 (second row and third column in K^3). The reader should confirm this result as was done for c_1 and c_2.

Saaty [1980] uses this analogy of total intensities to compute the **weights** of the criteria. The **dominance** of one criterion at the level of r-walks is the sum of its row elements divided by the sum of all elements in the matrix K^r. The dominance of one criterion at the level of 1-walks is the sum of its row elements divided by the sum of all elements in the matrix K.

The **total dominance** of a criterion over t r-walks, where $r=1,...,t$, is simply the average of the t dominance values. These total dominances, for $t \to \infty$, are what Saaty defines as the **weights** of the criteria. In matrix notation, the weights (and thus the total dominance) of the criteria, where w_i is the weight of criterion c_i, can be written as:

$$w = \begin{bmatrix} w_1 \\ : \\ w_m \end{bmatrix} = \lim_{t \to \infty} \frac{1}{t} \sum_{r=1}^{t} \frac{K^r e}{e^T K^r e} \quad e^T = [1,...,1].$$

Note that the numerator, $K^r e$, is the vector with the i-th element being the sum of the i-th row of the matrix K^r which corresponds to the total intensity of all r-walks for criterion c_i. The denominator, $e^T K^r e$, is the sum of all elements in the matrix K^r. The following relations simplify the numerical computation of the weights:

$$\lim_{t \to \infty} \frac{1}{t} \sum_{r=1}^{t} \frac{K^r e}{e^T K^r e} = \lim_{r \to \infty} \frac{K^r e}{e^T K^r e} = w.$$

It can be shown that the dominance (weight) of each criterion along all r-walks, as $t \to \infty$, is given by the solution of the **eigenvalue** problem:

$$Kw = \lambda_{max} w.$$

These two equations have the two following important implications. First, we can compute the weights of an inconsistent matrix K just as we would for a consistent matrix but with K^r ($r \to \infty$) instead of K. Second, we can compute the weights of the m criteria as the **eigenvector** of the maximum eigenvalue (λ_{max}). Numerical approximations with $r=5$ are often sufficient for $m \leq 5$. After w has been computed, λ_{max} can be computed from the relation $Kw = \lambda_{max} w$.

The computation of the weights of criteria whose relative preference intensities have been assessed on a ratio scale in a positive reciprocal-symmetric matrix K can be summarized as follows:

- If K is consistent (i.e., intensity transitive), then weight w_i of criterion c_i is the sum of the elements in row i divided by the sum of all elements in the matrix K.

- If K is inconsistent (i.e., not intensity transitive), then weight w_i of criterion c_i is the sum of the elements in row i divided by the sum of all elements in the matrix $\lim_{r \to \infty} K^r$.

Going back to the example of Figure III.1, for the inconsistent matrix K we get $w_1=0.61$, $w_2=0.24$, and $w_3=0.15$ which resulted from the matrix K^5. The corresponding maximum eigenvalue can be computed from the relation $Kw=\lambda_{max}w$ using any of the m equations:

$$\lambda_{max} = \frac{k_{i1}w_1 + ... + k_{im}w_m}{w_i}, \text{ for any } i=1,...,m.$$

For the above numeric example we get $\lambda_{max}=3.45$. The reader should use these values to verify $Kw = \lambda_{max}w$.

It should be noted that for a consistent matrix $K^{m \times m}$, we get $\lambda_{max}=m$ (order of the matrix). In the example above, for a consistent matrix, we would get $\lambda_{max}=3$. It can also be shown that $\lambda_{max} \geq m$.

Saaty introduces a **consistency index**, μ, which reflects the degree of consistency of the assessment:

$$\mu = \frac{\lambda_{max} - m}{m - 1}.$$

The difference between λ_{max} and m is used in the numerator because for a consistent matrix $K^{m \times m}$, λ_{max} is equal to m (the order of the matrix). Therefore, if the assessment by the decision maker is consistent, the consistency index is zero.

Table III.3 shows the average consistencies, $\bar{\mu}_m$, of 50 randomly generated reciprocal-symmetric matrices, for different orders of m [Harker, 1989]. The **consistency ratio**, CR, defined as $CR = \mu / \bar{\mu}_m$, is used to judge whether the consistency in the assessment should be accepted or not. As rule of thumb, CR should be 10 percent or less; otherwise, the assessment should be repeated, hoping that the consistency ratio decreases.

order of matrix: m	2	3	4	5	6	7	8	9	10
consistency: $\bar{\mu}_m$	0.00	0.58	0.90	1.12	1.24	1.32	1.41	1.45	1.49

Table III.3: Average measures of consistency, $\bar{\mu}_m$.

The matrix in Figure III.1 has a consistency ratio of 0.39, which is unacceptably high. One reason, however, for such a high value is that the matrix is not even ordinal transitive. If we make it ordinal transitive, by changing k_{12} from 5 to 2 and k_{13} from 2 to 7, the consistency ratio becomes 0.002. If we change in this new matrix k_{13} from 7 to 6, the assessment is even consistent.

However, ordinal transitivity does not guaranty a low CR value. For example, $k_{12}=1$, $k_{23}=9$, and $k_{13}=2$ are ordinal transitive, but we have $CR=30\%$, which is unacceptably high. To assure that ordinal (or even multiplicative) transitivity holds and that we get a small CR value, we would have to perform the assessments in a systematic way.

For example, to arrive at a perfectly consistent assessment (i.e., multiplicative transitivity holds) with as little effort as possible, we would assess m-1 pairs such that all other pairs can be derived transitively. For $m=5$ we could assess: k_{12}, k_{23}, k_{34}, and k_{45}. With these values, we could infer perfectly consistent assessments for all other k_{ij} values. For example, we could compute $k_{13}=k_{12}\times k_{23}$. These inferred assessments are proposed to the decision maker, hoping that s/he would not change the values too much, and thus keeping the CR value acceptably low.

There has been criticism in the literature regarding the eigenvector approach to resolve inconsistencies. Saaty and Vargas [1984] have addressed this issue by comparing the eigenvector method (EM) to the method of least squares (LSM) and to the method of logarithmic least squares (LLSM). The LSM computes the weight vector by minimizing the Euclidean metric:

$$\sum_{j=1}^{m}\sum_{i=1}^{m}\left(k_{ij}-\frac{w_i}{w_j}\right)^2.$$

The LLSM minimizes:

$$\sum_{j=1}^{m}\sum_{i=1}^{m}\left(\log k_{ij}-\log\left(\frac{w_i}{w_j}\right)\right)^2.$$

In all three cases we have a reciprocal intensity matrix, $K^{m\times m}$, with $k_{ij}=1/k_{ji}$. Saaty and Vargas discuss some interesting results [1984]. Obviously, if K is consistent, all three methods produce the same result. If K is ordinal transitive then the three methods preserve rank strongly. This means that if $k_{ih}\geq k_{jh}$ for all h, it follows that $w_i\geq w_j$.

More recently, Lootsma proposed alterations to the AHP with the system called REMBRANDT. He also proposed a logarithmic scale. A comparison of the AHP and REMBRANDT is discussed in Olson et al. [1995]. They conclude that the two approaches produce the same decision recommendations if the geometric mean is used for the aggregation of preferences in the AHP, while using the arithmetic mean produces different decision recommendations. A more in-depth discussion on the geometric and arithmetic mean follows in Chapter IX, Section 2.2 C.

2. Hierarchical Decomposition of Criteria

A complete paired comparison of all m criteria would be very work intensive. As mentioned earlier, it takes $(m^2-m)/2$ comparisons to do a complete paired assessment with m criteria. Therefore, it is more efficient to cluster the criteria into groups and to arrange the groups in a hierarchy, as proposed in the AHP. Then, paired comparisons are done within and across the groups. This approach can reduce the number of paired assessments significantly.

2.1 Structural Model

An example of a decomposition of the criteria is given in Figure III.2. The decision problem might be to choose the best landfill location in a community from a set of feasible locations (alternatives), where the goal is to minimize the negative impacts of the landfill. The criteria at the ends of the hierarchy are called the **root-criteria** and are used as the **evaluation-criteria** with which the alternatives are evaluated and the content goals defined (shaded in Figure III.2).

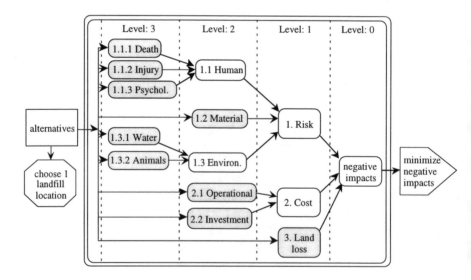

Figure III.2: Hierarchical decomposition (or clustering) of criteria.

The negative impacts are described with the three meta-criteria: risk, cost, and loss of land due to the landfill (Level 1). The third criterion, land loss, has not been

---

decomposed into sub-criteria. The criterion cost has been decomposed into operational and investment costs (Level 2). The criterion risk has been divided into risk to human, material, and environment (Level 2). The risk to human has been further decomposed into death, injury, and psychological risks (Level 3), and the environmental risk has also been further divided into risks to water and to animals (Level 3).

The root-criteria are 1.1.1, 1.1.2, 1.1.3, 1.2, 1.3.1, 1.3.2, 2.1, 2.2, and 3. What we are looking for are the weights of these nine evaluation-criteria, where the sum of the weights is one. To apply the paired assessment procedure in such a structure, an additional **axiom** must be introduced, saying that the preferences of elements are **independent** of the preference of elements at a lower level. If this independence cannot be assumed, Saaty [1980] says the system has **feedback**. He then suggests assessing the preferences in a **supermatrix**. We will address this issue in greater depth in Chapter IV, Section 3.2.

According to the structure of the criteria in Figure III.2, we can reduce the total number of paired assessments from $36=m(m-1)/2$ with $m=9$ to only 11 assessments. These are 3 at Level 1, 3+1 at Level 2, and 3+1 at Level 3. Thus, we could reduce the number of paired assessments by about 70 percent which is quite a significant reduction in effort, making descriptive approaches more appealing.

2.2 Formal Model

Pairwise comparisons of criteria are done for all clusters. In the example given in Figure III.2 we have 5 clusters, one at Level 1, two at Level 2, and also two at Level 3:

NI	1.	2.	3.	w
1.	1	3	4	.625
2.	1/3	1	2	.238
3.	1/4	1/2	1	.137

1	1.1	1.2	1.3	w
1.1	1	7	4	.705
1.2	1/7	1	1/3	.084
1.3	1/4	3	1	.211

1.1	1.1.1	1.1.2	1.1.3	w
1.1.1	1	7	9	.772
1.1.2	1/7	1	5	.174
1.1.3	1/9	1/5	1	.054

1.3	1.3.1	1.3.2	w
1.3.1	1	1/2	1/3
1.3.2	2	1	2/3

2	2.1	2.2	w
2.1	1	4	.80
2.2	1/4	1	.20

The weight of a criterion is computed by multiplying its relative weight with the relative weight of the parent element. For the example of Figure III.2, we get the following root weights (note that the weights of the root-criteria are in boxes and that they add up to one):

$$w_{1.1.1}^* = w_{1.1.1} w_{1.1} w_1 \qquad = 0.772 \times 0.705 \times 0.625 \quad \boxed{= \mathbf{0.340}}$$

$$w_{1.1.2}^* = w_{1.1.2} w_{1.1} w_1 \qquad = 0.174 \times 0.705 \times 0.625 \quad \boxed{= \mathbf{0.077}}$$

$$w_{1.1.3}^* = w_{1.1.3} w_{1.1} w_1 \qquad = 0.054 \times 0.705 \times 0.625 \quad \boxed{= \underline{\mathbf{0.024}}} \; \Sigma^\daleth$$

$$= 0.441 \quad = w_{1.1} w_1$$

$$w_{1.2}^* = w_{1.2} w_1 \qquad = 0.084 \times 0.625 \qquad\qquad \boxed{= \mathbf{0.052}} = w_{1.2} w_1$$

$$w_{1.3.1}^* = w_{1.3.1} w_{1.3} w_1 \qquad = 1/3 \times 0.211 \times 0.625 \quad \boxed{= \mathbf{0.044}} \quad \downarrow$$

$$w_{1.3.2}^* = w_{1.3.2} w_{1.3} w_1 \qquad = 2/3 \times 0.211 \times 0.625 \quad \boxed{= \underline{\mathbf{0.088}}} \; \Sigma^\daleth \;\; \downarrow$$

$$= \underline{0.132} \quad = w_{1.3} w_1 \quad \Sigma^\daleth$$

$$= 0.625 \quad = w_1$$

$$w_{2.1}^* = w_{2.1} w_2 \qquad = 0.80 \times 0.238 \qquad\qquad \boxed{= \mathbf{0.190}} \quad \downarrow$$

$$w_{2.2}^* = w_{2.2} w_2 \qquad = 0.2 \times 0.238 \qquad\qquad \boxed{= \underline{\mathbf{0.048}}} \quad \downarrow$$

$$\mathrel{\llcorner}\Sigma \qquad\qquad \rightarrow \qquad = 0.238 \quad = w_2$$

$$w_3^* = w_3 \qquad\qquad\qquad\qquad\qquad\qquad\qquad \boxed{= \underline{\mathbf{0.137}}} = w_3$$

$$\Sigma = \; 1.000$$

In Chapter IV we will show that the principle of paired assessments with a ratio scale can also be used to assess the relative preferences of alternatives, and in Chapter IX the importances of decision makers. There we will discuss the problems that can occur if the assumption of preferential independence across the levels does not hold. A potential shortcoming is that preference orders of a set of alternatives (or criteria) could get reversed if new alternatives (or criteria) are added. We will discuss ways to remedy this shortcoming, making the descriptive approach of preference assessment very valuable for practical purposes.

3. Aggregation of Criteria

3.1 Differentiation Power of Criteria

Criteria are used to evaluate the actions of a decision problem. There are two major aspects for choosing a specific criterion:

- The criterion is an important point of view
- The criterion differentiates the alternatives

For example, if cost is considered to be a very important criterion, but all alternatives turn out to have the same costs, then it is not a helpful criterion to rank the alternatives because it has no differentiation power. When choosing a criterion, we need to

---

consider not only whether it is important, but also whether it helps in differentiating the alternatives. For example, if the travel distances are evaluated in kilometers, nothing can be gained by evaluating them also in miles. If there is a (linear) relation between different criteria we say the criteria **correlate** with one another. If more than two criteria have a linear association, we speak of **multicollinearity**.

Figure III.3 shows such a situation. The normalized preference values of nine alternatives $\{a_1,...,a_9\}$ which have been evaluated with three criteria, c_1, c_2, and c_3, are represented in a three-dimensional graph with the axis n_1, n_2, and n_3.

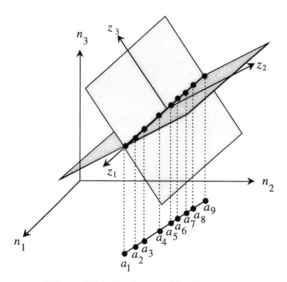

Figure III.3: Perfect multicollinearity.

The nine alternatives seem to lie exactly on a straight line which is a 1-dimensional sub-space of the 3-dimensional normalized preference space. This means that we have perfect multicollinearity and that instead of three criteria we would only need one criterion to visualize the same information about the differences among the alternatives. To reduce the three dimensions to one dimension we must perform a **rotation** of the original coordinate system such that one of the original coordinate axes is parallel to the straight line through the nine points. Because the remaining two coordinate axes do not give any additional information about the differences of the nine actions, we can omit them. Thus, dimension reduction consists of:

- rotation of the original coordinate system,
- omission of new dimensions with small differentiation power, and
- projection of the actions into the resulting plane.

The reduction of criteria calls for caution in two regards. First, we must draw careful conclusions about the original criteria. The fact that a reduction of the dimensions could be done by pure rotation means that there is a linear association between the dimensions. In such a case we say that there is **collinearity** between two dimensions or **multicollinearity** between the multiple criteria. How would we interpret this multicollinearity? There are two reasons for multicollinearity [Glantz and Slinker, 1990]. Assume the three criteria of Figure III.3 refer to the length of the alternatives in meters (c_1), in miles (c_2), and in yards (c_3). It is obvious that there are perfect linear relations among the three criteria. This cause of multicollinearity is called **structural multicollinearity**. It is simply a matter of the choice of the criteria. In other words, structural multicollinearity refers to the structural model, and it can be avoided by choosing only criteria that reflect conceptually different points of view.

Another possible reason why the nine alternatives in Figure III.3 lie on a straight line is a coincidence of the data. For example, assume three construction projects are evaluated with the three criteria costs, risks, and benefits, and that by some coincidence the numeric values are such that the risks are directly proportional to the costs and the benefits. We would then see perfect multicollinearity between the criteria costs, risks, and benefits. However, could we conclude that for any other construction project, costs, risks, and benefits have a linear relation? Certainly not! What we have detected here is **data-based multicollinearity**, which is due to the evaluation values of our alternatives and which cannot necessarily be generalized to all construction projects.

The second aspect where care must be taken when reducing the number of dimensions is the interpretation of the new dimensions in terms of the original criteria. For example, if the five original criteria are risk to life, risk to the environment, risk to material values, investment costs, and operational costs, and we can reduce these five criteria to two dimensions, then it would be meaningful if we could interpret these two dimensions as risk and cost dimensions.

It is often possible to interpret the first two new dimensions in terms of the original criteria, while it is more difficult to find meaningful interpretations for the third or higher new dimensions. More than three new dimensions are not sought because the most important purpose of dimension reduction is to visualize the alternatives in two or three dimensions for exploratory evaluation.

Figure III.4 shows the structural model of normalization and criteria reduction. The evaluation matrix E has been transformed into the normalized matrix N, and will be possibly weighted, N_w. A smaller set of new criteria is sought, if possible three or less, without losing too much of the differentiation power of the original criteria. Ideally, the new dimensions can be interpreted in terms of the original criteria.

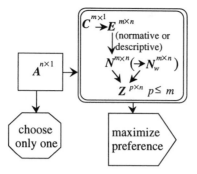

Figure III.4: Normalization of evaluation matrix and reduction of number of criteria.

3.2 Computational Aspects

Let N be the normalized $m \times n$ evaluation or preference matrix (several descriptive methods to normalize the evaluation matrix E will be discussed in Chapter IV; normative approaches will be discussed in Chapters V and VII),

$$
N^{m \times n} = \begin{bmatrix} n_{11} \cdots \cdots n_{1n} \\ \vdots \qquad \ddots \qquad \vdots \\ n_{m1} \cdots \cdots n_{mn} \end{bmatrix} .
$$

A measure of differentiation power is the unbiased **sample variance-covariance matrix** S, which is defined as follows, where s_{ik} is the **sample covariance** of criteria c_i and c_k, and s_{ii} is the **sample variance** of criterion c_i:

$$
S^{m \times m} = \begin{bmatrix} s_{11} \cdots \cdots s_{1m} \\ \vdots \qquad \ddots \qquad \vdots \\ s_{m1} \cdots \cdots s_{mm} \end{bmatrix} = \left\{ s_{ik} = \frac{1}{n-1} \sum_{j=1}^{n} (n_{ij} - \bar{n}_i)(n_{kj} - \bar{n}_k) \right\} .
$$

In matrix notation:

$$
S^{m \times m} = \frac{1}{n-1} \sum_{j=1}^{n} (n_j - \bar{n})(n_j - \bar{n})^T .
$$

The measure of differentiation power for a single criterion is the sample variance which is zero if all values are identical. In Figure III.3, the sample variance is maximal for criterion z_1 if its coordinate axis is parallel to the straight line through the nine alternatives. In this case, the variances for the other two criteria z_2 and z_3 are zero because their coordinate axes are orthogonal to that straight line. The sample covariance may be interpreted as a measure of the simultaneous differentiation power of two criteria.

An approach to reduce the number of criteria without losing too much of the differentiation power of the original criteria is the **principal components technique**. In mathematical terms, the original coordinate system (n_1, n_2, n_3) is rotated such that the three (still orthogonal) coordinate axes distribute as much as possible of the total differentiation power to a small number of principal components (new dimensions). Either before or after this rotation, a translation is performed which corresponds to a centralization of the alternatives around their means. Then, the n alternatives are projected into a plane which is defined by the most differentiating dimensions p ($<$ m). If this plane is of dimension $p \leq 3$, we can visualize the n actions. Figure III.5 shows an example of projecting a 3-dimensional evaluation into a 2-dimensional plane, also called a **bi-plot**.

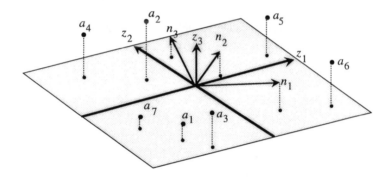

Figure III.5: Rotation and translation of coordinate system (n_1, n_2, n_3) into (z_1, z_2, z_3), and projection into the plane (z_1, z_2) which contains most of the information.

It can be shown [Morrison, 1976, p. 270] that with S being the unbiased sample variance-covariance matrix with eigenvalues $\lambda_1 \geq ... \geq \lambda_m \geq 0$, the vector l_i, belonging to the i-th principal component (new dimension) is the **eigenvector** of the i-th largest eigenvalue, λ_i, of S. If $\lambda_j \neq \lambda_k$ then the vectors l_j and l_k, belonging to the j-th and k-th principal components are automatically orthogonal; if $\lambda_j = \lambda_k$ then the components can be chosen to be orthogonal (there is an infinite amount of such orthogonal vectors; see Figure III.3). The result of the **rotation** gives therefore the i-th principal component as:

$$z_i = \hat{e}_i^T n = \hat{e}_{i1} n_1 + ... + \hat{e}_{im} n_m,$$

where $i = 1, ..., m$, $\hat{e}_i$ is the **eigenvector** that belongs to the **eigenvalue** λ_i of the matrix S, and $\lambda_1 \geq ... \geq \lambda_m \geq 0$. Moreover, it can also be shown that these orthogonal principal components are uncorrelated.

The **projection** of the actions in the plane (defined by the new dimensions) can be done such that the origin of the coordinate system coincides with the center of gravity (mean values) of the actions. This corresponds to a **translation** of the coordinate system which has no impact on the sample variance and therefore also not on the differentiation power. In fact, the translation could also be done prior to the rotation of the coordinate system. The values of the actions in the new m-dimensional coordinate system are computed as follows:

$$z_1 = \hat{e}_{11}(n_1 - \bar{n}_1) + \ldots + \hat{e}_{1m}(n_m - \bar{n}_m)$$

$$\vdots$$

$$z_i = \hat{e}_{i1}(n_1 - \bar{n}_i) + \ldots + \hat{e}_{im}(n_m - \bar{n}_m)$$

$$\vdots$$

$$z_m = \hat{e}_{m1}(n_1 - \bar{n}_1) + \ldots + \hat{e}_{mm}(n_m - \bar{n}_m)$$

If we omit all but two new dimensions, only the first two equations are used. To decide how many of the new dimensions should be kept, we must know how much of the total differentiation power they explain. The sample variance of the i-th principal component (new dimension) is λ_i and the total system variance is:

$$\lambda_1 + \ldots + \lambda_m = tr(S) = s_{11} + \ldots + s_{mm} = \sum_{i=1}^{m} Var(n_i) = \sum_{i=1}^{m} Var(z_i),$$

where $tr(S)$ is the sum of the diagonal elements of S. This means that the total system sample variance after the rotation (and translation) is the same as the total system sample variance before these transformations. The proportion of the total variance (differentiation power) explained by the i-th principal component is:

$$\frac{\lambda_i}{\lambda_1 + \ldots + \lambda_m}.$$

This means that we can compute the amount of the total sample variance that is explained by, let's say, the first two principal components as:

$$\frac{\lambda_1 + \lambda_2}{tr(S)}.$$

The difference between the total sample variance and the sum of the largest two eigenvalues, $tr(S)-(\lambda_1+\lambda_2)$, is the loss of information due to the reduction of the m dimensions to two dimensions. As a rule of thumb, we would like to reduce the dimensions to one or two, such that the amount of the total sample variance that is explained by these new dimensions is at least 70 percent of the total sample variance. Thus, we can project all actions (as well as the original m criteria) into the plane with these one or two dimensions, as illustrated in Figure III.5.

How would we go about computing numerically the principal components, $\hat{e}_1$, $\hat{e}_2$, ...? We discussed in Section 1.3 how to compute the largest eigenvalue, λ_1, and its corresponding eigenvector, w_1, of a matrix S. We used the relations:

$$Sw_1 = \lambda_1 w_1 \text{ and } \frac{S^r e}{e^T S^r e} \to w_1 \text{ as } r \to \infty.$$

With these relations we can compute the eigenvector, $w_1 \equiv \hat{e}_1$. The value in row i of w_1 is the sum of the elements in row i of the matrix S^r ($r \to \infty$) divided by the sum all elements of the matrix S^r. The corresponding eigenvalue λ_1 is computed from the equation $Sw_1 = \lambda_1 w_1$. The second principal component, $\hat{e}_2 \equiv w_2$, and corresponding eigenvalue, λ_2, are computed like the first ones, but with the matrix S_2 ($S \equiv S_1$):

$$S_2 = S_1 - \lambda_1 \hat{e}_1 \hat{e}_1^T.$$

If $(\lambda_1 - \lambda_2)/tr(S) < 0.8$, this procedure of computing principal components could be continued to for $\hat{e}_3$, $\hat{e}_4$, etc. However, it should be kept in mind that this numerical method of computing principal components is an approximation and it assumes that certain values converge. For some numerical values, this method might not be appropriate and more sophisticated approaches to compute eigenvectors should be used.

3.3 An Example of Dimension Reduction and Interpretation

Let's look at a numerical example to illustrate the discussions about dimension reduction. A decision maker must rank five potential landfill sites. The evaluation criteria used for this example are risk, zone, cost, landuse, and profit. The evaluation matrix ($E^{5 \times 5}$) shows the values of the alternatives; the '+' and '-' next to a criterion indicate whether higher values are preferred to lower values (+), or lower values to higher values (-).

$E^{5 \times 5}$:

	a_1	a_2	a_3	a_4	a_5
risk [10^{-6}] (-)	0.3	0.9	0.6	0.2	0.7
zone [class] (+)	II	I	II	III	II
cost [M$] (-)	15	17	20	8	10
landuse [km^2] (-)	30	42	18	16	47
profit [value] (+)	l	m	l	h	m

$N^{5 \times 5}$:

a_1	a_2	a_3	a_4	a_5
0.857	0.000	0.429	1.000	0.286
0.500	0.000	0.500	1.000	0.500
0.417	0.250	0.000	1.000	0.833
0.548	0.161	0.935	1.000	0.000
0.000	0.500	0.000	1.000	0.500

$N_w^{5 \times 5}$:

a_1	a_2	a_3	a_4	a_5
0.086	0.000	0.043	0.100	0.029
0.100	0.000	0.100	0.200	0.100
0.125	0.075	0.000	0.300	0.250
0.055	0.016	0.094	0.100	0.000
0.000	0.150	0.000	0.300	0.150

$S^{5 \times 5}$: ($\times 10^{-3}$)

1.693	2.500	2.322	1.277	0.857
2.500	5.000	5.625	2.098	3.750
2.322	5.625	15.306	-0.399	12.184
1.277	2.098	-0.399	2.006	0.172
0.857	3.750	12.184	0.172	15.750

Table III.4: Matrices for landfill location example.

The weight vector is: $w=[0.1,0.2,0.3,0.1,0.3]$. The evaluation matrix has been normalized ($N^{5\times5}$) and weighted, to yield $N_w^{5\times5}$, from which the sample variance-covariance matrix ($S^{5\times5}$) was computed (see Table III.4).

With this normalized and weighted evaluation matrix, N_w, a principal components analysis has been performed. The first eigenvalue is $\lambda_1=0.030$ and its corresponding eigenvector is $\hat{e}_1 =[0.102,0.268,0.684,0.019,0.671]^T$.

The second eigenvalue is $\lambda_2=0.006$ and its corresponding eigenvector is $\hat{e}_2=[0.435,0.652,0.109,0.415,-0.450]^T$. The amount of the total variance explained by these first two principal components is 91%; that is, only 9% of the differentiation power of the original five criteria is lost if we replace them by this two new dimensions. Therefore, we can use these first two components to visualize the differences among the alternatives (Figure III.6). The transformation equations to compute the new coordinates (centered around the mean) are:

• $z_1=0.102(n_1-0.051)+0.268(n_2-0.1)+0.684(n_3-0.15)+0.019(n_4-0.053)+0.671(n_5-0.12)$

• $z_2=0.435(n_1-0.051)+0.652(n_2-0.1)+0.109(n_3-0.15)+0.415(n_4-0.053)-0.450(n_5-0.12)$

Applying these two equations to all five 5-dimensional alternatives of the normalized and weighted matrix $N_w^{5\times5}$ leads to the graph of Figure III.6. The unit lengths of each of the five original criteria are also projected into the plane. The weight vector, which in the original system has been 'stretched' to unit length, is also projected into the plane. The projected coordinates of the alternatives have been multiplied by a factor 5 in comparison to the criteria and weight projections.

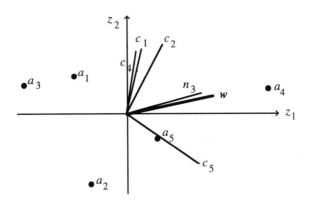

Figure III.6: Projections into plane of first two principal components.

Projecting the m-dimensional alternatives into two dimensions makes it possible to interpret the data and select the best alternative visually. The use of the principal components technique to analyze decision options has been proposed by the GAIA method (see also Chapter IV), for which the following interpretations of the two-dimensional graph are discussed [Brans and Mareschal, 1991]:

- **Differentiation power of criteria**: Longer projections of the original criteria (dimensions) in the new plane reflect higher differentiation power (higher variance s_{ii}). For our example, n_2 (5.0), n_3 (15.306), and n_5 (15.75) have the longest projections and seem thus to have the highest differentiation power of the criteria.

- **Similarity of criteria**: From S we see that s_{35} is large (12.184), which means that n_3 and n_5 describe almost the same preferences, and so their projections point in similar directions.

- **Independence of criteria**: From S we see that the covariance between n_4 and n_5 (s_{45}=0.172) is very small. This means that the two criteria are uncorrelated and describe mostly independent preference relations. In the graph this means that the projected axes are nearly orthogonal.

- **Conflict between criteria**: If two criteria have a large negative variance it means that they are conflicting, and the projections of these criteria point in opposite directions.

- **Decision power**: If the projected unit-length weight vector is long, the graph has strong decision power. The best alternatives lie in the direction of the projected weight vector. In fact, alternative a_4 is the dominant alternative prior to the normalization and, after the weighted normalization, it lies pretty much in the direction of the weight vector.

Brans and Mareschal [1991] note that in all real-world problems they have treated, the amount of the total variance that is explained by the first two principal components has always been larger than 60 percent and in most cases larger than 80 percent, even for 20 or more criteria. An important aspect in principal components analysis is the subjective interpretation of the new dimensions (criteria).

As mentioned earlier, these new criteria require an interpretation in terms of the original criteria. The components of the first eigenvector are all positive and of the same order of magnitude. We may therefore conclude that the first principal component appears to be essentially a weighted average of all criteria. The components of the second eigenvector are also of the same order of magnitude. The fifth component, however, is negative, leading us to conclude that the second principal component contrasts returns (profit) and investments (risks, zone, costs, and landuse). Looking at Figure III.6 we can identify four clusters with the following relations: $a_4 \succ a_5 \succ (a_1 \sim a_2) \succ a_3$ (see Problem 7 in Chapter IV).

4. Summary and Further Readings

Weights express the importance of different criteria for a descriptive preference elicitation approach, where the sum of all weights is one. A descriptive assessment of weights consists of a paired assessment of the relative preferences between two elements (criteria). The descriptive approach can lead to inconsistencies in the assessments.

A ratio-scale assessment which is reciprocal-symmetric and intensity-transitive is called consistent. The analytic hierarchy process (AHP) assumes that a human decision maker complies with reciprocal symmetry but not necessarily with intensity transitivity. The relative preference intensity of one element over another is expressed on a ratio scale, where 1 reflects equal preference and 9 very strong dominance.

The resolution of inconsistencies can be done with different approaches. However, the eigenvalue approach seems to be the soundest approach conceptually. The absolute weight vector can be obtained as the normalized row-sum vector of the assessment matrix raised to a high power.

To reduce the number of paired comparisons, a conceptual aggregation of criteria into hierarchical clusters may be done as proposed for the AHP. Then, the relative importance of the criteria can be computed within and across the clusters. An overview of the AHP can be found in [Golden et al., 1989], and information on the more recent concept of the Analytic Network Process (ANP) can be found in [Saaty, 1996].

Criteria can be aggregated through data-based or structural multicollinearity, using the principal components approach. If the amount of the sample variance which is explained by the first two principal components is high, the m-dimensional actions can be projected into a two-dimensional plane. These two new dimensions require appropriate interpretation. An overview of the principal components technique can be found in [Morrison, 1976] and [Johnson and Wichern, 1998]. The GAIA method uses the concept of principal components analysis [Brans and Mareschal, 1991]. Other approaches to bi-plot analysis can be found in [Gower and Hand, 1996].

5. Problems

1. Identify five elements referring to, for example, classes you are taking this semester, apartments, cars, dinners, or beverages. With these five elements, determine the weights with the approaches discussed in this chapter. Discuss the differences in the results and in the approaches.

2. Which of the following matrices are: a) consistent, b) ordinal transitive, and c) reciprocal-symmetric? Determine the weights, the consistency ratios of the three

criteria, and the corresponding maximum eigenvalue. Verify the results with the equation: $Kw = \lambda_{max}w$.

A	c_1	c_2	c_3
c_1	1	5	6
c_2	1/3	1	2
c_3	1/9	1/2	1

B	c_1	c_2	c_3
c_1	1	2	6
c_2	1/2	1	3
c_3	1/6	1/3	1

C	c_1	c_2	c_3
c_1	1	5	2
c_2	1/5	1	3
c_3	1/2	1/3	1

D	c_1	c_2	c_3
c_1	1	3	5
c_2	1/3	1	2
c_3	1/5	1/2	1

3. How would you modify the interactive strong preference ordering algorithm discussed in Chapter II, Section 5, so that it can be used with a ratio scale?

4. Check if the matrix below is ordinal transitive (where the criteria are arranged in decreasing preference order); use the method of Finan and Hurley discussed in Section 1.3 to determine a 4-criteria ordinal transitive intensity matrix K.

$$K = \begin{bmatrix} 1 & 2 & 5 & 4 \\ 1/2 & 1 & 4 & 6 \\ 1/5 & 1/4 & 1 & 3 \\ 1/7 & 1/6 & 1/3 & 1 \end{bmatrix}$$

5. Compute the largest eigenvalue λ_{max} with the eigenvector w and the relations $Kw = \lambda_{max}w$. Do it with all three equations for the numeric example discussed in Section 1.3.

6. Discuss the hierarchical decomposition of criteria in the example given in Figure III.2. Introduce more decompositions so that new roots are created, and discuss how you would compute the weights of these new roots.

7. For the nine actions in Figure III.3, $\{a_1,...,a_9\}$, describe what the results of a principal component analysis would be; that is, to how many dimensions (criteria) could the 3-dimensional problem be reduced, what would be the translation and rotation of the original coordinate system, and how much of the sample variance would be described by the new dimensions?

8. Johnson and Wichern [1998, p. 482] analyzed the weekly rates of return for the five stocks Allied Chemical, DuPont, Union Carbide, Exxon, and Texaco, listed on the New York Stock Exchange for the period January 1975 through December 1976. The data were represented in the evaluation matrix, $E^{5 \times 100}$. The first two principal components and their corresponding eigenvalues are:

- $\lambda_1 = 2.857$, $\hat{e}_1 = [0.464, 0.457, 0.470, 0.421, 0.421]^T$
- $\lambda_2 = 0.809$, $\hat{e}_2 = [0.240, 0.509, 0.260, -0.526, -0.582]^T$

These first two principal components account for 73% of the total sample variance. The authors derived from these results the following interpretations: The first component might be called a *general stock-market component*, because it is roughly an equally weighted sum, or 'index,' of the five stocks. The second component might be called an *industry component*, because it represents a contrast between the chemical stocks (Allied Chemical, DuPont, and Union Carbide) and the oil stocks (Exxon and Texaco). Comment on these findings and discuss how the authors derived their conclusions.

9. Describe how you would compute numerically the second principal component, $\hat{e}_2$, when the first principal component, $\hat{e}_1$, is known.

10. Discuss the differences between structural and sample-based multicollinearity. How does structural multicollinearity relate to the decomposition of criteria? Change the evaluation values of some actions in the problem given in Table III.4 and discuss what you would expect as results.

CHAPTER IV

DESCRIPTIVE ASSESSMENT - ALTERNATIVES AND RANKING

1. Structural Model of Descriptive Approaches

1.1 Basic Concepts

In this chapter, we address such problems as the selection of the best suited construction project, the prioritization of a set of feasible hospital locations, the ranking of a set of transportation policies, and the identification of promising risk abatement strategies. All of these decision problems involve binary decision variables, where $x_j=1$ if alternative a_j is chosen and $x_j=0$ otherwise.

The solutions to such decision problems consist of a complete or incomplete **preference order** of the alternatives under consideration. The type of preference order that is sought for the decision problem (e.g., complete strong or weak ranking, grouping of alternatives by feasibility and efficiency, or pseudo order) is stated in the structural goals. However, the solution of a decision problem is determined not only by the structural goals, but also by the evaluation of the alternatives, the subjective preferences, and the content goals. A descriptive preference elicitation approach is based on paired preference comparisons of the alternatives for each criterion. Accordingly, the definition of the binary relations is of major consideration in a descriptive problem solving method. The definition of these relations is done as part of the formal model. With the formal model and an appropriate resolution model, it is possible to find the desired solution(s) to the decision problem.

Most of the discussion in Chapter III about the assessment of the preferences (weights) of criteria also applies when we assess preferences for alternatives in a pairwise fashion. For alternatives, however, paired comparisons are done in two regards: to assess the dominance, strength, or **concordance** of each alternative over the others, and to assess the inferiority, shortcomings or **discordance** of each alternative over the others. Therefore, descriptive assessments aim not only at a preference ranking but also at a dis-preference ranking. The combination of the two ranks, or their aggregation, leads to the overall preference order of the alternatives.

Dominance and inferiority can also be defined as rules, stating, for example, when two alternatives are not comparable. Consequently, descriptive assessments focus more on the process of preference elicitation and preference ranking and less on rational or behavioral aspects of human decision making, as opposed to the axiomatic normative preference elicitation theories (Chapters V and VII).

The idea of dominance and inferiority of alternatives dates back to the 18-th century when the Marquis de Condorcet introduced the **majority rule** to aggregate preference assessments of multiple decision makers (Chapter IX). The majority rule says to prefer one alternative over another if the number of voters favoring it (concordance) is larger than the number of voters favoring the other alternative. The concept of paired comparisons and the statistical relevance of preference orders was formally addressed in the 1930s by Sir Maurice Kendall (Chapter IX). He proposed a procedure to test the **statistical significance** of rank-orders for different criteria, multiple decision makers, or scenarios.

The paired comparison approach with concordance and discordance measures has been further extended at the qualitative level by Bernhard Roy with the ELECTRE methods (Section 2.3). Almost thirty years ago, he introduced the concept of **outranking**, where the paired preference comparisons of alternatives can lead to an incomplete pseudo order, in which some alternatives are neither preferred nor indifferent to each other - they simply remain incomparable. Around 20 years ago, Thomas Saaty introduced a new dimension to descriptive preference elicitation by proposing a **ratio scale** for the paired preference assessments of criteria (Chapter III) and of alternatives (Chapter IV). Around ten years ago, Jean-Pierre Brans introduced the PROMETHEE method which is based on the **outranking flow** (Chapter IV) and combined it with the bi-plot representation, called GAIA (Chapter III). Around the same time, Roubens [1982] introduced the ORESTE method. In recent years, several combinations and extensions of these methods have been proposed, such as the **conflict analysis method** [van Huylenbroeck, 1995].

The descriptive approach is often seen as the *European school* of decision analysis, while the *American school* assesses preferences of alternatives in form of normative preference functions that capture the decision maker's multicriteria preference profile (Chapter V and VII). The descriptive dominance (concordance) measure describes the preference of an alternative over the others, while the inferiority (discordance) measure captures the shortcomings of the alternative. The descriptive outranking approach, as originally introduced by Roy, explicitly separates the 'good' and the 'bad' of the alternatives (relative to each other). An alternative outranks another, if its good parts are superior to the good parts of the other, and, at the same time, if the negative parts are not too strong.

Roy [1974] defines the outranking relation as a binary relation, $\succ$, in A (set of feasible alternatives), $a_j \succ a_k$, such that there are enough arguments to decide that a_j is better than a_k (or at least as good: $\succsim$) while there is no essential reason to refuse that statement.

Three aspects of outranking are worth noting. First, outranking is a relative (descriptive) approach because the assessment is based on paired comparisons. If new alternatives are added or some deleted, care must be taken with the preference order of the original set of alternatives. Second, paired comparison deals with ordinal (numeric and linguistic) and categorical scales. This mixed measurement implies some subjective interpretations for the assessments. Third, paired comparisons of m alternatives are represented in $K_i^{n \times n}$ matrices (as introduced in Chapter III); however, there is neither a rational nor behavioral basis as far as the preference structure of the decision maker and the preference aggregation principles are concerned. Despite the transparency of the descriptive methods and the intuitively sound approach of paired assessments, most of the proposed approaches encounter both theoretical and practical criticisms. A major theoretical problem is the possibility of rank reversals when adding near replicas of alternatives (Section 3), and a crucial practical shortcoming is the large number of necessary preference assessments.

1.2 Preference Aggregation: Basic Principles

Figure IV.1 shows the general concept of the structural model of descriptive assessment. It can be recognized as a descriptive approach because the decision maker must determine weights for the criteria (as discussed in Chapter III); normative approaches use scaling constants instead of weights (Chapter V and VII). With the evaluation matrix, $E^{m \times n}$, a paired preference assessment matrix, $K_i^{n \times n}$, $i=1,...,m$, is determined for each of the m criteria. The entries of the matrix $K_i^{n \times n}$, $k_{jk|i}$, stand for the (non-weighted) preference intensities of alternative a_j over alternative a_k for criterion c_i.

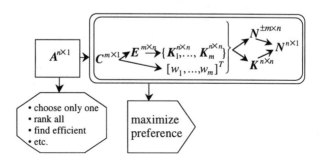

Figure IV.1: General concept of the structural model of descriptive approaches
(see Figure IV.2 for details).

The aggregation of these m quadratic preference matrices, together with the weights and discordance relations, leads either to a quadratic matrix, $K^{n \times n}$, or to two normalized evaluation matrices, $N^{+m \times n}$ (**dominance**) and $N^{-m \times n}$ (**inferiority**) which are of the same dimensions as the original evaluation matrix (see Figure IV.2).

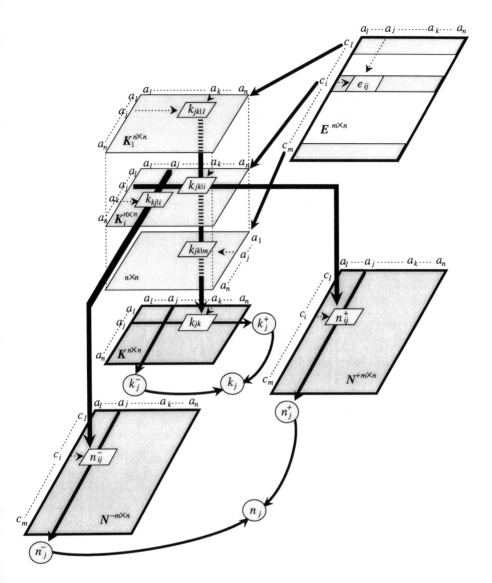

Figure IV.2: Preference aggregation across ($K^{n \times n}$) and within ($N^{+m \times n}$, $N^{-m \times n}$) criteria.

The weighted matrices are computed as $w_i \times K_i^{n \times n}$. However, we will not refer to K_w or N_w but simply to K and N (i.e., we omit the subscript 'w'). Moreover, most of the following discussion holds for both the $N^{+m \times n}$ and $N^{-m \times n}$ matrix; therefore, we will refer to them as $N^{\pm m \times n}$. The entry of the matrix $K^{n \times n}$, k_{jk}, stands for the weighted criteria-aggregated preference of alternative a_j over alternative a_k. The entry of the matrix $N^{\pm m \times n}$, $n_{ij}^{\pm}$, stands for the alternative-aggregated weighted preference ('+' for strength and '-' for weakness) of alternative a_j with respect to criterion c_i.

Both the $K^{n\times n}$ and the $N^{\pm m\times n}$ matrices (and this is one of the major differences from the normative approaches) are derived from paired comparisons of the alternatives under consideration for all criteria (see Figure IV.2). Therefore, the quadratic matrix, $K^{n\times n}$, is a result of a preference aggregation of the m matrices, $\{K_i^{n\times n}\}$, **across** the criteria, while the normalized matrix, $N^{\pm m\times n}$, is a result of a preference aggregation **within** the criteria (Figure IV.2).

It should also be noted that neither of the two matrices $N^{\pm m\times n}$ and $K^{n\times n}$ can be transformed into the other one. The two matrices contain more information than the original evaluation matrix, $E^{m\times n}$. This added information refers to the weights of the criteria and the subjective preferences over the alternatives for each criterion. Through the preference aggregation process (across and within the criteria), however, some of this information contained in the m matrices, $K_i^{n\times n}$, gets lost. Preference aggregation ($\circ$) within the matrices $K^{n\times n}$ and $N^{\pm m\times n}$ leads then to the overall preference of alternative a_j: $n_j = n_j^+ \circ n_j^-$, and $k_j = k_j^+ \circ k_j^-$.

2. Formal and Resolution Models

2.1 Ordinal and Interval Scales

For monotonically increasing preferences (e.g., benefits), the weighted preference value, $k_{jk|i}$, of alternative a_j over a_k with respect to criterion c_i, could, for example, be defined as:

$$k_{jk|i} = \begin{cases} w_i & \text{if } 5 < e_{ij} - e_{ik} \\ \dfrac{w_i}{2} & \text{if } 0 \le e_{ij} - e_{ik} \le 5 \\ 0 & \text{otherwise} \end{cases} .$$

This means that $k_{jk}=w_i$, if the benefit of alternative a_j more than 5 units larger than the one of alternative a_k, $k_{jk}=w_i/2$ if it is larger but not more than 5 units, and $k_{jk}=0$ if it is smaller than the one of a_k.

However, the preference of alternative a_j over alternative a_k is not necessarily a step function. It is reasonable to assume that there could be stronger or weaker preferences. The concept of **valued preferences** for one alternative **outranking** another was originally proposed by Roy and Bertier [1973]. It is based on the thought that preference (outranking) may be either present, not present, or a linear function between 1 (i.e., pure preference or outranking) and 0 (i.e., no preference or outranking).

Brans and Vincke [1985] propose different functions for the non-weighted preference (outranking) function, $k_{jk|i}$. Six types of the non-weighted preference function $k_{jk|i}$ are shown in Table IV.1. The larger the difference (Δ) between the evaluation values of two alternatives, the closer to 1 is $k_{jk|i}$. By difference (Δ) between two alternatives (or their measure of effectiveness) we do not necessarily mean numeric differences; differences may also be described qualitatively, as we will see in some of the following examples. The difference between two evaluations e_{ij} and e_{ik} for criterion c_i, $\Delta(e_{ij},e_{ik})$, is *positive* if $e_{ij} \succ e_{ik}$, *negative* if $e_{ik} \succ e_{ij}$, and *zero* if $e_{ij} \sim e_{ik}$. If we deal with numeric values, the difference can refer to a categorical, ordinal, interval, or ratio scale.

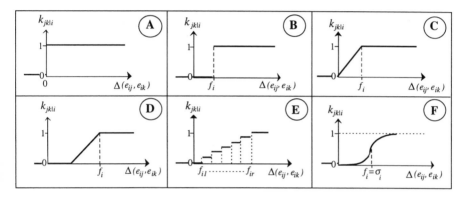

Table IV.1: Examples of non-weighted dominance functions $k_{jk|i}$.

For the PROMETHEE methods [Brans and Vincke, 1985], the threshold function, f_i, is a constant, while for the ELECTRE methods [Roy and Bertier, 1973] it is not. An indifference threshold function, g_i, and an inferiority (discordance) measure may also be introduced. The threshold function varies across the criteria but not within the criteria.

The non-weighted preference function of type A says that alternative a_j outranks (is more preferred than) alternative a_k for criterion c_i, if $e_{ij} \succ e_{ik}$. The non-weighted preference function of type B is a step function with the step occurring not at $e_{ij}=e_{ik}$ but at the value f_i, which must be defined by the decision maker. Type C shows a linear increase of the preference value between the minimum value (0) and the maximum (1). The preference values start growing at $\Delta(e_{ij},e_{ik})=0$, and the slope depends on the threshold value f_i. Type D of the non-weighted preference function is similar to the one used in the ELECTRE III method [Roy, 1978]. Type E is a multi-step function. This step function is conceptually similar to the approach used in the AHP. There, the "difference," $\Delta(e_{ij},e_{jk})$, is not an interval measure but a ratio measure: $k_{jk|i} \in \{1/9,1/8,...,8,9\}$. Type F corresponds to the normal distribution function with threshold value $f_i=\sigma_i$.

The type of preference function has an impact on the rank-order of the alternatives. For example, if we choose type B, we may get a pseudo order because transitivity for the indifference relation ($\sim$) might not hold anymore. Consider the following three alternatives $\{a_1, a_2, a_3\}$ which, for criteria c_i, have the following numeric evaluation values (gains): $e_{i1}=4.0$, $e_{i2}=4.8$, and $e_{i3}=5.7$. If we choose $f_i=1$, we get the following preference relations: $a_1 \sim a_2$, $a_2 \sim a_3$ but $a_3 \succ a_1$. Thus, the preference order is not transitive with regard to the indifference relation.

2.2 Aggregation of Preference Values

Different preference aggregation models that can be defined for the aggregation process illustrated in Figure IV.2. For example, the PROMETHEE method [Brans and Vincke, 1985] computes the overall weighted dominance measure k_j^+ and inferiority measure k_j^- of alternative a_j as:

$$k_j^+ = \frac{1}{n-1} \sum_{k=1}^{n} k_{jk}, \text{ where } k_{jk} = \sum_{i=1}^{m} w_i k_{jk|i} \, ,$$

$$k_j^- = \frac{1}{n-1} \sum_{k=1}^{n} k_{kj}, \text{ where } k_{kj} = \sum_{i=1}^{m} w_i k_{kj|i} \, .$$

The dominance and inferiority measures are divided by n-1 because each alternative is facing n-1 other alternatives. Each alternative, a_j, has therefore a two-dimensional overall preference measure associated with it: (k_j^+, k_j^-). This two-dimensional preference measure can be transferred into a one-dimensional preference measure as follows:

$$k_j = k_j^+ - k_j^- \, .$$

It should always be kept in mind that this transformation entails some loss of information, just as for any other preference aggregation.

The overall weighted dominance measure n_j^+ of alternative a_j, and its overall weighted inferiority measure n_j^- are defined as:

$$n_j^+ = \sum_{i=1}^{m} n_{ij}^+, \text{ where } n_{ij}^+ = \frac{1}{n-1} \sum_{k=1}^{n} w_i k_{jk|i} \, ,$$

$$n_j^- = \sum_{i=1}^{m} n_{ij}^-, \text{ where } n_{ij}^- = \frac{1}{n-1} \sum_{k=1}^{n} w_i k_{kj|i} \, .$$

Here too, the dominance and inferiority measures are divided by n-1 because each alternative is facing n-1 other alternatives. Each alternative has an overall preference

---

measure associated with it (n_j^+, n_j^-), which can be transferred into a one-dimensional preference measure as follows:

$$n_j = n_j^+ - n_j^-.$$

Again, the transformation of a two-dimensional measure into a one-dimensional measure means loss of information. With the chosen preference aggregation models to determine n_j^+, n_j^-, k_j^+, k_j^-, n_j, and k_j we can conclude the following relationships:

$$n_j^+ = k_j^+, \quad n_j^- = k_j^-, \text{ and thus } n_j = k_j.$$

To understand the equations introduced above, let's look at a numerical example. A decision maker has evaluated eight infrastructure projects, $\{a_1,...,a_8\}$, with four criteria, $\{c_1,...,c_4\}$. The results are summarized in the evaluation matrix (non-weighted score card), $E^{m \times n}$, below.

c_i: w_i	a_1	a_2	a_3	a_4	a_5	a_6	a_7	a_8
costs [$ 10^6$]: 0.4	3.0	2.5	6.2	1.3	4.5	3.2	1.9	5.3
risks [10^{-6}]: 0.1	4.0	1.2	0.3	2.8	1.7	4.2	5.1	0.6
value [-]: 0.2	high	med.	med.	low	high	low	med.	high
satisfaction [-]: 0.3	low	low	high	med.	low	med.	high	med.

It is interesting to note that a_6 is the only non-efficient (i.e., dominated) alternative. The Pareto optimal set is therefore $\{a_1,a_2,a_3,a_4,a_5,a_7,a_8\}$. The reader should check this.

A) Preference Aggregation with Type A Preference Function

The weighted preferences $(k_{jk|i})$ of the alternatives are assessed with the preference function of type A (that is, a_j outranks a_k for criterion c_i only if $e_{ij} \succ e_{ik}$) and are represented in a weighted preference matrix, $K^{8 \times 8}$, with entries k_{jk}:

	a_1	a_2	a_3	a_4	a_5	a_6	a_7	a_8	k_j^+	$k_j = k_j^+ - k_j^-$
a_1	0	.2	.6	.2	.4	.7	.3	.4	.400	-.057
a_2	.5	0	.4	.3	.5	.7	.1	.4	.414	-.029
a_3	.4	.4	0	.6	.4	.6	.1	.4	.414	-.072
a_4	.8	.7	.4	0	.7	.5	.5	.4	.571	.257
a_5	.1	.2	.6	.3	0	.3	.3	.4	.314	-.229
a_6	.3	.3	.4	.0	.7	0	.1	.4	.314	-.257
a_7	.7	.7	.4	.5	.7	.9	0	.7	.657	.414
a_8	.4	.6	.6	.3	.4	.3	.3	0	.414	-.029
k_j^-	.457	.443	.486	.314	.543	.571	.243	.443		

---

for the weighted $N^{+4\times8}$ matrix, we get:

	a_1	a_2	a_3	a_4	a_5	a_6	a_7	a_8
c_1: costs	.229	.286	0	.400	.114	.171	.343	.057
c_2: risks	.029	.071	.100	.043	.057	.014	0	.086
c_3: value	.143	.057	.057	0	.143	0	.057	.143
c_4: satisfaction	0	0	.257	.129	0	.129	.257	.129
n_j^+	.400	.414	.414	.571	.314	.314	.657	.414

and for the weighted $N^{-4\times8}$ matrix, we get:

	a_1	a_2	a_3	a_4	a_5	a_6	a_7	a_8
c_1: costs	.171	.114	.400	0	.286	.229	.057	.343
c_2: risks	.071	.029	0	.057	.043	.086	.100	.014
c_3: value	0	.086	.086	.171	0	.171	.086	0
c_4: satisfaction	.214	.214	0	.086	.214	.086	0	.086
n_j^-	.457	.443	.486	.314	.543	.571	.243	.443
$n_j = n_j^+ - n_j^-$	-.057	-.029	-.072	.257	-.229	-.257	.414	-.029

The computations can be checked as follows: $n_j^+ = k_j^+$, $n_j^- = k_j^-$, and thus $n_j = k_j$.

B) Preference Aggregation with Type B Preference Function

To illustrate the role of the preference function, $k_{jk|i}$, let's repeat the computations for the same numerical example with the type B preference function and the threshold value:

$$f_i = [\$3m, 2\times10^{-6}, 1\Delta, 1\Delta],$$

where 1Δ means that only the difference between high and low is noticeable (i.e., equal to 1). Thus, if one alternative provides, for example, medium satisfaction and another provides low or high satisfaction, then they are preferentially indifferent with regard to the satisfaction. In general, a_j outranks a_k for c_i if $e_{ij} \succ e_{ik} + f_i$. The weighted dominance matrix, $K^{8\times8}$, for this example is the following:

	a_1	a_2	a_3	a_4	a_5	a_6	a_7	a_8	k_j^+	$k_j = k_j^+ - k_j^-$
a_1	0	0	.4	.2	0	.2	0	0	.114	-.029
a_2	.1	0	.4	0	0	.1	.1	0	.100	.014
a_3	.4	.3	0	.1	.3	.1	.1	0	.186	-.043
a_4	0	0	.4	0	.4	0	.1	.4	.186	.072
a_5	.1	0	0	.2	0	.3	.1	0	.100	-.043
a_6	0	0	0	0	0	0	0	0	0	-.143
a_7	.3	.3	.4	0	.3	0	0	.4	.243	.172
a_8	.1	0	0	.3	0	.3	.1	0	.114	0
k_j^-	.143	.086	.229	.114	.143	.143	.071	.114		

for the weighted $N^{+4\times8}$ matrix, we get:

	a_1	a_2	a_3	a_4	a_5	a_6	a_7	a_8
c_1: costs	.057	.057	0	.171	0	0	.114	0
c_2: risks	0	.043	.057	.014	.043	0	0	.057
c_3: value	.057	0	0	0	.057	0	0	.057
c_4: satisfaction	0	0	.129	0	0	0	.129	0
n_j^+	.114	.100	.186	.186	.100	0	.243	.114

and for the weighted $N^{-4\times8}$ matrix, we get:

	a_1	a_2	a_3	a_4	a_5	a_6	a_7	a_8
c_1: costs	0	0	.229	0	.057	0	0	.114
c_2: risks	.057	0	0	.029	0	.057	.071	0
c_3: value	0	0	0	.086	0	.086	0	0
c_4: satisfaction	.086	.086	0	0	.086	0	0	0
n_j^-	.143	.086	.229	.114	.143	.143	.071	.114
$n_j = n_j^+ - n_j^-$	-.029	.014	-.043	.072	-.043	-.143	.172	0

We can now check the computations as follows: $n_j^+ = k_j^+$, $n_j^- = k_j^-$, and thus $n_j = k_j$.

With the type B preference function we see that many more preference values take on the value 0. This simply means that alternatives do not dominate each other that much. The resulting preference orders for type A and B functions are given in the table below. It should be noted that for n_j^+, k_j^+, n_j, and k_j higher values are preferred to lower values, while for n_j^- and k_j^- lower values are preferred to higher values.

Type A preference function	Type B preference function
rank order from both n^+ and k^+:	rank order from both n^+ and k^+:
$a_7 \succ a_4 \succ (a_2 \sim a_3 \sim a_8) \succ a_1 \succ (a_5 \sim a_6)$	$a_7 \succ (a_3 \sim a_4) \succ (a_1 \sim a_8) \succ (a_2 \sim a_5) \succ a_6$
rank order from both n^- and k^-:	rank order from both n^- and k^-:
$a_7 \succ a_4 \succ (a_2 \sim a_8) \succ a_1 \succ a_3 \succ a_5 \succ a_6$	$a_7 \succ a_2 \succ (a_4 \sim a_8) \succ (a_1 \sim a_5 \sim a_6) \succ a_3$
rank order from n and k is equal to rank order from n^- and k^-	rank order from n and k:
	$a_7 \succ a_4 \succ a_2 \succ a_8 \succ a_1 \succ (a_3 \sim a_5) \succ a_6$

We see that the choice of a preference function has an effect on the resulting preference order of the alternatives (although in a minor way for this example). Therefore, the preference function should be chosen with care.

2.3 Incomplete Preference Ranking of Alternatives

To illustrate the concept of dominance (concordance) and inferiority (discordance) of alternatives with incomplete preference orders, we discuss the example given by Roy and Vincke [1981] for the ELECTRE I method. A person wants to buy a car and considers seven cars (alternatives). The cars are evaluated with the following set of criteria: price (c_1); comfort (c_2); speed (c_3); and style (c_4). The table below shows the evaluation matrix and the weights of the criteria [Roy and Vincke, 1981]. It should be noted, that a_3 and a_5 are dominated alternatives. The reader should check this.

		a_1	a_2	a_3	a_4	a_5	a_6	a_7
min price:	c_1: $w_1 = 5/15$	300	250	250	200	200	200	100
max comfort:	c_2: $w_2 = 4/15$	E	E	A	A	A	W	W
max speed:	c_3: $w_3 = 3/15$	F	A	F	F	A	F	A
max style:	c_4: $w_4 = 3/15$	D	D	D	O	D	D	O

E: excellent, A: average, W: weak, F: fast, D: distinctive, O: ordinary

Discordance is defined in terms of when the outranking of one alternative by another is not allowed. This is the case for alternatives which have evaluation pairs (e_{ij}, e_{ik}) belonging to the set:

$$D_{\text{price}} = \{(100,300),(100,250)\} \text{ and } D_{\text{comfort}} = \{(E,W)\}.$$

Thus, even if alternative a_j would outrank alternative a_k for criterion c_i $(e_{ij} \succsim e_{ik})$ we would set $k_{jk|i}=0$. The non-weighted preference function of alternative a_j over alternative a_k for all four criteria, c_i, is thus defined as follows:

$$k_{jk|i} = \begin{cases} 1 & \text{if } e_{ij} \succsim e_{ik} \text{ and } (e_{ij}, e_{ik}) \notin D_i, \forall i \\ 0 & \text{otherwise} \end{cases}$$

which is type A function (Table IV.1), except for the '$\succsim$' preference relation instead of '$\succ$' and the discordance aspect. The weighted overall dominance value of alternative a_j over alternative a_k across all four criteria is:

$$k_{jk} = \sum_{i=1}^{4}(w_i k_{jk|i}) = \sum_{i: e_{ij} \succsim e_{ik}} w_i \ , \quad j,k = 1,...,7.$$

It should be noted that, like for the preference function in Section 2.3, we could have divided the sum by $n-1$. Anyway, after the aggregation of the preferences $k_{jk|i}$ across the criteria, we get the following matrix $K^{n \times n}$ with entries k_{jk}:

	a_1	a_2	a_3	a_4	a_5	a_6	a_7	k_j
a_1	-	10/15	10/15	10/15	10/15	6/15	6/15	52/15
a_2	12/15	-	12/15	7/15	10/15	3/15	6/15	50/15
a_3	11/15	11/15	-	10/15	10/15	10/15	10/15	62/15
a_4	8/15	8/15	12/15	-	12/15	12/15	10/15	62/15
a_5	8/15	11/15	12/15	12/15	-	12/15	10/15	65/15
a_6	11/15	11/15	11/15	11/15	11/15	-	10/15	65/15
a_7	0/15	3/15	0/15	8/15	8/15	9/15	-	27/15

From the k_j values in the table above we can derive the following preference order: $(a_5,a_6) \succ (a_3,a_4) \succ a_1 \succ a_2 \succ a_7$. However, Roy and Vincke [1981] propose an additional outranking condition in terms of a threshold k^*. The dominance of one alternative over another is only allowed if $k_{jk} \geq k^* = 12/15$. The dominance values k_{jk} of the matrix $K^{n \times n}$ fulfilling this condition are the framed entries in the table above.

The resulting preference relations can be represented in a $n \times n$ adjacency matrix, G, where each column and row represents an alternative. The matrix has an entry $g_{jk}=1$, if $a_j \succsim a_k$, and $g_{jk}=0$ otherwise. The resulting preference order is an **incomplete preference order** (Figure IV.3).

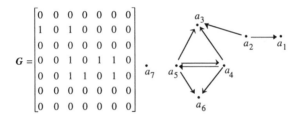

$$G = \begin{bmatrix} 0 & 0 & 0 & 0 & 0 & 0 & 0 \\ 1 & 0 & 1 & 0 & 0 & 0 & 0 \\ 0 & 0 & 0 & 0 & 0 & 0 & 0 \\ 0 & 0 & 1 & 0 & 1 & 1 & 0 \\ 0 & 0 & 1 & 1 & 0 & 1 & 0 \\ 0 & 0 & 0 & 0 & 0 & 0 & 0 \\ 0 & 0 & 0 & 0 & 0 & 0 & 0 \end{bmatrix}$$

Figure IV.3: Outranking relations of car buy problem [Roy and Vincke, 1981].

Roy and Vincke [1981] propose to find the subset E of A, such that each element of $A \backslash E$ is outranked by at least one element of E, and the elements of E do not outrank each other:

- $\forall a_k \in A \backslash E, \exists \, a_j \in E : a_j \succsim a_k$
- $\forall (a_j, a_k) \in E : a_j \not\succsim a_k$ and $a_k \not\succsim a_j$

In mathematical terms, the set E is called the **kernel** of the preference graph. For this example, the kernels are the following sets: $\{a_2, a_4, a_7\}$ and $\{a_2, a_5, a_7\}$, where a_4 and a_5 can be considered tied. Considering a_4 and a_5 tied means that the circuit in the graph has been reduced (eliminated). In order for two alternatives to be tied, not only must they be indifferent, but they also must be superior and inferior to the same alternatives. Let's assume that in the preference graph of Figure IV.3, a_3 could not be outranked by a_4 (i.e., the arc from a_4 to a_3 is removed). Then, the circuit (a_3, a_4) could not be reduced without introducing further assumptions.

If we compare the preference order based on the k_j values with the kernel, we see that a_7 performs quite differently. For the preference order based on the k_j values, a_7 is by far the least preferred alternative. With the k^* outranking condition, a_7 does quite well, because it is not outranked by any other alternative. On the other hand, a_6, which, together with a_5, had the highest k_j preference values, is now outranked by both and a_4 and a_5. This is so, because a_6 has five k_{jk} values which are 11/15. With the threshold value $k^*=12/15$, a_6 misses by 1/15 to outrank five other alternatives. This example shows how important it is to choose very carefully the outranking rules and the preference functions $k_{jk|i}$.

2.4 Ratio Scale

The Analytic Hierarchy Process (AHP), which was introduced in Chapter III for the assessment of the weights of criteria, is a descriptive preference elicitation approach. However, the AHP assesses alternative preferences as well as criteria preferences. This means that the alternatives are evaluated conditioned on the root-criteria. This approach is called the **relative AHP**, as opposed to the **absolute AHP**, where no paired comparisons of the alternatives is done.

To see the difference between relative and absolute assessment of preferences in the AHP, we will continue the example introduced in Chapter III (Section 2.1) but here we will use only the three meta-criteria risk, cost, and land use. We assume there are four landfill locations (a_1, a_2, a_3, a_4) to be ranked according to their preference. The weights of these three criteria were assessed in Chapter III as follows:

NI	risk	costs	landuse	w
risk	1	3	4	.625
cost	1/3	1	2	.238
landuse	1/4	1/2	1	.137

A) Relative Preference Assessment

For a relative assessment of the preferences of the four alternatives, we must do paired comparisons of these four alternatives with respect to three criteria which are at the bottom of the hierarchy. The results of these assessments are represented in the three quadratic matrices, $K_1^{4\times4}$, $K_2^{4\times4}$, and $K_3^{4\times4}$. The resulting relative importances (preferences) are then multiplied with the weights of the root-criteria to obtain the weighted relative preference of each alternative.

Let's assume that the decision maker assessed the following preferences, where instead of weights we speak now of **priorities**.

$K_1^{4\times4}$:

risk	a_1	a_2	a_3	a_4	priority
a_1	1	3	1/2	2	.256
a_2	1/3	1	1/7	1/2	.076
a_3	2	7	1	4	.531
a_4	1/2	2	1/4	1	.137

$K_2^{4\times4}$:

cost	a_1	a_2	a_3	a_4	priority
a_1	1	1/2	2	1/3	.157
a_2	2	1	3	1/2	.272
a_3	1/2	1/3	1	1/5	.088
a_4	3	2	5	1	.483

$K_3^{4\times4}$:

l-use	a_1	a_2	a_3	a_4	priority
a_1	1	1/2	2	5	.283
a_2	2	1	4	7	.518
a_3	1/2	1/4	1	2	.133
a_4	1/5	1/7	1/2	1	.066

The consistency ratio of all the assessments was below the required 10%, which means that we will accept the decision maker's inconsistencies. To compute the overall priorities of the four alternatives, we assume an additive preference structure. This means we compute the total priority n_j of alternative a_j as follows:

$$n_j = \sum_{i=1}^{3} w_i n_{ij},$$

where n_{ij} is the priority of alternative a_j for criterion c_i. With this preference aggregation procedure across the criteria we then get: $n_1=0.236$, $n_2=0.183$, $n_3=0.371$, and $n_4=0.210$ (note that $\Sigma_j n_j =1.0$). This implies the following preference order: $a_3 \succ a_1 \succ a_4 \succ a_2$, which completes the relative preference assessment with a ratio scale. As an exercise, the reader should check whether the four alternatives are Pareto optimal.

B) Absolute Preference Assessment

For an absolute assessment of the preferences of the four alternatives, we would introduce for each root-criterion an additional sub-level of criteria that captures intervals of the measurement units. For example, we could introduce for each criterion the three classes 'high' (H), 'medium' (M), and 'low' (L). Thus, we have to assess the relative importance of these three classes for each root-criterion. The resulting nine root-criteria would then be used to assess the alternatives. Because the root-criteria are expressed as linguistic values, describing sub-classes of the original criteria, each alternative takes on exactly one of the three sub-criteria for each original criterion. The following numerical example will clarify the principle of absolute preference assessment. Let's assume the decision maker provides the following assessments for the three sub-levels:

risk	H	M	L	weight
H	1	1/3	1/9	.066
M	3	1	1/7	.149
L	9	7	1	.785

cost	H	M	L	weight
H	1	1/2	1/8	.091
M	2	1	1/4	.182
L	8	4	1	.727

l-use	H	M	L	weight
H	1	1/4	1/6	.082
M	4	1	1/4	.236
L	6	4	1	.682

We can now insert these values into the matrix N and compute the total intensities of each alternative. The results are given in the table below.

$(N^{9\times4})^T$:

	risk .625			costs .238			land use .137			
	H .066	M .149	L 0.785	H .091	M .182	L .727	H .082	M .236	L .682	total intensity
$\Sigma=1$:	.041	.093	.491	.022	.043	.173	.011	.033	.093	
a_1		1			1				1	$p_1=.229$
a_2	1				1				1	$p_2=.177$
a_3			1	1				1		$p_3=.546$
a_4		1				1	1			$p_4=.277$

From the total preference intensities we can derive the following preference order: $a_3 \succ a_4 \succ a_1 \succ a_2$, which completes the absolute preference assessment of alternatives with a ratio scale. If we compare this preference order to the one determined with the relative assessment, we see that alternative a_1 and a_4 have switched places. This is not a flaw of the preference elicitation approach; it is simply a result of the decision maker's assessment.

While for paired assessments, adding or deleting an alternative can affect the preference order of the alternatives, we can see from this example that absolute preference assessments have no impact on the preference order of the alternatives. This means that with an absolute preference assessment alternatives can be deleted or added without affecting the preference order of the other alternatives. An absolute assessment of the alternatives is, however, always conditioned on the criteria. Thus, adding or deleting criteria might very well affect the total preference intensities of the alternatives.

This holds also for the assessment of the weights, as we discussed in Chapter III. Weights also have an absolute character. However, weights are conditioned on the higher level criterion. Also, if we add or delete criteria, we must reassess all weights which results in new weights for all criteria. In fact, by adding (or deleting) criteria we are changing the spectrum of our decision problem and thus also the meaning of each individual weight. Just because the names of the weights do not change does not mean that their meaning remains the same. To be correct, every criterion should have not only the own name, but also should carry the names of all the other criteria. Then, if the set of criteria is altered, all the non-changed criteria would change not only their weights but also their names.

2.5 Formalizing Structural and Content Goals

The structural goals of a decision model refer to the alternatives and content goals. They define the form of the solution, while the content goals define the content of the solution, such as the minimization of costs and risks. Structural goals state, for example, that only one out of n alternatives should be chosen, that all alternatives must be ranked, or that some of the content goals (e.g., constraints) are optional.

To select the r best of n alternatives, we define the structural and content goals as follows, where x_j is the decision variable for a_j:

- Structural goal: $\sum_{j=1}^{n} x_j = r$, where r is an integer, $x_j \in \{0,1\}$.

- Content goal: $\max_{a_j \in A} \sum_{j=1}^{n} x_j e(a_j)$.

The structural goal says that the sum of the binary decision variables is r. This implies that exactly r alternatives are chosen. The content goal says that the sum of the product of the binary decision variables with the corresponding evaluation of the alternatives should be maximal. These two goals lead to the selection of the r alternatives with the highest evaluation values. It should be noted that $e(a_j)$ is the evaluation value of choosing the alternative (i.e., $x_j = 1$), while it is assumed that $x_j e(a_j) = 0$ for $x_j = 0$. There are many real-world decision problems where we would also want to evaluate the renouncement of an alternative, $x_j e(a_j) \neq 0$, for $x_j = 0$. Renouncing to an alternative may have positive or negative consequences.

For example, if the choice is to find the alternative with the least negative effect, we would base our decision on the evaluations of the non-actions. Most alternatives have both positive and negative aspects, which means that $e(a_j | x_j = 0) \neq 0$. To account for these types of preferences, we could introduce a regret matrix, $R^{m \times n}$, expressing the regret for not taking an alternative.

Let us examine a structural model where both content and structural goals are given (Figure IV.4).

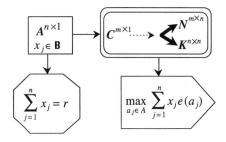

Figure IV.4: Structural model with formalized goals for choosing the r best alternatives.

If we set $r=1$, the solution will be the alternative with highest preference value. If we want to rank all n alternatives, $a_j \in A$, in ascending order (worst has rank 1 and best rank n) we would use $x_1, \ldots x_n$ as the decision variables, where n is the number of alternatives. In this case, x_j takes on positive integer rather than binary values. We then define the structural and content goals as follows:

- Structural goal: $\sum_{j=1}^{n} x_j = (n^2 + n)/2$; where $x_j \in \{1,2,\ldots,n\}$; $x_j \neq x_k, j \neq k$.

- Content goal: $\max_{a_j \in A} [\sum_{j=1}^{n} x_j e(a_j)]$.

The structural goal assures that the sum $x_1 + \ldots + x_n$ is equal to the sum of n ranks $1+2+\ldots+n = n(n+1)/2$. We have assumed that there are no ties, meaning that we have a complete strong preference order as discussed in Chapter II, Section 5. The content goal says that the ranks must be assigned to the alternatives such that the sum of the rank-weighted evaluations is maximal. Should there be ties between two or more alternatives, for example, $e(a_i)=e(a_k)$ for $i \neq k$, then these ties could be broken arbitrarily. We could also assign an average rank to the tied alternatives.

For the formalization of structural goals, many other logistic statements may also be used. Let $x_j = 1$ if action a_j is chosen and, $x_j = 0$ if action a_j is not chosen (the same holds for x_k). Table IV.2 shows several logistic relations for the selection of two actions ($x_j \in \mathbf{B}$):

Formal Model	$x_1+x_2 = 1$	$x_1+x_2 \leq 1$	$x_1+x_2 \geq 1$	$x_1-x_2 = 0$	$x_1-x_2 \leq 0$
Interpretation	either a_1 or a_2	at most one of the two	at least one of the two	either both or none	if a_1 then also a_2

Table IV.2: Logistic equations for choosing among two actions.

3. Sensitivity Analysis

We have seen that the selection of formalization and resolution models is fundamental and can have a major influence on the solutions. It is therefore important that the analyst and the decision maker understand and agree on the models. Once a specific model is constructed (or a method chosen), we would also like to know how stable the resulting preference order is. There are three kinds of instabilities: structural, functional, and numerical. **Structural instability** refers to possible **rank reversal** in the preference order of alternatives and criteria. If an alternative is added or deleted, the preference order of the other alternatives may change. Even for consistent assessments (i.e., intensity transitivity holds), structural instability can cause a reversal of the preference order of the alternatives.

Functional instability refers to **rank reversal** due to dependencies between criteria and alternatives. The additive model is based, among other things, on the assumption that preferences for criteria are independent of the alternatives.

Numerical instability means that rank reversal occurs as a result of changes in the weights of the criteria or the evaluation values of the alternatives. Structural and numerical instabilities are not necessarily flaws in the model. The analyst should simply understand when they can occur and how they can be prevented. For an inexperienced driver, a car can be more of a hazard than a benefit, while an experienced driver may be better able to appreciate the benefits of a car. The experienced driver can assess when to use a sophisticated off-road vehicle, when to use a truck, and when to use a small city car. The same is true for the different preference ordering methods proposed in the literature. All of them have their merits and limitations. It is not so much the choice of a method as the interpretations of the results in terms of the chosen method that make the difference in good decision support.

Structural and functional instabilities have been studied for the AHP. Some researchers accuse the AHP of generating random results [Dyer, 1990]. Others say that rank reversal results from misuse of the AHP, because it occurs if the weights of the criteria are not independent of the alternatives or if two alternatives are almost identical (near replicas). Others argue that rank reversal makes sense when new alternatives or criteria are added [Saaty and Vargas, 1984]. A famous example of rank reversal is given by Luce and Raiffa [1985], where a person at a new restaurant chooses salmon over steak, but when s/he hears that they also serve snails and frog legs, s/he prefers steak over salmon.

3.1 Structural Instability

Rank reversal due to addition of near replicas of alternatives (structural or outer dependence) is illustrated with the following example. This example was originally proposed by Belton and Gear [1983] with the duplication of an alternative, and later revised by Dyer [1990] with a near replica of an alternative. The evaluation matrix of four alternatives and the paired comparisons of the first three alternatives for the three criteria are given in the table below:

E	a_1	a_2	a_3	a_4
c_1	1	9	1	8
c_2	9	1	1	1
c_3	8	9	1	8

It should be noted that a_2 and a_4 are almost identical. It is assumed that the weights of the three criteria are the same; that is, each equals 1/3. The assessment of the priority weights of the first three alternatives is given below.

C_1	a_1	a_2	a_3	w_{1j}	C_2	a_1	a_2	a_3	w_{2j}	C_3	a_1	a_2	a_3	w_{3j}
a_1	1	1/9	1	1/11	a_1	1	9	9	9/11	a_1	1	8/9	8	8/18
a_2	9	1	9	9/11	a_2	1/9	1	1	1/11	a_2	9/8	1	9	9/18
a_3	1	1/9	1	1/11	a_3	1/9	1	1	1/11	a_3	1/8	1/9	1	1/18

By multiplying the criteria weights by the alternative's relative priorities and adding over all three criteria we get the following priorities for the three alternatives:

$$p_1 = 1/3 \times [1/11 + 9/11 + 8/18] = 0.45$$
$$p_2 = 1/3 \times [9/11 + 1/11 + 9/18] = 0.47$$
$$p_3 = 1/3 \times [1/11 + 1/11 + 1/18] = 0.08$$

Thus, the preference order of the first three alternatives is: $a_2 \succ a_1 \succ a_3$. If we repeat the assessment with all four alternatives, we get the following results:

C_1	a_1	a_2	a_3	a_4	w_{1j}
a_1	1	1/9	1	1/8	1/19
a_2	9	1	9	9/8	9/19
a_3	1	1/9	1	1/8	1/19
a_4	8	8/9	8	1	8/19

C_2	a_1	a_2	a_3	a_4	w_{2j}
a_1	1	9	9	9	9/12
a_2	1/9	1	1	1	1/12
a_3	1/9	1	1	1	1/12
a_4	1/9	1	1	1	1/12

C_3	a_1	a_2	a_3	a_4	w_{3j}
a_1	1	8/9	8	1	8/26
a_2	9/8	1	9	9/8	9/26
a_3	1/8	1/9	1	1/8	1/26
a_4	1	8/9	8	1	8/26

If we compute the total priority for each alternative we get the following results: $w_1 = 0.37$, $w_2 = 0.30$, $w_3 = 0.06$, and $w_4 = 0.27$.

Thus, the preference order of all four alternatives is: $a_1 \succ a_2 \succ a_4 \succ a_3$. Comparing this preference order with the preference order of only the first three alternatives ($a_2 \succ a_1 \succ a_3$) we see that the preference order between a_1 and a_2 is reversed. The rank reversal occurred because the evaluation values of a_1 and a_2 are very close for all three criteria. In both cases, a_2 is ranked higher than a_1 for two criteria. However, a_4 'takes more away' from a_2 than from a_1, which results in the rank reversal.

Saaty [1987] proposes a way to prevent rank reversal due to near replicas. An alternative a_j is said to be a **near replica** of an alternative a_k with respect to relative measurements if and only if: (i) their paired comparison judgment values $k_{jr|i}$ and $k_{kr|i}$ for criterion c_i satisfy the relation:

$$\left| \frac{k_{jrli}}{k_{krli}} - 1 \right| \le \delta_i, \ \delta_i > 0;$$

and (ii) their respective priorities w_{ij} and w_{ik} for criterion c_i satisfy the relation:

$$\frac{|w_{ij} - w_{ik}|}{\min(w_{ij}, w_{ik})} \le \varepsilon_i.$$

Although the choice of ε_i is arbitrary, Saaty [1987] uses experimental data to show that $\varepsilon_i \le 10$ percent appears to be a reasonable value to prevent rank reversal due to near copies of alternatives. The idea is based on the fact that human judgment used to make paired comparisons is insensitive to small changes in stimuli. The experiment by Saaty used small random perturbations in the k_{jkli} (± 0.5 with 50-50 chance), with which eigenvectors of matrices of size (n) 3 to 9 were generated. From the results of the experiment it could be concluded that the 10% upper bound for pooling near replicas seems to be reasonable. Using this 10% threshold for our example, we see indeed that a_2 and a_4 are near replicas.

3.2 Functional Instability

A system with feedback means that the preferences of criteria depend also on the preferences of the alternatives [Saaty, 1996]. In such a case, the problem cannot be represented anymore as a hierarchy as we did in Figure III.2 (Chapter III). In a system with feedback, rank reversal can occur even if all preference assessments are consistent (i.e., multiplicative transitivity holds). To clarify the following discussion we use the example by Dyer and Wendell [1985], where the evaluation matrix E for four alternatives and four criteria is the following:

E	a_1	a_2	a_3	a_4
c_1	1	9	8	4
c_2	9	1	1	1
c_3	1	9	4	8
c_4	3	1	5	5
$\Sigma = 70$	14	20	18	18

Let's assume that the evaluation value e_{ij} in the matrix E is identical to the subjective preference value k_{ij} ($e_{ij} \equiv k_{ij}$), and that the four criteria all have the same weight ($w_i = 0.25$ for four criteria). Then, the preference intensity k_j of a_j is the sum of the elements in column j of the evaluation matrix E divided by the sum of all elements in E: $k_j = (\Sigma_i e_{ij}) / (\Sigma_i \Sigma_j e_{ij})$.

With the numerical values of the matrix E, we get the following normalized preference intensities: a_1: 14/70; a_2: 20/70; a_3: 18/70; a_4: 18/70. The preference order for these four alternatives is therefore: $a_2 \succ (a_3 \sim a_4) \succ a_1$.

Let's now use the pairwise assessment approach with a ratio scale, and assume that the relative preference intensity of alternative a_j over alternative a_k for criterion c_i is $k_{jk|i} = e_{ij}/e_{ik} \equiv k_{ij}/k_{ik}$. Moreover, we assume that the decision maker is perfectly consistent ($k_{jk|i} \times k_{kl|i} = k_{jl|i}$).

With the values of the matrix E we now perform consistent paired preference assessments for the first three alternatives. The resulting preference intensities of the three alternatives per criteria, as well as the overall preference intensities of the three alternatives, are given below (the reader should confirm these results):

	a_1	a_2	a_3
c_1	1/18	9/18	8/18
c_2	9/11	1/11	1/11
c_3	1/14	9/14	4/14
c_4	3/9	1/9	5/9
overall preference intensities of alternatives:	0.320	0.336	0.344

With these overall preference intensities of the first three alternatives, we get the following preference order: $a_3 \succ a_2 \succ a_1$. If we repeat the assessment with all four alternatives, we get the following results:

	a_1	a_2	a_3	a_4
c_1	1/22	9/22	8/22	4/22
c_2	9/12	1/12	1/12	1/12
c_3	1/22	9/22	4/22	8/22
c_4	3/14	1/14	5/14	5/14
overall preference intensities of alternatives:	0.264	0.243	0.246	0.246

With these overall preference intensities for all four alternatives, we get the following preference order: $a_1 \succ (a_3 \sim a_4) \succ a_2$. If we compare the preference order of the first three alternatives with that of all four alternatives, we see that the ranks of the first three alternatives got reversed. But the preference order is also different from the one we derived directly from the evaluation matrix E: $a_2 \succ (a_3 \sim a_4) \succ a_1$ (without paired assessment).

The reason for this rank reversal lies in the interpretation of the evaluation matrix. This example by Dyer and Wendell deals with investment opportunities (alternatives) for each year (criteria), where the evaluation e_{ij} is the return due to investment opportunity a_j for the year c_i. The goal is to find the alternative that maximizes the return from the investment. One reason for rank reversal is that the criteria (time periods) should not have been assumed equally important (weights of

0.25), because the returns are not the same in each year [Harker and Vargas, 1987]. Thus, the weights of the criteria are not independent of the alternatives. In such a case, we have to use the **supermatrix** W (see table E):

$$
W \equiv
\begin{array}{c}
 \\
\begin{array}{c} c_1 \\ c_2 \\ c_3 \\ c_4 \\ a_1 \\ a_2 \\ a_3 \\ a_4 \end{array}
\end{array}
\begin{array}{cccccccc}
c_1 & c_2 & c_3 & c_4 & a_1 & a_2 & a_3 & a_4 \\
\hline
\multicolumn{4}{c}{} & \multicolumn{4}{c}{\text{relative weights}} \\
\multicolumn{4}{c}{\mathbf{0}} & \multicolumn{4}{c}{\text{of criteria given}} \\
\multicolumn{4}{c}{} & \multicolumn{4}{c}{\text{the alternatives}} \\
\hline
\multicolumn{4}{c}{\text{relative weights}} & \multicolumn{4}{c}{} \\
\multicolumn{4}{c}{\text{of alternatives}} & \multicolumn{4}{c}{\mathbf{0}} \\
\multicolumn{4}{c}{\text{given the criteria}} & \multicolumn{4}{c}{} \\
\end{array}
$$

$$
W =
\begin{array}{c}
\begin{array}{c} c_1 \\ c_2 \\ c_3 \\ c_4 \\ a_1 \\ a_2 \\ a_3 \\ a_4 \end{array}
\end{array}
\begin{array}{cccccccc}
c_1 & c_2 & c_3 & c_4 & a_1 & a_2 & a_3 & a_4 \\
& & & & 1/14 & 9/20 & 8/18 & 4/18 \\
& \mathbf{0} & & & 9/14 & 1/20 & 1/18 & 1/18 \\
& & & & 1/14 & 9/20 & 4/18 & 8/18 \\
& & & & 3/14 & 3/20 & 5/18 & 5/18 \\
\hline
1/22 & 9/12 & 1/22 & 3/14 & & & & \\
9/22 & 1/12 & 9/22 & 1/14 & & \mathbf{0} & & \\
8/22 & 1/12 & 4/22 & 5/14 & & & & \\
4/22 & 1/12 & 8/22 & 5/14 & & & & \\
\end{array}
$$

The number of comparisons in this supermatrix is $m(n^2-n)/2+n(m^2-m)/2$. With this matrix W we can do the same graph theoretic interpretation as we did with the matrix K concerning the path intensities in Chapter III. We compute a high power of W from which we can obtain the principal eigenvector as the preference intensities. Because of the structure of W, the exponent must be an odd number. Thus, we compute: $\lim_{r \to \infty} W^{2r+1}$:

$$
\lim_{r \to \infty} W^{2r+1} =
\begin{bmatrix}
& & & & 0.314 & 0.314 & 0.314 & 0.314 \\
& \multicolumn{3}{c}{0} & & 0.171 & 0.171 & 0.171 & 0.171 \\
& & & & 0.314 & 0.314 & 0.314 & 0.314 \\
& & & & 0.200 & 0.200 & 0.200 & 0.200 \\
0.200 & 0.200 & 0.200 & 0.200 & & & & \\
0.286 & 0.286 & 0.286 & 0.286 & & \multicolumn{3}{c}{0} & \\
0.257 & 0.257 & 0.257 & 0.257 & & & & \\
0.257 & 0.257 & 0.257 & 0.257 & & & & \\
\end{bmatrix}.
$$

The resulting preference intensities for the four alternatives are: 0.200 (a_1); 0.286 (a_2); 0.257 (a_3); 0.257 (a_4). The preference order for these four alternatives is therefore $a_2 \succ (a_3 \sim a_4) \succ a_1$, which corresponds to the preference order which we derived from the E matrix without paired comparisons.

3.3 Numerical Instability

Numerical instability affects the numeric values used in the analysis. Numeric values are used for the criteria weights (w_i), the evaluation matrix (e_{ij}), and the preference values (k_{jkli}). These values are exposed to different sources of numerical sensitivity which include: (1) measurement errors (stochastic or systematic), (2) limitations in human judgment (e.g., limitations in differentiating small differences), (3) problem uncertainty (assessment of scenarios), and modeling deficiencies (e.g., additive preference aggregation for non-additive preferences). All these sources of numerical sensitivity can cause numerical instability, which can lead to a reversal of the preference order. For each numeric value, it is possible to determine by inspection upper and lower bounds within which the preference order is not affected. These stability boundaries, however, depend on the fixed numerical values of all other parameters. Simultaneous variations of several parameters could lead to multi-dimensional stability boundaries.

Numerical instability is usually assessed by inspection. For example, one could determine for each numeric value upper and lower fluctuation limits and determine the preference orders for these three sets of data. If the orders differ a lot, suspicious parameters must be analyzed with greater scrutiny. Another approach to detect numerical sensitivity is simulation. Wolters and Mareschal [1995] discuss three types of sensitivity analysis to detect numerical instability for additive multicriteria decision models. The three methods are intended to detect (1) the sensitivity of a preference order to specific changes in the evaluations of all alternatives on certain criteria; (2) the influence of specific changes in certain criterion-scores of an alternative; and (3) the minimum modification of the weights required to make an alternative rank first.

4. Summary and Further Readings

Descriptive preference assessment methods are based on paired comparisons of the alternatives for all criteria. Preference aggregation may be done within or across criteria. Preference aggregation across criteria leads to a quadratic and weighted paired preference intensity matrix. Preference aggregation within criteria leads to a normalized weighted preference intensity matrix that has the same dimensions as the evaluation matrix.

The resulting matrices may be further aggregated to one preference intensity value for each alternative. The preference intensity of one alternative over another is a function of their dissimilarity. The non-weighted preference intensity is usually a value between zero and one. There are many different types of non-weighted preference functions. Preference aggregation is often assumed to be additive.

Inferiority is often defined in terms of rules, stating when dominance is not allowed. While dominance reflects the strong aspects of an alternative over the others, inferiority reflects the weak aspects. The 'sum' of dominances and inferiorities gives the resulting preference index of an alternative.

With the weighted preference values, preference orders for the feasible alternatives can be determined. The type of preference order is defined in the structural goals. The determination of the wanted preference order is done with an appropriate resolution model.

Descriptive approaches are relative. Near replicas of alternatives can lead to structural instability. Functional stability is affected by dependencies among criteria and alternatives. Numerical instability can occur due to changes in parameters.

A concise overview of descriptive preference aggregation methods can be found in Vincke [1992]. Vincke's book contains also many references to more in-depth descriptions of the ELECTRE, PROMETHEE, and GAIA methods, as well as real-world applications. Several articles on a scholarly debate about the analytic hierarchy process and the related problems of rank reversal can be found in *Management Science*, Vol. 36, No. 3, 1990. Potential and limitations of integrating interval and ratio scales in paired assessments are discussed in several articles in the *Journal of Multi-Criteria Decision Analysis*, Vol. 6, 1997.

5. Problems

1. Discuss the differences between absolute and relative descriptive preference assessments with a ratio scale and the problems that can occur.

2. Discuss the approach of paired assessment for a problem with multiple criteria and the concerns related to an additive preference aggregation function approach.

3. Discuss conditions under which the relations $k_{jk|i}+k_{kj|i}=1$ and $k_{jk|i}+k_{kj|i}=0$ hold.

4. Verify the results of the example in Section 2.2 by computing the matrices $K_i^{n \times n}$.

5. Repeat the example in Section 2.3 with the function A in Section 2.1. Compare the preference order with the two preference orders derived in Section 2.3.

6. Formalize the structural goal of choosing a_1 if both a_2 and a_3 are chosen.

7. With the formal and resolution models discussed in Section 2.2 (type A and B models) rank the alternatives of the example in Chapter III, Section 3.3. Compare the result of type A model with the graph in Figure III.6.

8. Verify the example of structural instability in Section 3.1.

9. Verify the example of functional instability in Section 3.2.

10. Summarize the concept of descriptive assessment. Especially discuss the role of axioms, the selection of a preference function $k_{jk|i}$ (does it reflect the decision makers preferences?), the assumptions for preference aggregation (if any), and whether all alternatives can be ranked from best to worst.

CHAPTER V

VALUES AND NORMATIVE CHOICE

1. The Structural Model

1.1 Conceptual Aspects

The **normative** approach to decision analysis assumes that the decision maker's subjective preference structure can be captured by an m-dimensional preference function $v(e_1,...,e_m)$, as opposed to the descriptive approach which is based on paired comparisons. This m-dimensional preference function will be called m-dimensional **value function**. Other authors, especially economists, call it **utility function**. However, we will reserve the term utility for subjective normative preference functions in the presence of uncertainty (Chapter VII).

The primary purpose of a value function is to rank alternatives $\{a_1,...,a_n\}$ according to their preferences with a **weak preference order**. For this purpose, an ordinal value measure v of the alternatives suffices, such that:

$$a_j \succeq a_k \Leftrightarrow v(a_j) \geq v(a_k).$$

For example, if we have evaluated the costs of two alternatives as $60 and $45, then the value function $v(\$60)=0.2$ and $v(\$45)=0.7$ is an **ordinal value function**. A value function is a subjective formalization of the preference structure for a specific decision maker. Although it is subjective, the decision maker must comply with a set of axioms. If s/he does, s/he is said to be **rational** and a value function exists which captures his/her subjective preference structure.

In this chapter we will also discuss an elicitation procedure to determine component value functions $v_i(e_i)$ for criterion c_i. The component value functions, as opposed to the descriptive approach, are not selected by the decision analyst (Chapter IV, Table IV.1). Instead, they are assessed in an interactive question-answer session.

In the descriptive approach as discussed in Chapters III and IV, we worked with binary decision variables. The normative approach can be used for both binary and real-valued decision variables. In fact, even if the value function is assessed with a finite set of discrete alternatives, appropriate interpolations can lead to a continuous m-dimensional value function, $v(a)$.

For $m=2$, the 2-dimensional value function can be represented graphically by its **indifference curves** in the e_1-e_2 plane. Indifference curves indicate actions for which the decision maker is preferentially indifferent; these actions therefore have identical preference values. Figure V.1 shows an example of a 2-dimensional value function with indifference curves $v=const_i$. The three actions a_1, a_2, and a_3 lie on an indifference curve, $v(a)=const_1$, and are thus preferentially indifferent:

$$v(a_1)=v(a_2)=v(a_3)=const_1, \text{ or, equivalently, } a_1 \sim a_2 \sim a_3.$$

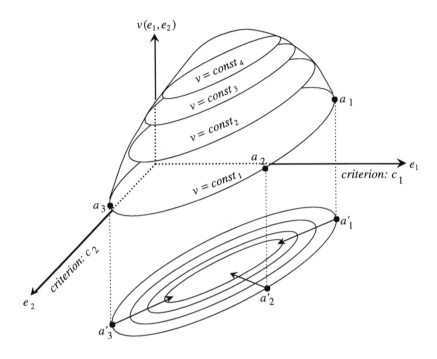

Figure V.1: Two-dimensional value function, $v(a):=v(e_1,e_2)$, with indifference curves.

The arrows in the e_1-e_2 plane indicate the directions of maximal value increase from the three actions. The value function in Figure V.1 seems to have one maximum. In the following discussions, however, we will address value functions which are monotonically increasing or decreasing. These value functions have indifference curves that are not closed, as the ones in Figure V.1 are.

It certainly would be convenient to determine an m-dimensional value function $v(a)$ in terms of m **component value functions** $v_i(e_i)$. However, we will discuss that this can only be done if **preferential independence** across the components holds (for reasons of consistency in the terminology as defined in Chapter I, we will refer to the components as *criteria* although the term *attributes* is mostly used in the literature). Because the concept of preferential independence has not been stressed for descriptive approaches, we will place more emphasis on it in this chapter. Figure V.2 shows the concept of deriving an m-dimensional value function from the m component value functions.

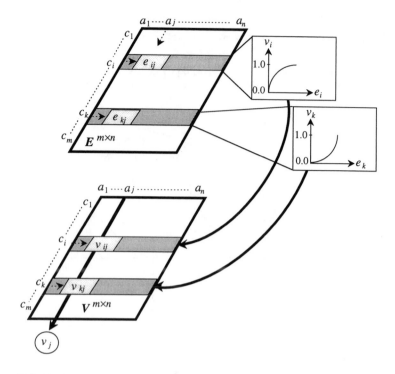

Figure V.2: Derivation of component value functions, $v_i(e_{ij})$, for all i, and preference aggregation across the criteria leading to the preference value v_j for alternative a_j.

1.2 The Structure

The structural model of Figure V.3 indicates that the evaluation matrix $E^{m \times n}$ gets transformed into a subjective normalized preference value matrix, $V^{m \times n}$ (assuming preferential independence holds). The component preferences are then further aggregated to the overall preference intensity values, v_j, of alternative a_j.

Even if we use only a finite set of actions to determine the value function, we can always interpolate to obtain a continuous preference value function. This means that we extend our discrete space of potential solutions to a continuous space with an infinite number of potential solutions (Figure V.1). The structural model of Figure V.3 is independent of the type of decision variables, $x_j \in \mathbf{R}$, $x_j \in \mathbf{Z}$, or $x_j \in \mathbf{B}$. When $x_j \in \mathbf{R}$, we might want to define the value function analytically in terms of e_1 and e_2; for example, $v(e_1, e_2) = 3(e_1)^2 - 2(e_2)^2$. The goal is to find the action with maximum value. For this purpose, the space of feasible solutions must be constrained with a relation between the effectiveness values for the two criteria. For example, we could specify a relation between e_1 and e_2: $e_1 = f(e_2)$. This relation between the two

---

Chapter V: Values and Normative Choice

effectiveness values can be graphed in the e_1-e_2 plane. The most preferred action is the one along that graph which has highest preference value in the value-field defined by $v(e_1,e_2)=3(e_1)^2-2(e_2)^2$.

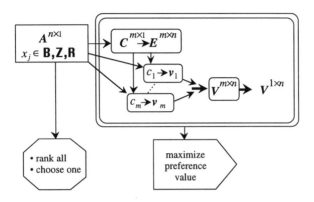

Figure V.3: Structural model of deriving normalized preference values.

We will discuss two normative approaches for preference assessment and aggregation, one in this chapter (without any scenario elements; i.e., decision making under certainty) and another in Chapter VII, with scenario elements (i.e., decision making under uncertainty). When uncertainty must be considered, preferences must also be aggregated across scenarios. This means that an approach which elicits subjective preferences in the presence of uncertainty must be employed, called **expected utility theory** or **multiattribute utility theory** (Chapter VII), as opposed to **multiattribute value theory**, which is the subject of this chapter.

For both approaches we will discuss first the formalization issues and then the resolution approach. Both theories have been extensively discussed in the literature. While many successful real-world applications have shown their benefit for decision making we will also discuss the criticisms concerning their applicability and behavioral assumptions.

2. The Formal Model of Value Theory

2.1. Motivation and Axioms of Value Theory

A decision maker, whose preference structure can be described by a value function is called **rational**. There have been many debates over whether or not a human can behave rationally but we will not elaborate further on this issue here. However, one of the major misconceptions concerning normative theories is to think that the value function imposes the choice on the human decision maker. Exactly the opposite is the case - the value function captures the decision maker's preference structure (if a value function can be determined). Should s/he ponder in greater depth the decision options and then decide differently than the value function suggests, then this would be a sign to revise the value function, and not a reason to blame the decision maker for not being rational or the approach for not computing 'correct' solutions. Thus, a normative approach does not tell how one should behave but how one can make best use of his/her resources given s/he complies with the requirements for a rational decision maker.

Rational behavior is based on a set of **axioms**. If the decision maker complies with these axioms of rationality, his/her preference structure can be captured by a normative value function. The three axioms for a rational decision maker are the following:

- **Completeness**: for any two actions, a_j and a_k, either $a_j \succ a_k$, or $a_k \succ a_j$, or $a_j \sim a_k$.

- **Transitivity**: if $a_i \succsim a_j$, and $a_j \succsim a_k$, then $a_i \succsim a_k$.

- **Substitution**: if $a_i \succ a_j$, and $a_i \sim a_k$ then $a_k \succ a_j$.

A direct consequence of these axioms is that the indifference curves in the e_1-e_2 plane never intersect (Figure V.4). Intersecting lines would be a violation of the axioms of rationality. Assume the lines $v=const_1$ and $v=const_2$ intersect at point X, points A and B are on $v=k_1$, and points C and D are on $v=k_2$. Then, with $v(C)>v(A)$, $v(B)>v(D)$, and $v(A)=v(B)$, we would conclude by substitution and transitivity that $v(C)>v(D)$ which, however, contradicts $v(C)=v(D)$.

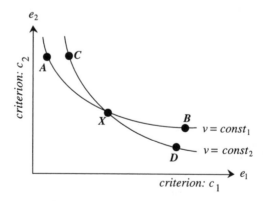

Figure V.4: Intersecting indifference lines violate the axioms of rationality.

A value function assigns numerical values to evaluation measures such that the most preferred evaluation measure gets the highest value (usually 1) and the least preferred the lowest value (usually 0). The values in-between have the same order as the actions' preference order. If the alternatives are ordered according to increasing (decreasing) preferences, we have a monotonically increasing (decreasing) value function.

An example of a monotonically increasing value function is the monetary gain. The higher the gain the more preferred is the outcome, tacitly assuming that there are lower and upper bounds. For such a monotonically increasing real-valued value function, $v(a)$, we have:

$$a_j \gtrsim a_k \Leftrightarrow v(a_j) \geq v(a_k) \Leftrightarrow e(a_j) \geq e(a_k).$$

An example of a monotonically decreasing value function is the monetary loss. The higher the loss, the lower the preference of the outcome. For such a monotonically decreasing real-valued value function, $v(a)$, we have:

$$a_j \gtrsim a_k \Leftrightarrow v(a_j) \geq v(a_k) \Leftrightarrow e(a_k) \geq e(a_j).$$

Often, however, preferences are not monotonic. For example, the optimal annual precipitation for agricultural purposes lies somewhere between the maximum and minimum historical precipitations. Precipitations smaller or higher than the optimum are rated less valuable. Figure V.5 shows different types of value functions.

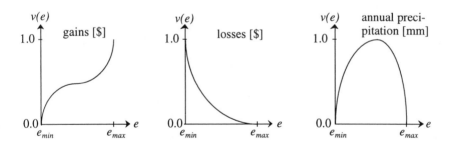

Figure V.5: Monotonically increasing (left), decreasing (middle), and non-monotonic (right) value functions.

The following discussions are limited to monotonically increasing value functions. Preferentially decreasing evaluations (Figure V.5, middle) can be transformed into preferentially increasing evaluations with the following transformation: $e^* = e_{worst} - e$, such that $e^*(e_{worst}) = e_{worst} - e_{worst} = 0$, $v(e_{worst}) = v(0) = v_{min} = 0$, and $v(e_{best}) = v_{max} = 1$. A single-peaked non-monotonic value function (Figure V.5, right) can be transformed into a monotonically increasing value function by mapping outcomes e_j above the peak to values below the peak e_k, such that $e_j \sim e_k$.

A subjective value function assigns numeric values (usually but not necessarily between zero and one) to the outcomes of the alternatives. If the scale is ordinal these numeric values must preserve the preference order of the alternatives. As was mentioned in Chapter I, ordinal scales are unique under strictly increasing transformations; thus, any strictly increasing transformation must preserve the preference order. Consequently, any strictly increasing transformation of an ordinal value function results in a **strategically equivalent** value function.

For example, a decision maker compares four construction projects in terms of monetary gains: $e(a_1)=\$10^7$, $e(a_2)=\$10^6$, $e(a_3)=\$10^8$, and $e(a_4)=\$10^5$. An ordinal value function is $v(e)=e$. The strictly increasing transformation $w=\log(v)$ gives the following result which preserves the preference order: $\log[v(a_1)]=7$, $\log[v(a_2)]=6$, $\log[v(a_3)]=8$, and $\log[v(a_4)]=5$. The preference order is still the same as before the transformation: $a_3 \succ a_1 \succ a_2 \succ a_4$.

2.2 Preferential Independence

To get an intuitive understanding of the concept of **preferential independence,** let's consider an example. If a person prefers ice-cream over cake in summer but cake over ice-cream in winter, then we would conclude that his/her preference for *desserts* is not preferentially independent of the *season.* We can see the options consisting of dessert and season as two-dimensional alternatives: $a_j:=(dessert_j,season_j)$. Thus, we can write for this example:

$$(ice\text{-}cream,\ summer) \succ (cake,\ summer) \nLeftrightarrow (ice\text{-}cream,\ winter) \succ (cake,\ winter),$$

which means that the statement 'ice-cream is preferred to cake' does not hold in general, but rather depends on the season.

However, if a person prefers strawberry over lemon (tastes) for any type of dessert (ice-cream, cake, etc.), then we conclude that the *taste* of the dessert is preferentially independent of the *type* of dessert. To generalize this for all tastes, we would have to check for all kinds of tastes (e.g., strawberry, lemon, chocolate, etc.). Preferential independence of *taste* over *type* of dessert can be expressed as:

$$(taste_j,type_\alpha) \succ (taste_i,\ type_\alpha) \Leftrightarrow (taste_j,\ type_\beta) \succ (taste_i,\ type_\beta).$$

If we can also assume that the type of dessert is preferentially independent of the taste, we would conclude that the two criteria are **mutually preferentially independent.** In the following section we will use a general notation to discuss preferential independence of two and more criteria (where criteria are also referred to as dimensions, components, or attributes).

A) <u>Two Criteria</u>:

If alternatives are evaluated with two criteria, c_1 (e.g., risk) and c_2 (e.g., cost), then c_1 is said to be preferentially independent of c_2 if the preference relation between any two alternatives a_j and a_i with varying levels of c_1 (x) and a common, fixed level of c_2 (α) is independent of that fixed level of c_2 (meaning that α can be replaced by any value β):

$$(x,\alpha) \succsim (y,\alpha) \Leftrightarrow (x,\beta) \succsim (y,\beta).$$

Conversely, criterion c_2 is said to be preferentially independent of c_1 if the preference relation between alternatives with varying levels of c_2 and a common, fixed level of c_1 is independent of that fixed level of c_1:

$$(\alpha,x) \succsim (\alpha,y) \Leftrightarrow (\beta,x) \succsim (\beta,y).$$

If c_1 is preferentially independent of c_2, and c_2 is also preferentially independent of c_1, then c_1 and c_2 are said to be mutually preferentially independent.

As an example let's assume that an agricultural engineer investigates the most preferred place to plant a certain crop. A place is characterized by the present humidity (c_1) and the optimal humidity of a soil (c_2), both measured in millimeters. The most preferred place is the one where the difference between present humidity and optimal humidity is smallest. That is, the larger the difference, the less preferred the place. Numerical values for four places are known: a_1: (3,1), a_2: (5,1), a_3: (5,6), and a_4: (3,6). By minimizing the differences between the present and the optimal humidity, we get the following preference relations:

$$a_1: (3,1) \succ a_2: (5,1), \text{ but } a_3: (5,6) \succ a_4: (3,6).$$

From these relations we can conclude the obvious result that the preference for the present humidity (c_1) depends on the optimal humidity (c_2), where for this numerical example the fixed levels at c_2 are 1 and 6. This means that if somebody asks the engineer how much s/he likes a certain amount of humidity, s//he would say that it depends on the soil type. Likewise, the preference for a certain soil type depends on the present humidity. That is, soils with high optimal humidity are preferred in humid regions, while soils with low optimal humidity are preferred in dry regions.

If we don't know if the preference is a function of the numeric outcomes, then we must elicit the decision maker's preference structure through an interview, by asking him or her preferences for different numeric outcomes. How to do this practically will be addressed as part of the resolution model.

Although this example might sound quite trivial, preferential independence is a very fundamental concept, especially when dealing with more than only two criteria. For example, if the criteria are not preferentially independent then it will not be possible to express the overall preference as an aggregation of the criterion-specific preferences. In such cases, preferences of combinations of criteria (called higher order effects) must also be considered as part of the multidimensional preference function.

B) <u>Three Criteria</u>:

For two criteria, we defined preferential independence by varying levels for one criterion with fixed levels for the other criterion. Thus, the second criterion can be seen as the **complement** of the first criterion. To define preferential independence of three criteria, c_1, c_2, c_3, we also use the concept of the complement. The criterion c_1 is said to be preferentially independent of its complement $\{c_2, c_3\}$ if the preference relation $\succeq$ between any two alternatives, a_j and a_k, with varying levels of c_1 and a common, fixed level of c_2 and c_3 (α, β) is independent of that fixed level.

$$(x, \alpha, \beta) \succeq (y, \alpha, \beta) \Leftrightarrow (x, \gamma, \delta) \succeq (y, \gamma, \delta).$$

If we have more than two criteria we therefore assess preferential independence not between criteria, but between sets of criteria. Thus, the criteria $\{c_2, c_3\}$ are said to be preferentially independent of $\{c_1\}$ if the preference relation between alternatives with varying levels of $\{c_2, c_3\}$ and a common, fixed level of $\{c_1\}$ is independent of that fixed level of $\{c_1\}$.

$$(\alpha, x, y) \succeq (\alpha, r, s) \Leftrightarrow (\beta, x, y) \succeq (\beta, r, s).$$

If $\{c_1\}$ is preferentially independent of $\{c_2, c_3\}$ and $\{c_2, c_3\}$ is also preferentially independent of $\{c_1\}$, then the two sets $\{c_1\}$ and $\{c_2, c_3\}$ are said to be **mutually preferentially independent**. With three criteria we thus get three pairs of sets for which we would like to assess mutual preferential independence: $\{c_1\}$ and $\{c_2, c_3\}$, $\{c_2\}$ and $\{c_1, c_3\}$, and $\{c_3\}$ and $\{c_1, c_2\}$.

For example, if we assess the preferences of traffic situations in terms of traffic density (c_1), travel speed (c_2), and time of day (c_3), we would have the following characteristics. First of all we should note that traffic density and time of day might be preferentially independent, although there might be a **correlation** between the two (high densities at rush hours, but low traffic densities are always better, regardless of the time of day). $\{c_1\}$ is preferentially independent of $\{c_2, c_3\}$ if, for example, lower traffic densities are always preferred, regardless of the travel speed and the time of day. This is obviously not the case. On the other hand, preference relations between any combinations of $\{c_2, c_3\}$ might be preferentially independent of any value of $\{c_1\}$. For example, low density and low speed is preferred to high density and high speed both at rush hours or at any other time of day.

C) <u>More than three Criteria</u>:

If we have m criteria, $C=\{c_1,...,c_m\}$, We would thus have to assess mutual preferential independence among any subsets $C_i \neq \emptyset$ of C. For example, to show that the second and fourth criteria are preferentially independent of their complement, we would have to show (with $m=5$):

$$(\alpha,x,\beta,y,\gamma) \succsim (\alpha,r,\beta,s,\gamma) \Leftrightarrow (\delta,x,\varepsilon,y,\zeta) \succsim (\delta,r,\varepsilon,s,\zeta).$$

In general, there are $\binom{m}{i}$ different subsets with i elements, $i=1,...,m-1$. For example, $\binom{m}{1} = m$ assessments of preferential independence must be made for the individual criteria $\{c_i\}$ with their complement $\{c_1,...,c_{i-1},c_{i+1},...,c_m\}$, and $\binom{m}{2}=m(m-1)/2$ assessments of preferential independence must be made for two criteria $\{c_i,c_k\}$ with their complement $\{c_1,...,c_{i-1}, c_{i+1},...,c_{k-1},c_{k+1},...,c_m\}$. The total number of paired assessments among all sets is therefore

$$\sum_{i=1}^{m-1}\binom{m}{i} = 2^m - 2.$$

This means that with 2 criteria we have to do 2 assessments of preferential independence, with 3 criteria 6, with 4 criteria 14, with 5 criteria 30, and with 6 criteria 62. In the context of the resolution model, we will discuss an approach to reduce this large number of assessments. If all subsets are mutually preferentially independent we simply say that the criteria $c_1,...,c_m$ are (mutually) preferentially independent.

2.3 Additive Value Functions

A special class of m-dimensional value functions are **additive value functions**. Let's assume that alternatives are evaluated in m dimensions $e(a):=(e_1,...,e_m)$, that preferential independence holds, and that we can determine for a decision maker his/her subjective component value functions, $v_i(e_i)$. The joint or aggregated value function over all criteria, $v(a)$, is said to be additive if it can be represented as the sum of the component value functions:

$$v(a) = \sum_{i=1}^{m} v_i(e_i).$$

A necessary, and for $m \geq 3$ sufficient, condition for an additive value function is that the criteria be mutually preferentially independent. To see why, assume we have an additive value function with the following preference relation:

- $v(x,y) \succsim v(r,s)$;
- because the value function is additive, we can rewrite this as:
- $v_1(x) + v_2(y) \succsim v_1(r) + v_2(s)$;
- we now set $y=s=\alpha$ and get:
- $v_1(x) + v_2(\alpha) \succsim v_1(r) + v_2(\alpha)$;
- by changing α to β and using the joint value function we get:
- $v(x,\alpha) \geq v(r,\alpha) \Leftrightarrow v(x,\beta) \geq v(r,\beta)$ which confirms that preferential

  independence is a necessary condition for an additive value function.

A more in-depth discussion of additional theoretical conditions that a value function must satisfy in order be additive can be found in [French, 1988, Chapter 4]. These conditions refer, among others, to the existence of a weak preference order ($\succsim$), the fact that all component value functions are finite, and the assumption that a continuous function can be derived.

In order for a 2-dimensional value function to be additive, the **Thompsen condition** must also be satisfied. The Thompsen condition says that if a decision maker agrees on the preference relations

$$(q,y) \sim (x,s) \text{ and } (r,y) \sim (x,t),$$

then s/he should also agree on

$$(r,s) \sim (q,t).$$

Figure V.6 illustrates the Thompsen condition. It says that given two indifferent actions (q,t) and (r,s), and a third action, (x,y), each two of the four resulting actions (by exchanging the components according to Figure V.6) also are indifferent.

$$\left. \begin{array}{c} (q,y) \sim (x,s) \\ (r,y) \sim (x,t) \end{array} \right\} \Rightarrow (r,s) \sim (q,t)$$

Problem 6 at the end of this chapter deals with a 2-dimensional value function where the criteria are mutually preferentially independent, but the Thompsen condition does not hold. Therefore, an additive value function for those two criteria would not reflect appropriately a decision maker's preference structure.

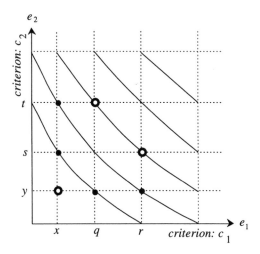

Figure V.6: Thompsen condition for 2-dimensional value function.

While value functions in general are unique up to a strictly increasing transformation (ordinal scale), it can be shown that additive value functions are unique up to a positive linear (affine) transformation. This means that $v(a)$ and $w(a)$ reflect the same preference order if and only if there is a positive linear (affine) transformation

$$v(a) = \sum_{i=1}^{m} v_i(e_i) \rightarrow w(a) = \sum_{i=1}^{m} w_i(e_i),$$

such that $w_i = \alpha v_i + \beta_i$, with $\alpha \geq 0$ (for a more in-depth discussion see [French, 1988, Chapter 4]). In other words, additive value functions refer to interval scales.

2.4 Linear Value Functions and Perfect Complements

Figure V.7 shows two important types of value functions: the **linear** value function (left) and the **perfect complement** value function (right).

A linear value function is an additive value function with constant tradeoffs between the evaluation values. In other words, for an m-dimensional linear value function

$$v(e_1,...,e_m) = k_1 e_1 + ... + k_m e_m,$$

where the k_i's are called the **scaling constants**, the following indifference relation

$$(e_1,e_2,...,e_i,...,e_j,...e_m) \sim (e_1,e_2,...,e_i+k,...,e_j\text{-}k\times k_i/k_j,...e_m)$$

holds for any $k \in \mathbf{R}$.

Let's set for all but the i-th and j-th criteria the evaluation values to the worst outcomes, such that $v_r(worst)=0$ ($r=1,...,m\backslash\{i,j\}$). Then, the linear value function becomes: $v(e_i,e_j)=k_ie_i+k_je_j$. From this value function we see that if we increase e_i by k, then we must reduce e_j by $k\times k_i/k_j$. That is, $k_ie_i+k_je_j=k_i(e_i+k)+k_j(e_j\text{-}kk_i/k_j)$.

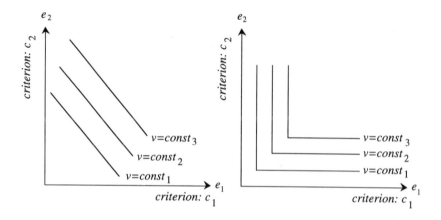

Figure V.7: Linear value function (left) and perfect complements (right).

It can be shown that linear value functions are unique up to a multiplicative (similarity) transformation. This means that if $v(a)$ and $w(a)$ reflect the same preference order if and only if there is a multiplicative (affine) transformation

$$v(a) = \sum_{i=1}^{m} v_i(e_i) \rightarrow w(a) = \sum_{i=1}^{m} w_i(e_i),$$

such that $w_i = \alpha v_i$ with $\alpha \geq 0$ (for an in-depth discussion see [French, 1988, Chapter 4]).

If we can assume that a decision maker has a linear 2-dimensional value function, we can visualize it in the 2-dimensional e_1-e_2 plane, as in Figure V.7: $v(e_1,e_2)=k_1e_1+k_2e_2$. We can rewrite this as: $e_2=v(e_1,e_2)/k_2-(k_1/k_2)e_1$. Thus, $-k_1/k_2$ is the slope of the line and $v(e_1,e_2)/k_2$ the intercept. Let's assume e_1 stands for gains in $\$10^5$ and e_2 for 100 cars traffic reduction. Then, to find the numerical values for any indifference line, we have to ask the decision maker for two indifferent actions. For example, s/he might express the following indifference: $(1,2) \sim (3,1)$. Inserting these values into the value function gives: $v=k_11+k_22$, and $v=k_13+k_21$. From these two equations we can compute: $k_1/k_2=1/2$ and $v/k_2=5/2$. We could chooses $v=1$, and get

$k_1=1/5$ and $k_2=2/5$. Because of the constant tradeoff ($k_1/k_2=1/2$), the linear value function is also called the **perfect substitute** value function.

Perfect complements (Figure V.7, right), as opposed to perfect substitute, means that the two criteria must grow simultaneously with a fixed proportion, in order to increase the preference value. The increase of only one criterion does not increase the overall value. For example, an agricultural engineer has a certain number of irrigators (c_1) and a certain amount of land to be cultivated in acres (c_2). Let's assume that the optimal ratio of irrigators and land to be cultivated is 1:2. Then, if the engineer gets six times as much land, s/he should want three times as many irrigators. The 2-dimensional value function of perfect complements is thus: $v(e_1,e_2)=\min(k_1e_1,k_2e_2)$. For our example, we have $v(e_1,e_2)=\min(2e_1,e_2)$.

Thus, if there are 10 irrigators and 16 acres, we have $v(10,16)=\min\{20,16\}=16$. This means that the engineer can give up two irrigators without losing any value. With 8 irrigators and 16 acres, increasing only the number of acres or only the number of irrigators does not increase the overall preference value; both must grow at a ratio of 1:2, which is the line connecting the corner points of the indifference curves in Figure V.7, right.

2.5 Value Functions Over Time Streams

A commonly used approach in policy decision making is **cost-benefit analysis**. The basic assumption is that all criteria can be expressed in monetary units and that the overall value is computed with a linear value function. Capital investment decisions are usually based on financial consequences over a certain number of time streams. Let's assume that a decision maker is comparing two infrastructure projects (a_1 and a_2) over five years, where negative values refer to investments and positive to returns:

$\$10^5$	1	2	3	4	5
a_1	-4	-2	1	3	6
a_2	-8	-4	3	6	10

If e_{ij} is the outcome of project a_j for the i-th time stream, then, the **net present value** (NPV) of a_j is defined as:

$$NPV(a_j) = \sum_{i=1}^{m} \frac{e_{ij}}{(1+r)^{i-1}} = \sum_{i=1}^{m} \rho^{i-1} e_{ij},$$

where r is the constant **discount rate**, and $\rho=(1+r)^{-1}$ is called the **discount factor**. If we see the projects over the m time streams (where for our example $m=5$) as an m-dimensional value function, and the conditions for a linear value function hold with constant tradeoff ρ between any two consecutive time periods, then the linear value function over the time streams can be represented as:

$$v(e_1,...,e_m) = NPV(a) = \rho e_1 + \rho^{-1} e_2 + ... + \rho^{m-1} e_m.$$

Therefore, the NPV can be used to order projects (alternatives a_j and a_k) according to their preference:

$$a_j \succcurlyeq a_k \Leftrightarrow NPV(a_j) \geq NPV(a_k).$$

Returning to the numeric example for the two projects, and assuming a discount rate of $r=10\%$, we get:

- $NPV(a_1) = -4 -2/1.1 + 1/1.1^2 + 3/1.1^3 + 6/1.1^4 = 1.36$
- $NPV(a_2) = -8 -4/1.1 + 3/1.1^2 + 6/1.1^3 + 10/1.1^4 = 2.18$

We would therefore conclude that project a_2 is more preferred than a_1.

2.6 Interpretation of 2-Dimensional Value Functions

We will use the 2-dimensional value function to make some interpretations of value functions that could be extended to more than two dimensions. Figure V.8 shows an indifference curve of a value function, $v(a)$.

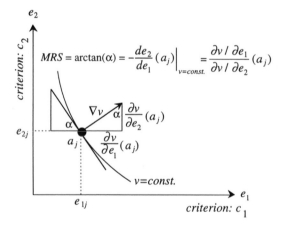

Figure V.8: Direction of maximum value increase and marginal rate of substitution.

If $v(a)$ has continuous partial derivatives $\partial v / \partial e_1$ and $\partial v / \partial e_2$, then the direction of maximal value increase from action a_j: (e_{1j}, e_{2j}) is the **gradient** of $v(a)$ at (e_{1j}, e_{2j}), where e_1 and e_2 are unit vectors in direction of e_1 and e_2, respectively:

$$\nabla v(e_{1j}, e_{2j}) = \frac{\partial}{\partial e_1} v(e_{1j}, e_{2j}) e_1 + \frac{\partial}{\partial e_2} v(e_{1j}, e_{2j}) e_2.$$

The partial derivatives, $\partial v / \partial e_i$, are called **marginal values**.

For example, let:

- $v(e_1, e_2) = 3(e_1)^2 - 2(e_2)^2$, and a_j: (2,-4),

then the direction of maximal increase of $v(a)$ from a_j is:

- $\nabla v(2, -4) = 6 \times 2 \times [1,0]^T - 4 \times (-4) \times [0,1]^T = [12,16]^T$.

Moreover, it is also known that the direction of maximal increase of the value function is orthogonal to the tangent in a_j, and that the absolute value of the gradient, $|\nabla v(e_1, e_2)|$, is the maximum value of the directional derivative at (e_{1j}, e_{2j}).

The **marginal rate of substitution** (MRS) reflects how much the decision maker needs to be compensated on e_2 in order to give up one unit on e_1, or, how much s/he is willing to give up on e_1 if s/he is offered one unit of e_2. The MRS is also called the **tradeoff value**. The MRS can be computed from the total differential which is defined as:

$$dv = \frac{\partial v}{\partial e_1} de_1 + \frac{\partial v}{\partial e_2} de_2.$$

Because tradeoff means to gain on one criterion by giving up on another, so that overall nothing changes, we must set $dv=0$. Thus, setting the total differential equal to zero, we get:

$$MRS \equiv -\frac{de_2}{de_1}\bigg|_{v=const.} = \frac{\partial v / \partial e_1}{\partial v / \partial e_2}.$$

For example, let's take the:

- **implicit** value function: $v(e_{1j}, e_{2j}) = 3(e_1)^2 - 2(e_2)^2$,
- and look at the indifference line: $19 = 3(e_1)^2 - 2(e_2)^2$, and the action a_j: (3,2).

We then get the **explicit** function:

- $e_2 = [0.5(3e_1^2 - 19)]^{0.5}$.
- Thus, $de_2 / de_1 = (3/2)e_1[0.5(3e_1^2 - 19)]^{-0.5}$.
- For the point (3,2), we then get: $MRS=-9/4$.

If we compute the MRS with the total differential, we get:

- $\partial v / d e_1 = 6e_1$, $\partial v / d e_2 = -4e_2$.
- Thus, we get: $MRS = -6e_1/4e_2$, and at a_j, $MRS = -9/4$.

For an ordinal value function we can show that the MRS is invariant under strictly increasing transformations. For example, let's assume that we have a value function:

- $v(e_{1j}, e_{2j}) = (e_1 e_2)^{0.5}$.
- Then: $\partial v / \partial e_1 = 0.5(e_2/e_1)^{0.5}$, and $\partial v / \partial e_2 = 0.5(e_1/e_2)^{0.5}$,
- which gives $MRS = (\partial v / \partial e_1)/(\partial v / \partial e_2) = e_2/e_1$.

If we use the:

- strictly increasing transformation, $w = \log(v)$,
- we have $w = 0.5\log(e_1) + 0.5\log(e_2)$.
- Thus, we get $\partial w / \partial e_1 = 0.5/e_1$, and $\partial w / \partial e_2 = 0.5/e_2$.
- Therefore, $MRS = (\partial w / \partial e_1)/(\partial w / \partial e_2) = [(1/e_1)/(1/e_2)] = e_2/e_1$,

which is identical to the MRS prior to the transformation.

For the additive value function $v(e_{1j}, e_{2j}) = 3(e_1)^2 - 2(e_2)^2$, we can show that the MRS is invariant under positive linear transformations. The reader should confirm that $MRS = -9/4$ in point (3,2) still holds using, for example, the transformation $w = 4v + 5$. A linear value function has a constant MRS; for the 2-dimensional example in Section 2.4, we have $MRS = k_1/k_2$.

2.7 Interpretation of Component Value Functions

In general, we can distinguish three different shapes for a monotonically increasing component value graph (Figure V.9).

A **convex** value graph ($d^2 v / de^2 < 0$) stands for a decision maker who prefers an increase of low outcomes more than an increase of high outcomes by the same amount (δ):

$$v(e_{worst} + \delta) - v(e_{worst}) > v(e_{best}) - v(e_{best} - \delta).$$

A **concave** value graph ($d^2 v / de^2 > 0$) stands for a decision maker who prefers an increase of high outcomes more than an increase of low outcomes by the same amount (δ):

$$v(e_{worst} + \delta) - v(e_{worst}) < v(e_{best}) - v(e_{best} - \delta).$$

Finally, a **linear** value graph ($d^2v/de^2 = 0$, $dv/de = const.$) stands for a decision maker who is indifferent between an increase of high and low outcomes by the same amount (δ):

$$v(e_{worst}+\delta) - v(e_{worst}) = v(e_{best}) - v(e_{best}-\delta).$$

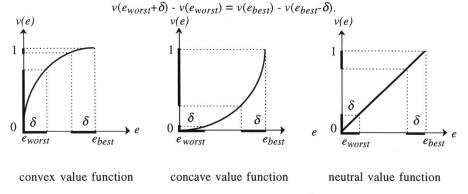

| convex value function | concave value function | neutral value function |

Figure V.9: Convex, concave, and linear value functions, $v(e)$.

3. The Resolution Model

3.1 The General Approach

The process of assessing a subjective multicriteria additive value function is illustrated in Figure V.10.

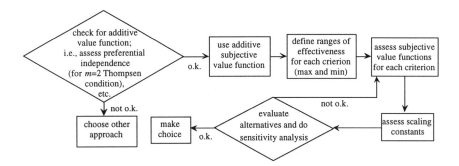

Figure V.10: Process of assessing a subjective multicriteria additive value function.

The alternatives may be either discrete (e.g., selection of most appropriate policy out of a finite set of feasible policies) or continuous. In the former case, a finite number of alternatives, the criteria, and the effectiveness values of the alternatives with respect to all criteria are provided in an evaluation table, $E^{m \times n}$. The decision maker must be aware of the axiomatic concept of the normative approach and s/he must accept the normative character of the modeling paradigm. The decision maker can act as an individual with personal preferences and interests, or s/he can be asked to take the view of a group of people or a segment of society. For example, the decision maker could make the subjective assessments as a representative of an environmental interest group or as the authority which has to decide on a public transportation project.

3.2 Assessment of Mutual Preferential Independence

The first step in the elicitation of an m-dimensional value function is to assess mutual preferential independence of the criteria. This means that for any subset of criteria it must be shown that it is preferentially independent of its complement (and vice versa for mutual preferential independence). That is, first the individual criteria are checked with their complements, then pairs, then triples, etc. If many criteria are involved, the assessment of preferential independence can become quite work intensive. We discussed earlier that for m criteria we would need to do 2^m-2 assessments. However, it can be shown that fewer assessments suffice to assure mutual preferential independence of the m criteria.

In fact, the following theorem can be shown ([Gorman, 1968], [Keeney and Raiffa, 1993, Chapter 3.6.3]): if any two subsets C_1 and C_2 of the set of criteria $C=\{c_1,...,c_m\}$, where C_1 and C_2 overlap but neither is contained in the other and $C_1 \cup C_2 \neq C$, are each preferentially independent of their complement, then the following sets of criteria

A) $C_1 \cup C_2$,
B) $C_1 \cap C_2$,
C) (C_1-C_2) and (C_2-C_1), and
D) $(C_1-C_2) \cup (C_2-C_1)$

are each preferentially independent of their respective complement. The operator '-' between two sets, C_1-C_2, means to remove all those elements from C_1 which are also contained in C_2. This result reduces the necessary checks for preferential independence to $m-1$ assessments.

From this theorem it can be derived that if every pair of criteria is PIoC then the criteria are mutually preferentially independent (see Problem 5). For example, let $C=\{c_1,c_2,c_3,c_4,c_5\}$; this should require 4 (=5-1) assessments of preferential independence. With five criteria there are 10 pairs of criteria. However, we will now show that we need to assess preferential independence for only four pairs. Let these four pairs be $\{c_1,c_2\}$, $\{c_2,c_3\}$, $\{c_3,c_4\}$, and $\{c_4,c_5\}$. Preferential independence holds if the decision maker agrees with the following relations, for example for $\{c_1,c_2\}$: $(x,y,\alpha,\beta,\gamma) \succsim (r,s,\alpha,\beta,\gamma) \Leftrightarrow (x,y,\delta,\varepsilon,\phi) \succsim (r,s,\delta,\varepsilon,\phi)$. If this can be assumed, we know that $\{c_1,c_2\}$ is preferentially independent of its complement (PIoC).

Using relation D for the two sets $C_1=\{c_1,c_2\}$ and $C_2=\{c_2,c_3\}$ gives $(C_1-C_2)=\{c_1\}$ and $(C_2-C_1)=\{c_3\}$; thus, $(C_1-C_2)\cup(C_2-C_1)=\{c_1,c_3\}$. Concluding PIoC of the other six pairs of criteria from the four pairs we assessed with the decision maker is done as follows:

- with $\{c_1,c_2\}$ and $\{c_2,c_3\}$ being PIoC applied to D we conclude $\{c_1,c_3\}$ is PIoC,
- with $\{c_2,c_3\}$ and $\{c_3,c_4\}$ being PIoC applied to D we conclude $\{c_2,c_4\}$ is PIoC,
- with $\{c_3,c_4\}$ and $\{c_4,c_5\}$ being PIoC applied to D we conclude $\{c_3,c_5\}$ is PIoC,
- with $\{c_1,c_3\}$ and $\{c_3,c_4\}$ being PIoC applied to D we conclude $\{c_1,c_4\}$ is PIoC,
- with $\{c_2,c_3\}$ and $\{c_3,c_5\}$ being PIoC applied to D we conclude $\{c_2,c_5\}$ is PIoC,
- with $\{c_1,c_4\}$ and $\{c_4,c_5\}$ being PIoC applied to D we conclude $\{c_1,c_5\}$ is PIoC.

Using the fact that if all pairs of criteria are PIoC, then the criteria are preferentially independent, we know that we are done. We see indeed that we could reduce the originally necessary 2^m-2 (2^5-2=30) PIoC assessments to m-1 (5-1=4) PIoC assessments of pairs of criteria.

To perform these assessments of preferential independence of the criteria, we use the effectiveness values from the evaluation matrix. This implies that preferential independence considers the range of the evaluation values of the alternatives under investigation. Therefore, if we add or delete alternatives in a decision problem, we should also check if there is an impact on the preferential independence of the criteria. Finally, it should not be forgotten to check Thompsen's condition for when we have only two criteria.

3.3 Elicitation of 2-Dimensional Value Functions

Assuming that preferential independence and Thompsen's condition hold, we can use an additive value function. The process of eliciting a 2-dimensional additive value function should now be obvious from our discussion so far.

Assume that a decision maker is considering a transportation project which will be evaluated in terms of annual gains (c_1, evaluated in \$1,000) and annual time savings (c_2, evaluated in hours). We would then choose arbitrary points x and y with value $v_1(x)=v_2(y)=0$; for example, $v_1(4)=v_2(22)=0$. We then choose an additional point, e.g., $e_1=6$, such that $v_1(6)=1$. With these three points and with the characteristics of an additive value function, we get for any project (i,j):

$$v(i,j) = v_1(i) + v_2(j).$$

The 2-dimensional additive value function in Figure V.11 has been assessed by further asking the decision maker to identify the time saving y, such that $(6,22) \sim (4,y)$. Let's assume s/he assesses $y=23$. Then we have: $v_2(23)=v_1(6)=1$. Because of the additive nature of the value function it can be concluded that $v(6,23)=v_1(6)+v_2(23)=2$, (Figure V.11, left).

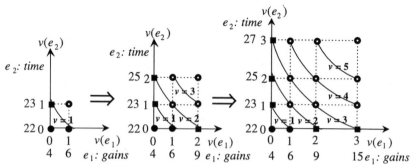

Figure V.11: Elicitation of a 2-dimensional additive value function
(●: chosen, ■: assessed, and ○: concluded values).

The decision maker is then asked to assess gains x and time savings y, such that: $(x,22) \sim (4,y) \sim (6,23)$. Let's assume s/he assesses $x=9$ and $y=25$. Then, we have: $v(9,22)= v(4,25)=v(6,23)=2$. From this we conclude $v(9,23)=v(6,25)=3$, and $v(9,25)=4$ (Figure V.11, middle).

We then go on by asking the decision maker to assess gains r and time savings s, such that: $(r,22) \sim (9,23) \sim (6,25) \sim (4,s)$. Let's assume s/he assesses $r=15$ and $s=27$. Then, we have: $v(15,22)=v(9,23)=v(6,25)=v(4,27)=3$. From this we conclude that $v(15,23)=v(9,25)=v(6,27)=4$, $v(15,25)=v(9,27)=5$, and $v(15,27)=6$ (Figure V.11, right).

3.4 Elicitation of Component Value Functions

If preferential independence holds, the component value function is independent of the levels of the other components. There are different ways to assess a component value function. The first step is to normalize the outcome ranges of the value function. Usually, the interval $[0,1]$ is chosen, but any other interval could also be used, for example, $[0,100]$. The best effectiveness value $e_{i,best}$ for criterion c_i is assigned a preference value of 1, and the worst, $e_{i,worst}$, a value of 0. Thus, for c_i, we have $v_i(e_{i,best})=1$, and $v_i(e_{i,worst})=0$.

With the **mid-value-splitting technique** [Torgerson, 1958], we ask the decision maker for the mid-value effectiveness $e_{i,0.5}$, where $v_i(e_{i,0.5})=0.5$ (Figure V.12). This means we ask the decision maker to identify the effectiveness value $e_{i,0.5}$, such that an improvement from $e_{i,worst}$ to $e_{i,0.5}$ is indifferent to an improvement from $e_{i,0.5}$ to $e_{i,best}$:

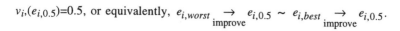

$v_i, (e_{i,0.5})=0.5$, or equivalently, $e_{i,worst} \xrightarrow[\text{improve}]{} e_{i,0.5} \sim e_{i,best} \xrightarrow[\text{improve}]{} e_{i,0.5}.$

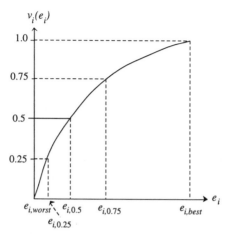

Figure V.12: Principle of mid-value-splitting technique.

This procedure is continued to determine $e_{i,0.75}$, using $e_{i,0.5}$ and $e_{i,best}$, $e_{i,0.25}$, using $e_{i,worst}$ and $e_{i,0.5}$, etc., until the value curve can be approximated through these points.

As an example let's assume two different communities (decision makers) are faced with a transportation project, where one criterion is cost of the project. The criteria are assumed to be preferentially independent. Best and worst costs have been identified (the same for both communities), and the value function ranges from $v(worst)=0$ to $v(best)=1$. Let's assume that one community is very rich and the other very poor. For the rich community, the worst outcome might not be too big a financial burden and the best outcome might be easily affordable. For the poor community, the worst outcome might cause major financial problems and the best outcome be just about affordable. Thus, the rich community is fairly neutral about costs, meaning that the mid-value lies in the middle of the effectiveness interval, $e_{i,0.5}=(e_{i,worst}-e_{i,best})/2$. Such a neutral value function can be obtained by the normalization function, n_{ij}, which we encountered in Chapter II, Section 3.2, B, where v_{ij} stands for the normalized evaluation value, e_{ij} (alternative a_j and criterion c_i):

$$v_{ij} = \frac{e_{ij} - e_{i\,worst}}{e_{i\,best} - e_{i\,worst}}.$$

Note that the neutral value function is linear. The poor community, on the other hand, prefers more decreases of high costs than decreases of low costs. This attitude towards project costs results in a convex value function (see Figure V.9).

Kirkwood and Sarin [1980] have shown that the following **exponential value function** is a reasonable approach for monotonically increasing value functions (i.e., e_{ij} gets transformed into the preference value v_{ij}):

$$v_{ij} = \frac{e^{\rho e_{ij}} - e^{\rho e_{i,worst}}}{e^{\rho e_{i,best}} - e^{\rho e_{i,worst}}}.$$

For $\rho \rightarrow \pm 0$, we have a linear value function $v_{ij} = (e_{ij} - e_{i,worst})/(e_{i,best} - e_{i,worst})$. Two characteristics of this function should be noted: (1) the shape of the function for a fixed ρ depends on the range defined by e_{best} and e_{worst} (e.g., gains expressed in $\$10^6$ or in $\$$ result in different shapes of v_{ij}); (2) for monotonically increasing preferences (e.g., gains, where higher outcomes are preferred to lower ones), $\rho > 0$ results in a concave, and $\rho < 0$ in a convex value function, while for monotonically decreasing preferences (e.g., costs, where lower outcomes are preferred to higher ones) $\rho > 0$ results in a convex and $\rho < 0$ in a concave value function. Figure V.13 shows several graphs for selected values of ρ between 0 (worst) and 10 (best).

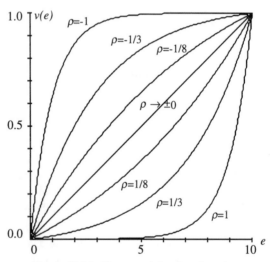

Figure V.13: Exponential value function.

3.5 Assessment of Scaling Constants

If we do not assess the additive value function directly, as we discussed for two criteria, but through the component value functions, scaled from zero to one, then we have the following additive value function over the m criteria:

$$v(a_j) = v(e_{1j},...,e_{mj}) = k_1 v_{1j}(e_{1j}) + ... + k_m v_{mj}(e_{mj}),$$

Chapter V: Values and Normative Choice

where $v_i(e_{ij})$ is the component value for criteria c_i with the scaling constant k_i, and with $\sum_{i=1}^{m} k_i = 1$, $v_{i,best}=1$, and $v_{i,worst}=0$.

To discuss the process of determining the scaling constants, let's introduce the following notation. The alternative $a_{\{x,y\}}$ is a (fictive) alternative, for which the best effectiveness value for criteria c_x and c_y are $e_{x,best}$ and $e_{y,best}$, and the worst effectiveness value is $e_{i,worst}$, for all other criteria. Thus, using the definition of the additive value function, we get, for example,

$$v_x(a_{\{x\}}) = k_1 \times 0 + ...+ k_x \times 1 + ...+ k_m \times 0 = k_x.$$

In the same way, we also get $v(a_{\{x,y\}})=k_x+k_y$.

The first step in assessing the scaling constants k_i is to rank the (fictive) alternatives $a_{\{1\}},...,a_{\{m\}}$; that is, the alternatives a_i which perform best for criterion c_i and worst for all other criteria. This ranking corresponds to the ranking of the scaling constants k_i. For example, if we have four criteria, we could get the following preference ranking: $a_{\{3\}} \succ a_{\{4\}} \succ a_{\{1\}} \succ a_{\{2\}}$. This ranking implies: $k_3 > k_4 > k_1 > k_2$. At this point we know the order of the scaling constants but not their numerical values.

To obtain the numerical values of the scaling constants we would ask the decision maker to compare two alternatives, for example, $a_{\{x\}}$ and $a_{\{y\}}$. Let's assume that s/he assesses $a_{\{x\}} \succ a_{\{y\}}$. Then, we deteriorate $a_{\{x\}}$ by reducing the effectiveness value at the level of criterion c_x to e_{x*}, such that $a_{\{x*\}} \sim a_{\{y\}}$. We now have:

$$k_x v_x(e_{x*}) = k_y.$$

The value $v_x(e_{x*})$ can be determined from the component value function of criterion c_x; thus, we know the proportional relationship between the two scaling constants. Comparing other pairs of k_i's, and considering that $\sum_{i=1}^{m} k_i = 1$, we can compute the numerical values for all k_i's.

For example, let's assume we have determined the following relationships for the example above $(k_3 > k_4 > k_1 > k_2)$: $k_4/k_3=0.3$, $k_1/k_3=0.2$, $k_2/k_3=0.1$, and $k_1+k_2+k_3+k_4=1$. We now must solve the following system of linear equations:

$$\begin{bmatrix} 0 & 0 & -.3 & 1 \\ 1 & 0 & -.2 & 0 \\ 0 & 1 & -.1 & 0 \\ 1 & 1 & 1 & 1 \end{bmatrix} \times \begin{bmatrix} k_1 \\ k_2 \\ k_3 \\ k_4 \end{bmatrix} = \begin{bmatrix} 0 \\ 0 \\ 0 \\ 1 \end{bmatrix}.$$

The result is: $k_3=0.625$, $k_4=0.1875$, $k_1=0.125$, $k_2=0.0625$. Here too, as in any procedure where we ask the decision maker for his/her preferences, the assessment of the scaling constants is not a unique procedure. For example, we could have asked the decision maker for any other preferences such that we get four equations to compute the four scaling constants.

With these scaling constants and the component value functions we now have a multi-dimensional additive value function. However, we must always remember that the assessments of preferential independence, the shape of the component value functions, and the scaling constants refer to a specific problem and decision maker. Consequently, if alternatives are altered or if new decision makers get involved in the problem solving process, the assessments must be repeated.

3.6 Solution Search

Value functions can be used to find optimal decisions from a set of discrete or continuous decision options. If we use discrete decision options, such as the selection of a transportation policy, we proceed in the same way as we did with the normalized and weighted evaluation matrix $N^{m \times n}$ in Chapter IV. Continuous decision options require a different resolution model to search for the most preferred solution.

A) <u>Convex Sets and Efficient Frontier</u>

The way we defined a monotonically increasing conditional value function to be convex implies that if we connect any two points on the value function, the graph lies above the connecting line defined by the two points. On the other hand, if it lies below this connecting line, the component value function is concave. Similarly, a set of actions, $A=\{a_j=(e_{1j},e_{2j})\}$, in the e_1-e_2 plane is said to be convex, if any straight line between two actions a_j and a_k is completely within A. Figure V.14 shows examples of **convex sets** and sets which are by definition not convex.

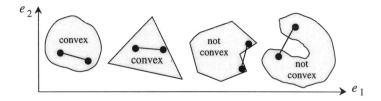

Figure V.14: Examples of convex sets.

In addition to defining convexity for 1-dimensional graphs and 2-dimensional sets, we can also define convexity of 2-dimensional indifference lines. An indifference line, $v(e_1,e_2)=const$ through the actions a_j and a_k is convex if all points a_x on the straight line connecting the two actions a_j and a_k either all have higher or all have lower values than a_j and a_k. Figure V.15 shows a convex and a non-convex set.

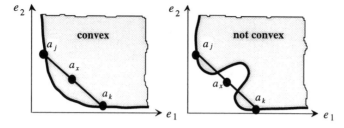

Figure V.15: Convex (left) and not convex (right) indifference curves.

Indifference curves contain often unbalanced alternatives. This means that an alternative scoring low on the first criterion and high on the second can be indifferent to an alternative scoring high on the first and low on the second. If the indifference lines are convex, we can see that between any two such unbalanced alternatives, there are alternatives which are more balanced and more preferred (higher value). If we assume in Figure V.15 (left) that a_j: (e_{1j}, e_{2j}) and a_k: (e_{1k}, e_{2k}) lie on a convex indifference curve, then the alternative a_x: $((e_{1j}+e_{1k})/2, (e_{2j}+e_{2k})/2))$ has more balanced effectiveness values and is more preferred than both a_j and a_k. This implies that balancing the effectiveness values results in higher preference values. For example, balanced meals are better than unbalanced ones, or balancing the length and width of a rectangle (given constant enclosure-length) results in a greater area.

Figure V.16 shows several examples of **efficient frontiers**. Top left shows the efficient frontier of a non-convex discrete set, top right shows the efficient frontier of a non-convex continuous set, bottom left shows an efficient frontier of a convex continuous set, and bottom right shows an efficient frontier of a non-convex continuous and non-connected set. If we also know the direction of growth of the value function, we can determine the most preferred alternative.

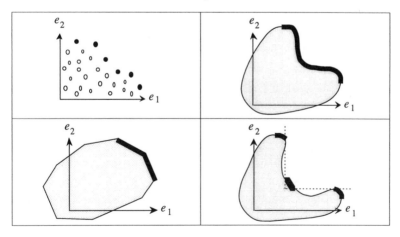

Figure V.16: Examples of efficient frontiers.

B) Evaluation of Explicit versus Search in Set for Implicit Alternatives

When the decision options are finite and represented explicitly (e.g., a set of infrastructure projects), we simply proceed as we did in Chapter IV, with the normalized evaluation matrix $N^{m \times n}$. We could also try to reduce the dimensions as discussed in Chapter III and try to find a 2-dimensional graphical representation of the m-dimensional decision problem, where m is the number of criteria.

If the alternatives are defined implicitly in terms of a set such as in Figure V.16, we have to search through this set to find the most preferred solution. A graphical approach is to overlay the value function (or its indifference curves); the alternative on the efficient frontier with greatest preference value is then the most preferred one. Thus, we are dealing with an optimization problem: to maximize the preference value subject to the constraint that the most preferred alternative must be contained in the set of feasible solutions. This optimum is defined by the tangency point (see Figure V.17, left).

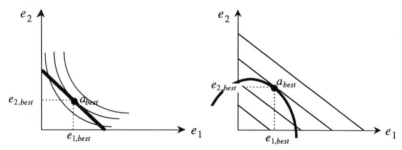

Figure V.17: Tangency method to determine the best alternatives.

The tangency approach works also for a linear value function and a non-linear constraint line (see Figure V.17, right). However, what we are searching for (in the 2-dimensional case) are two evaluation values e_1 and e_2 within the set of feasible pairs of evaluation values such that the value function is maximized (for monotonically decreasing value functions we would minimize the value function). This means that we can see the evaluation values as decision variables x_1 and x_2.

This makes especially sense if we specify the meanings of the value function and the constraint. For example, let c_1 be the expected savings and c_2 the future safety of a transportation project. The value function $v(e_1,e_2)=e_1+e_2$ could stand for the long term return, meaning that savings and safety contribute equally to the long term return. The goal would be to maximize the long term return. We can further assume that safety and savings cost money; that is, the safer and more cost saving a transportation system is, the more expensive it is to realize. The costs (C) of a transportation system, in terms of its amount of savings (e_1) and safety (e_2), could be $C=0.2(20-e_1)^2-e_2$.

What we have done now is the following: we have used the original two criteria (savings and safety) to define their impact on other criteria (return and costs). Because we search for the optimal outcomes, e_1 and e_2 (in terms of our original two criteria), it makes sense to see them as our decision variables, x_1 and x_2. Thus, the solution we are looking for consists of two basic actions: x_1 which is the amount of savings and x_2 which is the amount of safety. The two together define the optimal alternative. In summary, we can state our decision problem as follows:

- maximize return: $\quad v = x_1 + x_2$
- such that the costs: $\quad C = 0.2(20 - x_1^2) - x_2$

C) Real-Valued Solutions

The optimal solution in an m-dimensional convex set can be found by setting all partial derivatives of the value function equal to zero (first-order condition which suffices for convex sets). This gives m equations for m variables (the m coordinates of the best alternative). If we add a constraint, we have an additional equation but no new variable.

A mathematical method for solving such (non-linear) **constraint-optimization** problems is the **Lagrange multiplier method** (Joseph Louis Lagrange, 1736-1814, French mathematician). Suppose we want to maximize $v(x_1,...,x_m)$, subject to a constraint: $g(x_1,...,x_m)=0$. We use the equation

$$\mathcal{L} = v(x_1,...,x_m) + \lambda g(x_1,...,x_m),$$

where λ is a new variable, called the **Lagrange multiplier** (if we have n constraints, we add n new equations, each with a different multiplier). It should be noted that when the constraint holds, we have $\mathcal{L}=v$.

Thus, we apply the first-order condition to the function $\mathcal{L}$ and compute: $\partial \mathcal{L} / \partial x_i = \partial v / \partial x_i + \lambda(\partial g / \partial x_i)=0$, for $i=1,...,m$, and $\partial \mathcal{L} / \partial \lambda = g = 0$. This gives $m+1$ equations for the m decision variables $x_1, ... , x_m$ and λ. If we solve for λ, we get:

$$\lambda = \frac{\partial v / \partial x_i}{-\partial g / \partial x_i}.$$

If v is a benefit function and g a cost constraint, we get [Nicholson, 1995]:

$$\lambda = \frac{\text{marginal benefit of } x_i}{\text{marginal cost of } x_i}.$$

This means that λ can be seen as a **shadow price**, in the sense that much can be gained for v if we relax the constraint g for when λ is large, while not much can be gained if λ is small.

As an example, let's assume that a decision maker wants to decide on a transportation system by considering the criteria cost (c_1) and reliability (c_2). The decision maker requires a constant tradeoff (MRS) between reliability and cost of 2/3 (i.e., his/her value function is linear). Experience shows that there is a relation between cost and reliability: $x_2=0.2(20-x_1{}^2)$. What are the costs and reliability of the most preferred transportation system?

Solution: The decision maker seems to have a constant MRS of 2/3. This implies a linear value function of $v(x_1,x_2)=2x_1+3x_2$, which s/he wants to maximize, subject to the constraint $g=0=0.2(20-x_1{}^2)-x_2$. Figure V.18 shows the situation with the optimal solution.

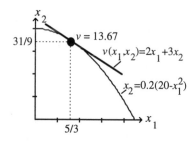

Figure V.18: Constraint-optimization problem.

The Lagrange multiplier method then gives: $\mathcal{L}=2x_1+3x_2+\lambda[0.2(20-x_1{}^2)-x_2]$. Setting the partial derivatives equal to zero gives: $\partial \mathcal{L}/\partial x_1=2-0.4\lambda x_1=0$, $\partial \mathcal{L}/\partial x_2=3-\lambda=0$, and $\partial \mathcal{L}/\partial\lambda=0.2(20-x_1{}^2)-x_2=0$. From the second equation we get $\lambda=3$, and inserting this into the first equation gives $x_1=5/3$. Inserting this into the third equation (which is the constraint) gives $x_2=31/9$. Thus, the optimal solution is $v=(5/3,31/9)=13.67$.

This means that because $\lambda=3$, we know that if we relax the constraint from $g=0$ to $g=k$ we gain $3\times k$ units on v. The reader should verify this by repeating the computations with $g=k=0.2(20-x_1{}^2)-x_2$.

3.7 Sensitivity Analysis

Sensitivity analysis is used to assess the robustness of the evaluations of the alternatives, and the resulting rankings of the alternatives under investigation. Much of what has been said in Chapter IV about sensitivity analysis also holds here. The robustness depends also on the value functions for each criterion and on the scaling constants. If small variations in these parameters result in a different ranking of the alternatives (or a different choice of the most preferred alternative), then the model is sensitive. The sensitivity of a parameter is checked by increasing and decreasing its value until the original results change (e.g., until the best alternative is not best anymore or until rank reversal occurs). This should be done for all component value functions and scaling factors. While the component value functions can be altered at will, the scaling factors, when changed, must still sum up to one. Thus, scaling factors can not be altered one at the time.

Sensitivity analysis can be done not only by changing the parameters of the additive value function, but also by varying the point of view of the decision maker. The assessment of the value function is done by a decision maker taking a specific point of view. S/he can take a more conservative or progressive point of view, or can try to represent different interest groups, such as environmental groups, citizens, industry, etc. Changing the points of view to assess the robustness of the results has a different purpose than changing the parameter values. Sensitivity analysis by altering the points of view provides important insights. First, the decision maker can assess which interest groups might support his/her point of view. Second, s/he can see where other interest groups might place most emphasis. Both findings would provide helpful information for the negotiation process in conflict resolution.

4. Summary and Further Readings

A rational decision maker complies with a set of axioms about his/her subjective preference structure. Based on the axioms and the assessments by the decision maker, an ordinal subjective value function can be derived. Decision analysts refer to preference theory under certainty as value theory, while economists call it utility theory. We use the term value theory. For a more in-depth treatment of value theory, see Keeney and Raiffa [1993], and for a basic introduction to the economic approach see Nicholson [1995]. These sources also cover utility theory, which will be discussed in Chapter VII. A more in-depth mathematical approach of rational decision making can be found in French [1988]. For an introduction to practical cost-benefit analysis see [Sugden and Williams, 1978].

If the criteria are preferentially independent (and for $m=2$ the Thompsen condition holds), the value function is additive. An additive value function is invariant under positive linear transformations. A linear value function has a constant marginal rate of substitution (tradeoff). Linear value functions are unique up to a multiplicative transformation.

The additive value function is assessed with the decision maker in the following steps: (i) assessment of mutual preferential independence of the m criteria in $m-1$ assessments (for $m=2$ the Thompsen condition must also be checked), (ii) identification of ranges for all criteria (best and worst evaluation values), (iii) assessment of conditional value functions for each criterion (e.g., with the mid-value splitting technique), (iv) assessment of scaling constants, and (v) evaluation of alternatives and sensitivity analysis (for parameters and for decision maker's points of view).

A component value function is either concave, convex, or linear. The gradient in a multidimensional value function is orthogonal to the indifference curves and points in the direction of maximum value growth.

In constraint-optimization problems, the most preferred alternatives are the ones on the efficient frontier with maximum preference value. They can be computed with the Lagrange multiplier method. In Chapter X we will discuss other computational methods for when both value and constraint functions are linear.

5. Problems

1. Describe decision problems where one decision maker has a convex, another a linear, and a third a concave value function for a specific criterion.

2. A subjective multicriteria value function returned the value of 0.6 for one alternative and the value of 0.3 for another alternative. The decision maker can conclude that the first is better than the second. However, can s/he also conclude that the first is twice as good as the second?

3. Show for two criteria ($m=2$) that mutual preferential independence is a necessary condition for an additive value function. Hint: insert the additive form of the value function into the definition of preferential independence and replace the fixed evaluation for the second criterion with another fixed evaluation.

4. Using the axioms of rational behavior discuss why a rational decision maker's indifference curves in the e_1-e_2 plane never intersect.

5. a) For five criteria, use the theorem mentioned in Section 3.12 to show that if every pair of criteria is PIoC then all criteria are mutually preferentially independent.

   b) Discuss how you would go about assessing preferential independence of six criteria. How many assessments are necessary and what are they?

6. Verify Thompsen's condition with the example in Figure V.7.

7. Let's assume that the decision maker has assessed the following subjective values for the nine 2-dimensional alternatives:

   $v(6,6) = 2,$ $v(6,7) = 3,$ $v(6,8) = 5$
   $v(7,6) = 3,$ $v(7,7) = 6,$ $v(7,8) = 7$
   $v(8,6) = 5,$ $v(8,7) = 8,$ $v(8,8) = 10$

Show that preferential independence holds, and verify that Thompsen's condition does not hold. Discuss the consequences.

8. Verify the example discussed in Figure V.18.

9. Discuss the concepts of convex sets and curves, and of efficient frontier and best alternative. Look at Figures V.14-16 and discuss additional examples.

10. A decision maker's value function (e.g., returns) for acquiring a certain number of acres of construction land in urban areas (x_1) and in rural areas (x_2) is: $v(x_1,x_2)=(x_1x_2)^{0.5}$. Assume that one acre urban land costs \$100 and one acre rural land \$25. If an organization has \$200 to spend, how many acres rural and urban land should it buy? Compute the Lagrange multiplier, λ, and make an interpretation of it.

CHAPTER VI

CHOICES UNDER UNCERTAINTY

1. Decision Making Under Complete Uncertainty

1.1 Structural Model

Decision making under uncertainty means that the alternatives are evaluated for different **scenarios**. A scenario was defined in Chapter I, Section 2.4, as a condition of the system under investigation, upon which all assessments and evaluations depend, but which is not under the control of the decision maker. For example, one could assess the benefit of a public transportation system for the four scenarios (1) high demand and strong economy, (2) low demand and strong economy, (3) high demand and weak economy, and (iv) low demand and weak economy.

Scenarios are described in terms of **partitions** of the total space of uncertainty. In this example we have two partitions: the demand (high or low) and the economy (strong or weak). A partition is therefore a collection of disjoint **states**, which describe completely the total space of uncertainty. This means that the demand is either high or low, and that no other type of demand can occur. The same holds for the economy; it is either strong or weak, and no other type of economy can be expected. States within a partition are therefore a **mutually exclusive** and **collectively exhaustive** description of the space of uncertainty. A scenario is any combination of the states. For two partitions, each with two states, we have a total of four scenarios. Consequently, the scenarios are also disjoint and form a mutually exclusive and collectively exhaustive description of the space of uncertainty. On the other hand, states from different partitions need not be mutually exclusive; for example, high demand (state from partition demand), and strong economy (state from partition economy) could be dependent on each other.

In this context we will make a distinction between complete uncertainty and informed uncertainty. **Complete uncertainty** means that we do not have enough data to quantify the chance of occurrence of the different states and resulting scenarios. **Informed uncertainty** refers to situations where we do have enough data to quantify the chance of occurrence of the different states and resulting scenarios. Decision making under informed uncertainty is also called **decision making under risk**. However, we will not make any distinction between the two types of uncertainty unless it is necessary in the context of the discussion.

A structural model illustrating the situation for problem solving under complete uncertainty is given in Figure VI.1. It is assumed that the subjective preferences with respect to all criteria have been aggregated to a subjective preference value (u), such that each alternative has one overall preference value for each scenario. This could have been done with any of the descriptive approaches discussed in Chapter IV or with the approach of value theory discussed in Chapter V. However, we will see in Chapter VII that the appropriate approach for assessing and aggregating preferences in the presence of uncertainty is utility theory. The exact meaning of a utility measure and the difference to a value measure (as introduced in Chapter V) will be discussed in Chapter VII. For the time being, we will use the term utility as a subjective preference measure, where higher values are preferred to lower ones.

Figure VI.1 shows that if we have v (5) scenarios, we must determine for each of the n (4) alternatives v overall utility values. The formal models which are inserted into the structural model in Figure VI.1 will be discussed in the next section. Obviously, we are dealing with complete uncertainty because the chances of occurrence (probabilities) for the five scenarios are not quantified.

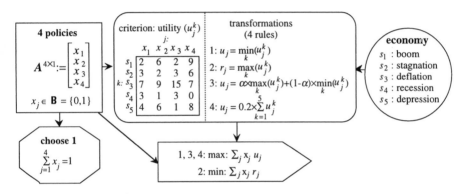

Figure VI.1: Structural model for complete uncertainty with five scenarios.

If the alternatives are evaluated with multiple criteria, it is possible that not all criteria are affected by uncertainty in the same way. For example, the costs of a policy could depend on the economic development, while the environmental impacts might not. This means for the structural model that not all criteria nodes are influenced by the same scenario nodes. In the following sections, however, we will confine our attention to the case where all evaluation criteria are affected by the same scenarios.

1.2 Formal Model

Problem solving under complete uncertainty means that the chance of occurrence of the different scenarios cannot be quantified. However, for each scenario the alternatives can be evaluated in terms of the chosen criteria. In this section we will only consider the cases with one criterion (e.g., gains), or where the multicriteria evaluations have been aggregated over all criteria to an overall preference or utility value (Figure VI.1). We will also assume that we have a monotonically increasing preference function, where higher utilities are preferred to lower ones.

The literature proposes four classical choice rules for solving single-criterion decision problems under complete uncertainty with explicit alternatives (Figure VI.1). These approaches are introduced and compared to each other in the following discussion. Let's assume that a decision maker must choose one out of four policies (alternatives) on the basis of their utilities (higher values are preferred to lower ones). Five economic scenarios have been identified (Figure VI.1). The score table below shows the four alternatives (a_j) and their utilities for each scenario ($u^k(e_j) \equiv u_j^k$).

u	a_1	a_2	a_3	a_4
s_1	2	6	2	9
s_2	3	2	3	6
s_3	7	9	15	7
s_4	3	1	3	0
s_5	4	6	1	8

An alternative (in our example a policy) a_j which for all scenarios s_k $(k=1,...,v)$ is weakly preferred to another alternative a_i and strongly preferred for at least one scenario is called **first-order stochastically dominant**:

- $a_j^k \succeq a_i^k$ for all k, and
- $a_j^k \succ a_i^k$ for at least one k.

This definition is identical to the one for dominant alternatives (introduced in Chapter I with respect to criteria) if we replace *scenario* with *criterion*. We can thus also say that the stochastically non-dominated alternatives are **(stochastically) efficient**. As we can see in the numeric example above, all alternatives are efficient and we need additional rules (or decision criteria) to decide which alternative is the most preferred one.

A) Wald's MaxMin Rule

A conservative decision maker bases his/her decision on the minimum utility, called **security level**, of the four alternatives, rather than on the highest possible utility. This decision maker agrees with the saying, 'a bird in the hand is better than two in the bush.' This preference principle is reflected in the **MaxMin** decision rule (first rule in Figure VI.1); it avoids worst-case scenarios, regardless of their chance of occurrence. In other words, it suggests taking the alternative whose smallest utility is the largest among all alternatives' smallest utilities:

$$\max_{j=1}^{n}[\min_{k=1}^{v}(u_j^k)].$$

Applying this rule to our decision problem, we find the minimum utilities for each alternative in the table below in gray. The highest of these minimum utilities is the one for the first alternative, a_1. Thus, we conclude that the best alternative is a_1.

u	a_1	a_2	a_3	a_4
s_1	2	6	2	9
s_2	3	2	3	6
s_3	7	9	15	7
s_4	3	1	3	0
s_5	4	6	1	8

B) Savage's MinMax Regret Rule

While Wald's rule suits a conservative decision maker, Savage's **MinMax Regret** decision rule is appropriate for a decision maker who wants to minimize possible regrets from having made the wrong decision (second rule in Figure VI.1). This means that Savage's rule suits a progressive decision maker. To evaluate the alternatives with Savage's rule, a regret table is constructed which shows the possible regrets for each alternative under each scenario. For example, if the decision maker chooses the first alternative when the second scenario occurs, s/he would miss 3 units of utility in regard to the fourth alternative. The optimal alternative is the one which minimizes the possible regrets:

$$\min_{j=1}^{n}[\max_{k=1}^{v}(r_j^k)].$$

When applying this decision rule to our decision problem, the following regret table results with the maximum regret for each alternative in gray. The alternative with minimum regret is the second alternative, a_2.

r	a_1	a_2	a_3	a_4
s_1	7	3	7	0
s_2	3	4	3	0
s_3	8	6	0	8
s_4	0	2	0	3
s_5	4	2	7	0

C) Hurwicz's Optimism-Pessimism Index

Wald's and Savage's decision rules are based on extreme values, the former on the minimum of possible utilities and the latter on the maximum of possibly missed utilities. Hurwicz proposes to look at the extreme outcomes for each alternative a_j: the highest, $\max_k u_j^k$, and the lowest, $\min_k u_j^k$. The overall measure of effectiveness used to judge an alternative should vary between these two extreme values (third rule in Figure VI.1). A more optimistic decision maker would place the measure of effectiveness closer to the maximum, and a more pessimistic decision maker would place it closer to the minimum. Hurwicz's **Optimism-Pessimism** decision rule says to maximize the weighted sum of highest and lowest outcomes:

$$\max_{j=1}^{n}\left(\alpha \max_{k=1}^{v} u_j^k + (1-\alpha)\min_{k=1}^{v} u_j^k\right), \quad 0 \le \alpha \le 1$$

If $\alpha=0$, then Hurwicz's decision rule is identical to Wald's decision rule; if $\alpha=1$, then the decision rule is the most optimistic rule possible; that is, a MaxMax approach which says to choose the alternative with the highest possible utility. Applying this rule to our decision problem with $\alpha=0.5$, we find the best alternative to be the third alternative, a_3.

u	a_1	a_2	a_3	a_4
s_1	2	6	2	9
s_2	3	2	3	6
s_3	7	9	15	7
s_4	3	1	3	0
s_5	4	6	1	8
$0.5\max_{k=1}^{v} u_i^k + (1-0.5)\min_{k=1}^{v} u_i^k:$	4.5	5.0	8.0	4.5

D) Laplace's Principle of Insufficient Reasoning

Wald's and Savage's decision rules were based on one utility value and Hurwicz's rule on two. The next step is to consider all utility values for the different scenarios. However, the chances of occurrence of the scenarios are unknown. Laplace assumes that all scenarios are equally likely; the decision is based on the average utility (fourth rule in Figure VI.1). This means that Laplace's **Principle of Insufficient Reasoning** decision rule considers all available information (i.e., all u_j^k) and makes some reasonable guesses about the missing information (i.e., the chance of occurrence of the scenarios). Thus, the decision rule says to choose the alternative a_j such that:

$$\max_{j=1}^{n} \left(\frac{1}{n} \sum_{k=1}^{v} u_j^k \right).$$

When applying this rule to our decision problem, the best alternative is the fourth alternative, a_4.

u	a_1	a_2	a_3	a_4
s_1	2	6	2	9
s_2	3	2	3	6
s_3	7	9	15	7
s_4	2	1	3	0
s_5	4	6	1	8
$\frac{1}{n} \sum_{k=1}^{v} u_j^k$:	3.6	4.8	4.8	6.0

1.3 Resolution Model

The numerical example for our decision problem was deliberately chosen so that each of the four decision rules would result in a different decision (optimal alternative). This is not always the case. Every decision rule is based on different subjective points of view that a decision maker can adopt. These include conservatism (Wald), no regret (Savage), variations from conservatism to optimism (Hurwicz), and average utility (Laplace). These rules might have some similarities to the approaches discussed so far. However, in the previous chapters, we were dealing with multiple criteria instead of multiple scenarios. As we discussed in Chapter I, there are in fact some similarities among criteria, scenarios, and decision makers. They all represent different points of view for a problem at hand. Consequently, concepts like dominant and efficient alternatives can be applied to criteria, scenarios, and decision makers.

2. Decision Making Under Risk

2.1 Structural Model

In decision making under informed uncertainty, also called decision making under risk, we have enough data to quantify the chances of occurrence of the scenarios. Scenarios are often defined in terms of partitions and their states. This is possible because every partition encompasses the total space of uncertainty. But partitions, unlike scenarios, represent only fragmented information about an uncertain situation. Moreover, the chance of occurrence of the states of one partition might be dependent

---

on the states from other partitions. These dependencies among partitions are illustrated in the structural model by arrows.

Figure VI.2 shows an example of a decision problem under uncertainty, with four partitions of the total space of uncertainty. The partition *weather* is divided into three states (sun, rain, snow), the partition *equipment* into two (new, old), the partition *worker* into two (novice, expert), and the partition *system safety* into three (safe, critical, unsafe).

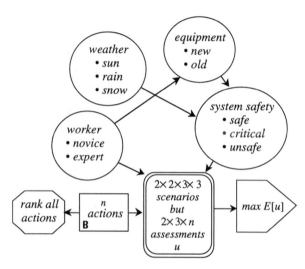

Figure VI. 2: Structural model for decision making under risk.

The uncertainties of the states and the resulting 36 (3×2×3×2) scenarios can be quantified with probabilities. The structural model says that the uncertainty (probability) of the equipment (new, old) is conditioned (dependent) on the qualification of the *worker* (novice or expert). This means that we have to make four probability assessments for the partition *equipment*. We also see that the *system safety* is directly influenced by the *weather* and the *equipment*.

Finally, we see that the evaluation of the alternatives (assessment of the utilities) must be done for six out of the resulting 36 scenarios. These six scenarios consist of all combinations that result for the two *worker* states (novice, expert) and the three *system safety* states (safe, critical, unsafe). Because the utility is influenced by 6 scenarios, we might also say that we have six **evaluation scenarios**. However, the joint uncertainty is made up of 36 scenarios, which are derived from of all combinations of the states of the four partitions. With the six evaluation scenarios, expected utilities are computed; the content goal is to maximize the expected utility. The discussion about the structural model might appear somewhat abstract at this point. However, as soon as we introduce the formal model, the above statements will become immediately clear.

2.2 Formal Model: Concepts of Probability Theory

The most common way to quantify uncertainty is with the concept of probability theory. The concept of probability theory dates back to the 17th century, the time of Blaise Pascal (1623-1662) and Pierre Fermat (1605-1665), when uncertainty referred to gambling situations. A century later, the works of Karl Gauss (1777-1855) and Pierre Laplace (1749-1827) laid the basis for today's probability theory. Two schools of thoughts prevail in probability theory: the objectivists and the subjectivists. The **objectivists** see probability as a measure which is independent of human judgment and thus constant from person to person. They can further be divided into **logical** (classicists) and **empirical** (frequentists) objectivists. The frequentists see probability as a limit of long-run relative frequency, a concept which was introduced by Denis Poisson (1781-1840) in 1837.

The **subjectivists** see probability as an individual's measure of belief or confidence in a statement. This type of probability varies from person to person. This does not mean that one can manipulate probability values to his/her favor; rather, subjectivists expect the analyst to act as rationally and consistently as possible. These characteristics, however, have been criticized by behavioral scientists. Human biases in probability assessments have been found to be a major concern, especially for extremely rare events where no historical data is available. Subjective probabilists are also called **Bayesians** because they revise beliefs with Bayes' theorem which will be discussed later on in this section.

A) Types of Probability

The axioms of Kolmogorov are accepted by both objectivists and subjectivists. In fact, they have been introduced to overcome perceptual and mathematical problems of defining probability as the limit of a long-run frequency. Rather than attempting to define probability, Kolmogorov's approach is to regard probability as a generic term which complies with certain axioms [Bronstein and Semendjajew, 1981].

The first axiom, which says that the probability of an event, state, or scenario is a number between zero and one, is based on the concept that for a large number (n) of trials, the relative frequency (number of occurrences of the event divided by n) fluctuates with increasingly small variation around a certain value between zero and one.

To better understand the following definitions, let's look at an example. The outcomes of rolling a die (1, 2, 3, 4, 5, and 6) are the possible basic events (states or scenarios) in the uncertainty space. Let's introduce two partitions A and B, where A refers to the size of the outcome (< 2 or ≥ 2), and B to the kind of outcome (even or odd). It should be noted that these two partitions are each described by two disjoint states, (a_1, a_2) for A and (b_1, b_2) for B. These two partitions with their two states define thus four scenarios: (a_1, b_1), (a_2, b_1), (a_1, b_2), and (a_2, b_2).

Let's assume that we have a fair die (all six outcomes are equally likely). The types of outcomes are given below (left). In the long-run (if we repeatedly roll the die), the proportions of the frequencies of the four scenarios are given below in the table in the middle. From these frequencies we can derive the probabilities of the four scenarios (table below on the right).

	$a_1 :< 2$	$a_2 : \geq 2$		a_1	a_2		$p(a_1)$ 1/6	$p(a_2)$ 5/6
b_1: even	$\varnothing$	$\{2,4,6\}$	b_1	0	3	$p(b_1)=3/6$	0	3/6
b_2: odd	$\{1\}$	$\{3,5\}$	b_2	1	2	$p(b_2)=3/6$	1/6	2/6
types of outcomes			frequency of scenarios			probabilities of scenarios		

The axioms of probability theory as defined by Kolmogorov can then be stated as follows. Let S be the total uncertainty space of all scenarios, a_i a state of partition A, b_j a state of partition B (where $A=B$ is possible), $p(a_i)$ the probability of state a_i and $p(b_j)$ the probability of state b_j. The three **axioms** of Kolmogorov are:

1. $p(a_i)$, $p(b_j) \geq 0$, for all i,j,
2. $p(S) = 1$,
3. If $(a_i \cap b_j) = \varnothing$, then $p(a_i \cup b_j) = p(a_i) + p(b_j)$,

where $(a_i \cap b_j)$ means that the two states a_i and b_j occur simultaneously, and $(a_i \cup b_j)$ means that a_i or b_j or both occur. For the example above, we have $(a_1 \cap b_1) = \varnothing$ (i.e., there is no outcome which is even and, at the same time, smaller than 2), and therefore $p(a_1 \cup b_1) = p(a_1) + p(b_1) = 1/6 + 3/6 = 4/6$. On the other hand, $(a_1 \cap b_2) = \{2,4,6\} \neq \varnothing$. Thus, $p(a_1 \cup b_2) \neq p(a_1) + p(b_2)$. Moreover, all states within a partition are mutually exclusive and therefore $p(a_i \cup a_j) = p(a_i) + p(a_j)$. For example, for the partition of the year into the four seasons, the probability that a day falls into summer or winter or spring is $1/4 + 1/4 + 1/4 = 3/4$ (assuming all seasons have the same number of days).

For any partition A, which by definition consists of mutually exclusive and collectively exhaustive states a_i, we therefore have $\Sigma_i \, p(a_i) = 1$. For the resulting scenarios s_k, we also have $\Sigma_k \, p(s_k) = 1$. The four partitions *weather, equipment, worker,* and *system safety* (Figure VI.2) produce 36 mutually exclusive and collectively exhaustive scenarios.

Let's assume that we have two partitions A and B, each with a certain number of partitions, $a_i \in A$ and $b_i \in B$. Then, we can distinguish the following types of probabilities:

- **Marginal** probabilities: $p(a_i)$ and $p(b_j)$.
- **Conditional** probabilities: $p(a_i|b_j)$, and $p(b_j|a_i)$.
- **Joint** probabilities: $p(a_i \cap b_j) \equiv p(a_i, b_j)$.

For the example of rolling a die we see that the joint states correspond to the scenarios, and the joint probabilities are the probabilities of the scenarios. For example, $p(a_2,b_2)=2/6$. The conditional probability of state a_1, given that state b_2 occurred, is $p(a_1|b_2)=(1/6)/(1/6+2/6)=p(a_1 \cap b_2)/p(b_2)=1/3$. The way we computed this conditional probability motivates the following definition:

$$p(a_i \cap b_j) = p(a_i|b_j)p(b_j) = p(b_j|a_i)p(a_i).$$

In words: the joint probability of two states (from different partitions) is equal to the conditional probability multiplied by the marginal probability of the conditional state. At this point we must extend our discussion beyond two partitions. As in Figure VI.2, there usually are more than two partitions of the uncertainty space. A combination of two or more partitions can be seen as a new partition.

For example, assume we have three partitions, demand D: (d_{high}, d_{low}), economy E: (e_{strong}, e_{weak}), and technology T: (t_{new}, t_{old}). Then, we could define conditional probabilities, e.g., as $p(d_{high}|e_{strong}, t_{new})$, where the two joint states (e_{strong}, t_{new}) form a new state. The collection of these new states forms a new partition of the uncertainty space.

From these definitions and relations we can derive the **multiplication law** for n states from different partitions. Let $\{z_i, ..., z_n\}$ be a set of states from different partitions. The joint probability of these n states is then (proof by induction using the definition of joint probability introduced above):

$$p(z_1, z_2, z_3, ..., z_n) = p(z_1)p(z_2|z_1)p(z_3|z_1, z_2)...p(z_n|z_1, z_2, z_3, ..., z_{n-1}).$$

B) Probabilistic Independence

A state a_i is **probabilistically independent** of another state, b_j, if $p(a_i|b_j)=p(a_i)$. Thus, we get for two probabilistically independent states $p(a_i \cap b_j)=p(a_i)p(b_j)$. It should be noted that if a_i is independent of b_j, then b_j also is independent of a_i. This means that probabilistic dependence is symmetric, as opposed to preferential dependence of criteria (Chapter V). If a_i is probabilistically independent of b_j, then $p(a_i)p(b_j)= p(a_i \cap b_j)=p(b_j|a_i)p(a_i)=p(b_j)p(a_i)$. Thus, b_j is also probabilistically independent of a_i.

For the example of rolling a die we see that the partitions $size$ of outcome (A) and $kind$ of outcome (B) are not probabilistically independent because $p(a_i|b_j) \neq p(a_i)$. However, if we modify the definition of A to outcomes < 3, we get:

	a_1 :< 3	a_2 :≥ 3		a_1	a_2		$p(a_1)$ 2/6	$p(a_2)$ 4/6
b_1: even	{2}	{4,6}	b_1	1	2	$p(b_1)=3/6$	1/6	2/6
b_2: odd	{1}	{3,5}	b_2	1	2	$p(b_2)=3/6$	1/6	2/6
	types of outcomes			frequency of scenarios			probabilities of scenarios	

As we see now, the frequencies for a_1 (1) and a_2 (2) are independent of the states b_1 and b_2. For example, we can confirm this by checking $p(b_2|a_1)p(a_1)=p(b_2)p(a_1)=$ 3/6×2/6=1/6. Thus, these redefined partitions A and B are (mutually) probabilistically independent. The concepts of **mutual exclusiveness**, **collective exhaustiveness**, and **probabilistic independence** can be summarized as in Figure VI.3.

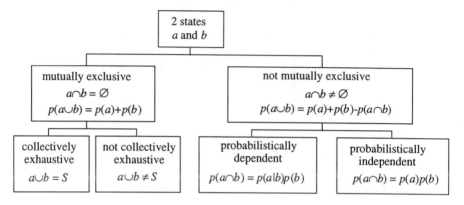

Figure VI.3: Probability concepts.

C) Arc Reversal (2-dimensional)

The results that we discussed so far can be generalized in terms of Bayes' theorem.

Bayes' Theorem: Let A and B be two partitions of the uncertainty space S. Then, with $p(b_j) > 0$, we have:

$$p(a_i|b_j) = \frac{p(b_j|a_i)p(a_i)}{\sum_i p(b_j|a_i)p(a_i)}$$

Note that: • $p(b_j|a_i)p(a_i) = p(b_j \cap a_i)$, and

• $\sum_i p(b_j|a_i)p(a_i) = \sum_i p(b_j \cap a_i) = p(b_j).$

In words: The conditional probability $p(a_i|b_j)$ of a state a_i, given another state b_j, equals the joint probability $p(b_j \cap a_i)$ divided by the probability of state b_j, $p(b_j)$.

When do we use Bayes' theorem? Bayes' theorem can be used to reverse conditional relations, called **relevances**, between partitions. Let's assume we have two partitions, $A=\{a_1,a_2\}$ and $B=\{b_1,b_2\}$, where B is conditioned on A. For example, look at the relation between *worker* and *equipment* in Figure VI.2. For these two partitions, we assess the following probabilities:

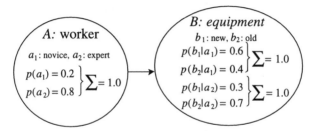

To reverse the conditional relation, we use Bayes' theorem, for example:

$$p(a_1|b_2) = \frac{p(b_2|a_1)p(a_1)}{p(b_2|a_1)p(a_1) + p(b_2|a_2)p(a_2)} = \frac{0.4 \times 0.2}{0.4 \times 0.2 + 0.7 \times 0.8} = 0.125$$

note: $\sum_{i=1}^{2} p(b_2|a_i)p(a_i) = p(b_2)$ or $p(b_2) = \frac{p(b_2|a_1)p(a_1)}{p(a_1|b_2)} = \frac{0.4 \times 0.2}{1/8} = 0.64$

and get **arc reversal**:

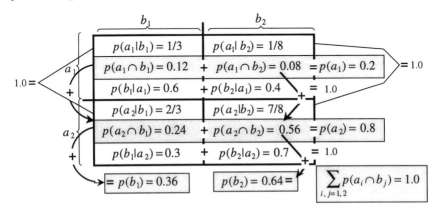

A more illustrative representation of the marginal, conditional, and joint probabilities is to see each partition as a dimension in the uncertainty space. Thus, we would represent the probabilities for these two partitions in a 2-dimensional table:

Table VI.1: 2-dimensional partition of the uncertainty space.

The relations in Table VI.1 are obvious and computed with Bayes' theorem. Assume that the marginal probabilities $p(a_1)$, $p(a_2)$, $p(b_1)$, and $p(b_2)$ are the lengths along the axes of the rectangle (Table VI.1). The joint probabilities for this example are not the areas which are defined by these lengths. The reader should explain why, and change the numeric values so that they are.

D) Arc Reversal (3-dimensional)

Let's now assume that we have three partitions: *traffic density* (high, low), $A=\{a_1,a_2\}$; *accident type* (collision, run-off-road), $B=\{b_1,b_2\}$; and *weather* (rain, sun), $C=\{c_1,c_2\}$; and that their probabilities have been defined as in the left structural model of Figure VI.4. The structural model shows that the likelihoods of the states in B depend on the occurrences of the states of A and C. It should be noted that the diagram shows that A and C are (mutually) probabilistically independent. The reader should explain why.

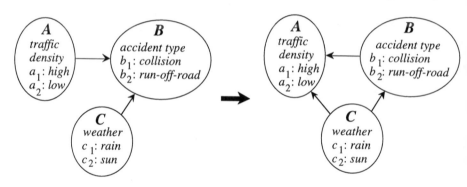

Figure VI.4: Arc reversal and inheritance of predecessor's arcs.

The marginal and conditional probabilities are assumed to have been assessed as follows:

$p(A)$:   • $p(a_1) = 0.2$, $p(a_2) = 0.8$

$p(B|A,C)$:   • $p(b_1|a_1,c_1) = 0.3$, $p(b_2|a_1,c_1) = 0.7$
                • $p(b_1|a_2,c_1) = 0.6$, $p(b_2|a_2,c_1) = 0.4$
                • $p(b_1|a_1,c_2) = 0.8$, $p(b_2|a_1,c_2) = 0.2$
                • $p(b_1|a_2,c_2) = 0.1$, $p(b_2|a_2,c_2) = 0.9$

$p(C)$:   • $p(c_1) = 0.4$, $p(c_2) = 0.6$

Given this structural model and the corresponding probabilities (approximated by frequencies), we might want to compute the conditional probability of the traffic density. This means that we need to reverse the arc between A and B. However, if we do this, we must add a new arc between C and A as shown in the right structural model of Figure VI.4. The reason for this is that B and C are probabilistically dependent. Thus, if A is conditioned on B, it also must be conditioned on C.

As a first step in the computation, we draw the 3-dimensional representation of the probabilities, as we did for the 2-dimensional example, but this time only with the joint probabilities. Given the marginal probabilities $p(a_i)$ and $p(c_k)$, and the conditional probabilities $p(b_j|a_i,c_k)$, $i,j,k=1,2$, we can compute the joint probabilities as follows:

$p(A,B,C)$:
- $p(a_1,b_1,c_1) = p(b_1|a_1,c_1)\,p(a_1)\,p(c_1) = 0.3{\times}0.2{\times}0.4 = 0.024,$
- $p(a_2,b_1,c_1) = p(b_1|a_2,c_1)\,p(a_2)\,p(c_1) = 0.6{\times}0.8{\times}0.4 = 0.192,$
- $p(a_1,b_2,c_1) = p(b_2|a_1,c_1)\,p(a_1)\,p(c_1) = 0.7{\times}0.2{\times}0.4 = 0.056,$
- $p(a_2,b_2,c_1) = p(b_2|a_2,c_1)\,p(a_2)\,p(c_1) = 0.4{\times}0.8{\times}0.4 = 0.128,$

$$\Sigma = 0.400 = p(c_1)$$

- $p(a_1,b_1,c_2) = p(b_1|a_1,c_2)\,p(a_1)\,p(c_2) = 0.8{\times}0.2{\times}0.6 = 0.096,$
- $p(a_2,b_1,c_2) = p(b_1|a_2,c_2)\,p(a_2)\,p(c_2) = 0.1{\times}0.8{\times}0.6 = 0.048,$
- $p(a_1,b_2,c_2) = p(b_2|a_1,c_2)\,p(a_1)\,p(c_2) = 0.2{\times}0.2{\times}0.6 = 0.024,$
- $p(a_2,b_2,c_2) = p(b_2|a_2,c_2)\,p(a_2)\,p(c_2) = 0.9{\times}0.8{\times}0.6 = 0.432,$

$$\Sigma = 0.600 = p(c_2)$$
$$\Sigma = 1.000$$

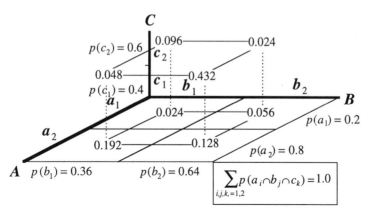

Figure VI.5: Marginal and joint probabilities over three partitions with two states each.

From this representation of the probabilities in Figure VI.5 we can see that the partitions A and C are probabilistically independent. For example, $p(a_1|c_1)= p(a_1,c_1)/p(c_1)=(0.024+0.056)/(0.4)=0.2$ which is in fact $p(a_1)$.

If we recall Bayes' theorem and the definition of conditional probabilities, we see that for any number of partitions, there is one requirement on the formal level when any arc reversal is performed:

- The original **joint probability distribution** does not change and the new structural model must contain all the information to compute it.

From this requirement, we can conclude two necessary conditions, one for the structural model and the other for the arc reversal procedure:

- **Structural model**: the arcs in the structural model never form cycles.

- **Arc reversal** between two nodes *A* and *B*: both *A* and *B* inherit each others' predecessors.

The first condition is obvious if we think about how we assess the probabilities for a given structural model. The probability of every successor (e.g., *equipment* in Figure VI.2) is assessed conditioned on the predecessors (e.g., *worker* in Figure VI.2). Thus, if we had cycles, we would have to evaluate a partition that we already did evaluate, and we might contradict the previously defined joint probability distribution.

The second necessary condition is illustrated in Figure VI.6 which shows several examples of arc reversal between *A* and *B*.

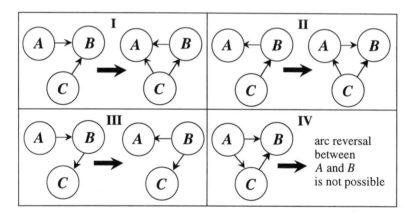

Figure VI.6: Arc reversal between *A* and *B* and inheritance of predecessors.

Example I in Figure VI.6 is the one we just discussed. Example II shows that partition *A* inherits also the predecessors of partition *B*. This might be intuitively less obvious than the first case, so the reader should check this with a numerical example. Example III shows that if both partitions have no predecessors, there will be no inheritances at all. Finally, example IV shows that arc reversal is not possible if it would cause cycles. Again, the arc reversals are done in such a way that the joint probability distribution is not altered and the information to determine it is preserved.

Reversing the arc between *A* and *B* in Figure VI.4, therefore gives the following results:

$p(A|B,C)$: • $p(a_1|b_1,c_1) = p(a_1,b_1,c_1)/[p(b_1|c_1)p(c_1)] = 0.024/[0.54 \times 0.4] = 0.111$
  • $p(a_2|b_1,c_1) = p(a_2,b_1,c_1)/[p(b_1|c_1)p(c_1)] = 0.192/[0.54 \times 0.4] = 0.889$
  • $p(a_1|b_2,c_1) = p(a_1,b_2,c_1)/[p(b_2|c_1)p(c_1)] = 0.056/[0.46 \times 0.4] = 0.303$
  • $p(a_2|b_2,c_1) = p(a_2,b_2,c_1)/[p(b_2|c_1)p(c_1)] = 0.128/[0.46 \times 0.4] = 0.696$
  • $p(a_1|b_1,c_2) = p(a_1,b_1,c_2)/[p(b_1|c_2)p(c_2)] = 0.096/[0.24 \times 0.6] = 0.667$
  • $p(a_2|b_1,c_2) = p(a_2,b_1,c_2)/[p(b_1|c_2)p(c_2)] = 0.048/[0.24 \times 0.6] = 0.333$
  • $p(a_1|b_2,c_2) = p(a_1,b_2,c_2)/[p(b_2|c_2)p(c_2)] = 0.024/[0.76 \times 0.6] = 0.053$
  • $p(a_2|b_2,c_2) = p(a_2,b_2,c_2)/[p(b_2|c_2)p(c_2)] = 0.432/[0.76 \times 0.6] = 0.947$

$p(B|C)$: • $p(b_1|c_1) = p(b_1,c_1)/p(c_1) = (0.024+0.192)/0.4 = 0.54$
  • $p(b_2|c_1) = p(b_2,c_1)/p(c_1) = (0.056+0.128)/0.4 = 0.46$
  • $p(b_1|c_2) = p(b_1,c_2)/p(c_2) = (0.096+0.048)/0.6 = 0.24$
  • $p(b_2|c_2) = p(b_2,c_2)/p(c_2) = (0.024+0.432)/0.6 = 0.76$

$p(C)$: • $p(c_1) = 0.4$
  • $p(c_2) = 0.6$

The reason for addressing the principle of arc reversal is that the data about probabilities or frequencies might not be available in the form of our (conceptual) structural model. Therefore, we need to reverse arcs to get the information that matches our problem. Schachter [1986] calls this approach to information acquisition **thinking backwards**. For example, if historical data shows that in 80% of all truck accidents (A) the truck was speeding (S), 5% of all truck trips lead to an accident, and 40% of all trucks are speeding, we have $p(S|A)=0.8$, $p(A)=0.2$, and $p(S)=0.4$. However, what we are interested in is the probability that a speeding truck ends up in an accident, $p(A|S)$. This can be computed through arc reversal as follows:

$$p(A|S) = p(S|A) \times p(A)/p(S) = 0.8 \times 0.05/0.4 = 0.1.$$

The use of arc reversal as part of solving a decision problem under uncertainty will be addressed in more depth in Chapter VIII.

Howard [1989] discusses an approach for getting fragmented information from people that is based on the arc reversal principle. One asks the decision maker to assess the uncertainty of redundant **relevance diagrams**, called **knowledge maps**. By reversing arcs and comparing the results, the analyst can check the consistencies of the decision maker's assessments.

As far as the evaluation of the alternatives is concerned, we only need to know which partitions point into the criteria node in the structural model. These partitions define the evaluation scenarios which are relevant for the assessment of the alternatives. Once we have assessed the alternatives, we must think of decision rules. One set of decision rules was discussed at the beginning of this chapter. However, those four rules were based on the assumption that there was not enough information to assign probabilities to the scenarios. That is, we were dealing with decision making under complete uncertainty. If probabilities can be determined (decision making under risk), the alternatives are often ranked according to the **expected risk**.

Risk management deals with random events that can be described quantitatively. The most frequently used concepts in risk management are given in Table VI.2.

Term	Symbol	Definition
Distribution function *(df)* or cumulative distribution function *(cdf)*	$F(x)$	Probability that the variable takes a value less than or equal to x $F(x) = \Pr[X \le x]$ $F[x] = \int_{-\infty}^{x} f(t)dt$
Probability density function *(pdf)*	$f(x)$	$\int_{X_L}^{X_U} f(x)dx = \Pr[X_L < X \le X_U]$ $f(x) = \dfrac{d}{dx} F(x)$
Probability function *(pf)* for discrete variables	$f(x)$	$f(x) = \Pr[X = x] = p(x)$
Survival function	$S(x)$	Probability that the variable takes a value greater than x. $S(x) = \Pr[X > x] = 1 - F(x)$
Hazard function (failure rate)	$h(x)$	$h(x) = \dfrac{f(x)}{S(x)} = \dfrac{f(x)}{1 - F(x)}$
Mean (first moment about the origin, expected value)	m, $E[x]$	$m = E[x] = \int_{-\infty}^{\infty} x\,dF(x)$ $m = E[x] = \sum_{i=1}^{n} x_i p(x_i)$, discrete variable
Variance (second moment about the mean)	σ^2, $V[x]$	$\sigma^2 = \int_{-\infty}^{\infty} (x - m)^2 dF(x)$ $\sigma^2 = \sum_{i=1}^{n} (x_i - m)^2 p(x_i)$, discrete variable
Standard deviation	σ	
Mode		point with maximum value in the *pdf* or *pf*
Median		root of $F(x) = 0.5$
Information content or entropy; entropy is maximal for the uniform distribution	I	$I = -\int_{-\infty}^{\infty} f(x) \log_2(f(x))dx$ $I = -\sum_{i=1}^{n} p(x_i) \log_2 p(x_i)$, discrete variable

Table VI.2: Probabilistic concepts [Evans et al., 1993].

The concepts of value theory as discussed in Chapter V did not involve any consideration of uncertainty. A suitable approach for assessing and aggregating preferences under uncertainty will be discussed in Chapter VII. We will then discuss that preferences can be aggregated across scenarios with the expected value approach.

2.3 Decision Rules

Financial risk management deals with uncertainty that refers both about gains and losses. In risk management of technological systems, on the other hand, uncertainty refers generally only to losses; the benefits of a technological system are assumed to be known. In the following discussion we will confine attention to the risks of technological systems, where the main goal is to assure the safety of the system and the environment surrounding the system. Safety goals for technological systems are defined for (and from the perspective of) different actors. The decision options are risk reduction measures which cost a certain amount of money to implement.

The task is to decide on an alternative to reduce the present risk of a technological system. Often, only the expected damage is used as a measure of risk. However, other measures, such as the entropy (a measure of the information content, see Table VI.2), can reflect aspects that could be as relevant as the expected value. In the following, we will discuss the expected value and the survival function as measures of risk for decision making about technological systems.

A) Individual Risk

Risk is measured in a certain unit and it refers to a certain time period. For example, risk could be measured as the annual probability of death, p. Then, the expected value of the resulting damage is computed as: $r=E[\text{damage}]=p\times1+(1-p)\times0$, where 1 stands for a fatal event and 0 for survival. Safety goals for **individual risk** are established as hard constraints in terms of the expected value.

The first nation to introduce a legally binding quantitative approach to decision making about technological risks was The Netherlands. Dutch safety regulations require that the exposure to a technological system does not increase the annual probability of death by more than 10^{-6} [VROM, 1989]. This safety goal is based on the premise that young people between six and twenty years have an individual death probability of 10^{-4} per year, and that a technological system should not increase this value by more than 1%. This safety goal does not consider any cost-benefit tradeoff. That is, the individual risk of 10^{-6}/year cannot be exceeded, regardless of the financial implications.

B) Collective Risk

The sum of individual risks is called **collective risk**. Collective risk refers to society as a whole and is therefore anonymous. For example, assume that n people are exposed to some hazard sources, where the persons face independent risks (probabilities of death) of r_i. Then, the collective risk (probability of death) of the n persons is:

$$R = 1 - \prod_{i=1}^{n}(1-r_i) \approx \sum_{i=1}^{n} r_i,$$ for small values of r_i; e.g., $r_i = 10^{-7}$ and $n=10$.

The collective risk is therefore the sum of independent individual risks, given that the individual risks are very small. This is so because the terms of higher order (e.g., $r_i \times r_j$) are negligibly small. The collective risk is also the sum of the individual risks in case that the risks are dependent (e.g., airplane passengers facing the risk of a crash).

Even if every individual is safe enough (i.e., the annual risk of every individual is smaller than 10^{-6}), the collective risk might be perceived as too high. Actors interested in reducing the collective risk are the government who invested in education, companies who might experience loss of image in case of fatal accidents, and insurance companies who want to reduce the overall damage claims.

Clearly, safety reductions beyond the individual safety goal must be based on a **risk-cost tradeoff**. The collective risk should therefore be reduced only if the risk reduction is cost-effective. To define cost-effectiveness, the collective risk must be transferred into monetary units. This is done by defining the **life-saving cost** (LSC).

The LSC could be based on the financial impact that the loss of a person poses to society, or on the losses which a third party has to pay for insurance claims. Values for LSCs differ across groups of individuals, just as life insurance rates vary across age groups. However, subjective or emotional LSCs are higher than economic LSCs. They are between $1 and $10 million. This means, for example, that we (e.g., society) would be willing to invest up to $10 million to reduce the expected annual number of anonymous fatalities on the national highways.

The risk reduction alternatives $(a_1(R_1,C_1), a_2(R_2,C_2), ...)$ are inserted into a Cost-Risk diagram (see Figure VI.7). The **Pareto optimal** or efficient (non-dominated) alternatives define the **efficient frontier**, and the best alternative is the efficient alternative that performs best in terms of the risk-cost tradeoffs.

The optimal alternative, from the point of view of a single system, is the one on the efficient frontier with $\Delta R_i / \Delta C_i = -LSC^{-1}$, which is a_5 (Figure VI.7) This is so, because for each dollar invested ($\Delta C = C_j - C_0$) on risk reduction we want to reduce the risk ($\Delta R = R_j - R_0$) by $\Delta C \times LSC^{-1}$.

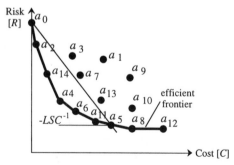

Figure VI.7: Optimal risk reduction measures.

Chapter VI: Choices under Uncertainty

However, if we take the point of view of an administrator of multiple (k) systems who has a limited amount of money (C_{tot}) to invest for risk reduction on those systems, s/he is facing the following problem:

(1) $min \ \Sigma_i \ R_i$

(2) $\Sigma_i \ C_i = C_{tot}$

where R_i is the risk and C_i the corresponding cost for system s_i. With the **Lagrange multiplier method**, we minimize:

$$\mathcal{L} = \Sigma_i \ R_i + \lambda(\Sigma_i \ C_i - C_{tot}),$$

Taking the partial derivatives, and setting the equations equal to zero gives:

- $\partial \mathcal{L}/\partial C_i = dR_i/dC_i + \lambda = 0$, thus, $\lambda = -dR_i/dC_i = LSC^{-1}$
- $\partial \mathcal{L}/\partial \lambda = \Sigma_i \ C_i - C_{tot} = 0$

This means that the risk reduction alternatives chosen for each system must result in the same **marginal rate of substitution**: $MRS=LSC^{-1}=-dR_i/dC_i=\lambda$. The optimal alternatives (choice) can be found with the tangent method as illustrated for three systems in Figure VI.8.

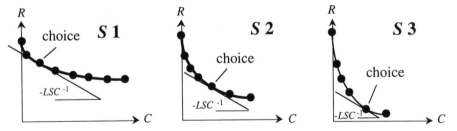

Figure VI.8: Constant MRS alternatives.

Let's look at a numerical example to understand what we have done. Three systems (S_i) are given with their risk-cost functions: $R_1=1/C_1$ (S_1), $R_2=2/C_2$ (S_2), and $R_3=3/C_3$ (S_3). Let's assume that we have $LSC=2$; that is, we are willing to invest 2 monetary units (e.g., \$2 millions) to save one anonymous and probabilistic life. With the above equations we get: $2=1/C_1^2$, $2=2/C_2^2$, and $2=3C_3^2$. This gives: $C_1=0.707$, $C_2=1.000$, $C_3=1.225$, and $C_{tot}=2.932$. Inserting these three cost values into the three risk-cost functions gives the remaining risks: $R_1=1.414$, $R_2=2.000$, and $R_3=2.449$. The total remaining risk of the three systems is: $R_{tot}=5.863$.

If we would allocate the same amount of money ($C_{tot}=2.932$) evenly to the three systems ($C_1=C_2=C_3=C_{tot}/3=0.977$), we would get: $R_1=1.023$, $R_2=2.046$, and $R_3=3.070$. The total remaining risk of all three systems would now be $R_{tot}=6.139$ which is higher than the one computed with the approach illustrated in Figure VI.8.

C) Group Risk

Safety goals which have been defined by national authorities for technological systems for licensing purposes are not necessarily based on collective risk but on **group risk**. Group risk is also composed of the sum of individual risks. The difference is that collective risk is an anonymous risk, while group risk refers to an identifiable group of people. For example, villages, organizations, industrial areas, and train passengers are seen as groups, often with identified individuals. Because of this personal note, safety goals for group risk are defined with hard constraints as for individuals and not with a risk-cost tradeoff as for the anonymous society.

Group safety goals are expressed in terms of a **survival function**, $S(N)$, where N is the number of fatalities in case of an accident. A survival function states the probability of an event that causes N or more fatalities (at least N). Therefore, group safety goals prescribe the maximum allowed probability with which a technological system will cause N or more fatalities, and this for all possible values of N.

The Dutch safety goals take into account the decision maker's degree of **risk aversion** or negative attitude toward technological risk [VROM, 1989]. A risk neutral decision maker makes choices according to expected damages. For example, 10 fatalities with a chance of 10^{-5}, and 10^5 fatalities with a chance of 10^{-9} have the same expected damage of 10^{-4} fatalities. However, the difference between these two risk profiles lies in the variance. For the first risk profile, the variance is 10^{-3}, while for the second it is 10. A risk averse decision maker would thus want to minimize both expected value and variance.

This means that a measure for subjectively perceived risks should consider both expected number of fatalities (N) and the variance (σ^2). Since the variance is a function of N^2, we would define as the acceptable risk (R_{acc}) not the product of the probability of more than N fatalities $(S(N))$ and the number of fatalities (N), $R_{acc}=S(N)\times N$, but the product of probability of more than N fatalities $(S(N))$ and the squared number of fatalities (N^2), $R_{acc}=S(N)\times N^2$.

In fact, the Dutch safety goal for group risk is defined in terms of the survival function, $S(N)$, as [VROM, 1989]:

$$S(N)\times N^2 \leq 10^{-3}.$$

This means, for example, that if the number of expected deaths (N) increases by factor 10, the probability of such an event (S) must decrease by factor 100 (instead of factor 10) to be indifferent between the two risks: $R_1=(N,S) \sim R_2=(10\times N, S/100)$. The risks of a system are depicted in a logarithmic N/S diagram (Figure VI.9).

The Dutch safety goal is based on the idea of 10 fatalities occurring with a probability of less than 10^{-5} per year (i.e., 10^{-6} per individual); that is, $log(10^{-5})+2\times log(10)=-5+2=-3$. For the safety goals defined above we get:

$$log(S) + 2\times log(N) \leq -3.$$

From the point (1,-5) in the logarithmic *N/S* diagram, a line with slope of -2 defines safe and unsafe systems (see Figure VI.9). Any hazard with a survival curve above this line is not acceptable and must be reduced. Hazards that are 1% of this threshold line are negligible. The area between these two threshold lines is called the **ALARA** zone, meaning that the risks should be reduced to as low as reasonably achievable. This could mean that the life-saving-cost concept could be employed for survival functions lying between these two threshold lines.

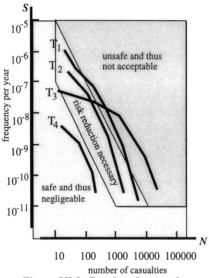

Figure VI.9: Dutch safety goals.

The risk of a technological system can be characterized by its survival function. Figure VI.9 shows different technological systems (T_i) and their corresponding survival functions. T_4 is the only system with negligible risks. T_1 and T_3 are unsafe, while the safety of T_2 needs to be improved according to the ALARA principle. In 1995, the group safety goal has been changed, saying that all not unsafe systems must comply with the ALARA principle; that is, even those which used to be considered safe.

It should be noted that the risk management literature often uses frequencies instead of probabilities. The survival function $S(N)$ is then replaced by a frequency function, unfortunately called $F(N)$, which should not be confused with the distribution function $F(x)$. In an **F/N diagram**, hazards with an annual frequency larger than 1.0 can also be depicted. For example traffic systems (air, road, rail) can have annual accident frequencies larger than 1.0. The annual probability, if it is measured as the long-run frequency, would then be larger than 1.0, which is meaningless by definition. However, the risk-cost approach discussed for probabilities also works for frequencies. The probabilistic approach can be seen as a normalized frequency approach. This is indeed the case if the probabilities of the different scenarios have been determined from a normalized histogram.

2.4 Aggregation of Linguistic Variables

We have discussed in previous chapters several approaches to aggregate effectiveness values across multiple criteria. Risk managers must assess the safety of a system with respect to its potential negative impacts. For large technological systems, the extent of an accident is often evaluated with linguistic values, such as *disruption, accident, disaster*, and *catastrophe*. Several factors (criteria) determine the classification of an event. These are, among others, the measurable damages, including fatalities, injuries, areas of contaminated land, amount of property loss, etc. Each of these factors contributes to how an event would be classified. For example, one fatality in a public transport accident could be classified as an accident, while an event causing a preventive evacuation of a larger urban area could be assessed as a disaster.

Various approaches have been proposed for aggregating linguistic values. They are based on the concept of **fuzzy set theory** and are called aggregation operators on fuzzy sets. The concept of fuzzy set theory was introduced by Lofti A. Zadeh [1965] as a complement to ordinary crisp set theory. Fuzzy set theory is not based on axioms derived from common sense, like probability theory. It is primarily a number crunching concept that has been applied very successfully in dynamic systems.

A **linguistic variable** is characterized by a fuzzy set, which consists of measurable outcomes (effectiveness values) and a normalized membership function over these values. Effectiveness values with membership function value 1.0 state that the effectiveness value belongs very strongly to the set defined by the linguistic variable.

Figure VI.10 shows three linguistic variables describing age groups in form of fuzzy sets and membership functions. At first glance, they look similar to probability density functions. However, the concept of fuzzy sets does not require that the area under the membership function be 1.0.

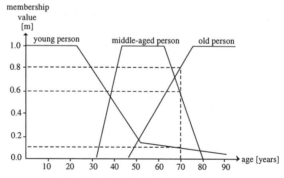

Figure VI.10: Membership functions for different linguistic variables.

From Figure VI.10 we can see that a person can belong simultaneously to different age groups. For example, a seventy year old person is young with a membership value of 0.1, middle-aged with a membership value of 0.6, and old with a membership value of 0.8.

The age of a person can depend on multiple factors (criteria), such as physical fitness, mental freshness, etc. Therefore, we have to define how to compute the union (disjunction: $\cup$) and the intersection (conjunction: $\cap$) of multiple fuzzy values. Different approaches have been proposed in the literature (Table VI.3).

Union (or: $\cup$) and Intersection (and: $\cap$), where a and b are two fuzzy values	Range	Author
$\cup$: $1-\max\left[0,\left((1-a)^{-p}\right)+(1-b)^{-p}-1\right]^{1/p}$ $\cap$: $\max\left(0, a^{-p}+b^{-p}-1\right)^{-1/p}$	$p \in (-\infty, \infty)$	Schweizer and Sklar [1961]
$\cup$: $\dfrac{a+b-(2-\gamma)ab}{1-(1-\gamma)ab}$ $\cap$: $\dfrac{ab}{\gamma+(1-\gamma)(a+b-ab)}$	$\gamma \in (0, \infty)$	Hamacher [1978]
$\cup$: $1-\log_S\left[1+\dfrac{(S^{1-a}-1)(S^{1-b}-1)}{S-1}\right]$ $\cap$: $\log_S\left[1+\dfrac{(S^a-1)(S^b-1)}{S-1}\right]$	$S \in (0, \infty)$	Frank [1979]
$\cup$: $\min\left(1,\left(a^w+b^w\right)^{1/w}\right)$ $\cap$: $1-\min\left(1,\left((1-a)^w+(1-b)^w\right)^{1/w}\right)$	$w \in (0, \infty)$	Yager [1980]
$\cup$: $\dfrac{a+b-ab-\min(a,b,1-\alpha)}{\max(1-a,1-b,\alpha)}$ $\cap$: $\dfrac{ab}{\max(a,b,\alpha)}$	$\alpha \in (0,1)$	Dubois and Prade [1980]
$\cup$: $\dfrac{1}{1+\left[\left(\frac{1}{a}-1\right)^{-\lambda}+\left(\frac{1}{b}-1\right)^{-\lambda}\right]^{-1/\lambda}}$ $\cap$: $\dfrac{1}{1+\left[\left(\frac{1}{a}-1\right)^{\lambda}+\left(\frac{1}{b}-1\right)^{\lambda}\right]^{1/\lambda}}$	$\lambda \in (0, \infty)$	Dombi [1982]

Table VI.3: Fuzzy aggregation operators [Yager and Filev, 1994].

The Swiss safety regulation extends the Dutch approach to risk management by adopting the concept of aggregation of linguistic values [BUWAL, 1991]. Damages are determined with nine different criteria. These are: death, injuries, evacuations, alarm factor, loss of large animals, area of damaged ecosystem, area of contaminated soil, area of contaminated ground water, and material losses. Some of the membership functions are shown in Figure VI.11.

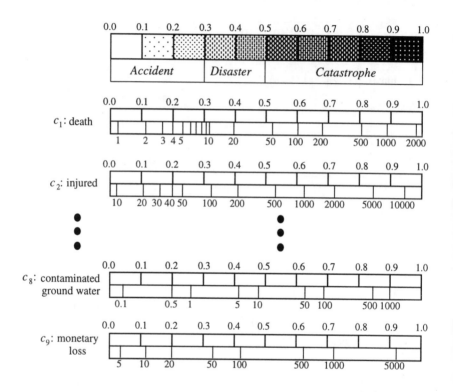

Figure VI.11: Damage indicators used in the Swiss safety regulation [BUWAL, 1991].

The safety regulation then suggests aggregating the nine fuzzy values with Yager's approach (see Table VI.3). For nine values, and n_i being the value for criteria c_i, we get for the aggregated value

$$n_{tot} = \min\left[1, \left(\sum_{i=1}^{9}(n_i)^w\right)^{1/w}\right],$$

where $w \in [1, \infty]$. For $w=1$ we get $n_{tot}=\min[1, \Sigma_i(n_i)]$; for $w=\infty$, we get $n_{tot}=\max(n_i)$.

The F/N diagram is replaced by an F/n_{tot} diagram. Safety goals are also defined with a survival function, $F(n_{\text{tot}})$.

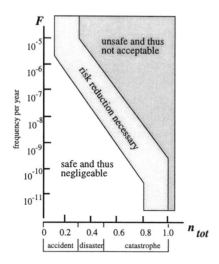

Figure VI.12: Swiss safety goals [BUWAL, 1991].

2.5 Resolution Model

The resolution model consists of the evaluation of all potential risk reduction measures, including the zero-alternative (no safety measure at all). If all individuals who are exposed to a particular technological system are safe enough under the present conditions, the zero-alternative will be implemented. If this is not the case, the most cost-effective alternative that reduces the risks below the safety threshold should be chosen. It should be noted that the individual system perspective and the multiple systems perspective do not necessarily concur on which alternatives should be implemented (Figures VI. 7 and 8).

The resolution model for the Dutch safety goals is summarized in Figure VI.13. Given that individual and group safety is assured, collective risk aspects are addressed. This means that the group risk is further reduced if the LSC justifies the reduction. However, if we apply the LSC principle to linguistic variables (such as in the Swiss safety regulation approach), we would first have to define a tradeoff between membership value and cost. This might be difficult to do because the numeric membership value n_{tot} does not disclose how many factors contributed to it and to what extent. It seems, however, that in most practical cases there is only one factor which contributes the most to the overall membership value.

Technological systems must often be ranked according to their risks. The reason to focus on rankings (relative assessments) is that absolute risk assessments are often hard to determine because of the high degree of uncertainty. However, rankings are also not always easy to determine based on the graphs in the F/N diagram. For example, how would we go about to rank the four systems given in Figure VI.9. It is obvious that T_4 is the safest system because it dominates (stochastically) all other systems. T_1 and T_3 could be compared based on the expected damage (see Problem 7).

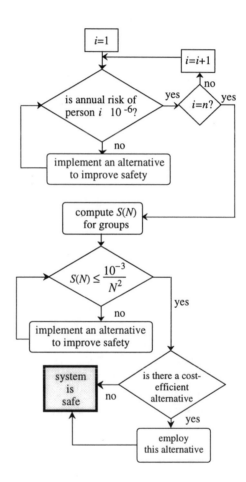

Figure VI.13: Resolution model for the Dutch safety regulations (up to 1995).

3. Summary and Further Readings

Four decision rules for single-criterion choice under complete uncertainty are proposed. These rules, which are based on different subjective points of view that decision makers adopt in their choice process, include conservatism (Wald), no regret (Savage), variations from conservatism to optimism (Hurwicz), and average (Laplace). A guide for managing uncertainty in quantitative risk and policy analysis can be found in Morgan and Henrion [1990].

Uncertainty is described by a set of mutually exclusive and collectively exhaustive scenarios. Scenarios are constructed with partitions of the uncertainty space. Partitions consist of disjoint states which give a description of the whole space of uncertainty. States of different partitions are related to each other by marginal, conditional, and joint probability distributions. The evaluation scenarios are described in terms of the different partitions and their states.

Two schools prevail in probability theory: the objectivists (logical or empirical probabilities) and the subjectivists (Bayesian reasoning). Probability theory is based on the axioms of Kolmogorov. Bayes' theorem is used to reverse arcs (conditional probabilities) in the structural model. Arc reversal does not affect the joint probability of the scenarios. Consequently, a structural model can never have cycles. To preserve all information given by the original structural model, arc reversal might generate new arcs between partitions. Arc reversal as part of solving a decision problem under uncertainty will be addressed in more depth in Chapter VIII.

Practical risk management of technological systems uses safety goals for individuals, groups and society. Individual risk should not exceed a predetermined level. Group safety is based on the stochastic dominance principle in the F/N diagram (survival function). Safety for society is assessed with an economic life-saving-cost approach. A technological system can be safe, conditionally safe, or unsafe. A conditionally safe system must be made save with a reasonable effort. This means that the risks should be reduced to as low as reasonably achievable (ALARA principle). Conditionally safe means that the system will be made safe if the life-saving-cost (marginal rate of substitution) justifies the effort. More on the issue of life-saving-costs can be found in Fischhoff et al. [1981].

Linguistic values operate on fuzzy, rather than crisp sets. Special operators have been defined for the union and intersection of fuzzy sets. Safety goals can also be defined with linguistic values, similar to numeric values. For an introduction to fuzzy set theory see Yager and Filev [1994] or Dubois and Prade [1980].

4. Problems

1. Discuss the preference functions for the four decision rules under complete uncertainty using the concepts of Figure V.9.

2. A decision maker adopts Wald's MaxMin rule to find the alternative that maximizes net profits. However, not one but two (or even more) alternatives turn out to be optimal. How would you choose the best among these optimal alternatives?

3. Revise the four decision rules under complete uncertainty for when the ordinal value function is decreasing instead of increasing. Give an example of a decreasing ordinal value function.

4. Define a new decision rule for decision making under complete uncertainty, as discussed in Section 1, by combining Wald's and Savage's rules. For example, use an optimism-pessimism index or something like the principle of insufficient reason. How does your rule compare to the four decision rules discussed in Section 2.1?

5. Modify the ELECTRE I outranking method (Chapter IV, Section 2.3) to a decision making approach under complete uncertainty for one criterion (preference value). Discuss the analogy between the two approaches. How would you proceed if you had multiple criteria under complete uncertainty?

6. A technical component (e.g., a computer) has the highest probability of failure during its initial stage of usage and after it has been used for an extended period of time (e.g., five years). Sketch the *pdf* for such a component and discuss reasonable measures of risk (e.g., expected value).

7. What (geometrically) is the expected value in the *F/N* diagram. Hint: use the definition $E[x] = \int x \, dF(x)$, where $F(x)$ is the distribution function. Discuss possible consequences if you rank the systems according to the expected number of fatalities or according to their *F/N* function.

8. Convince yourself of Bayes' theorem with the number of accidents given in the table below for the two partitions humidity and season. Compute first the joint probabilities $p(s,h)$ and then the conditional probabilities $p(s|h)$. From these joint probabilities use Bayes' theorem to reverse the conditional relation between s and h. Are the season and the humidity probabilistically independent?

---

season (s)	humidity (h)	
	dry (d)	wet (w)
April-September (a)	10	15
October-March (m)	20	55

9. Extent the 2-element union and intersection aggregation operators for fuzzy values (Table VI.3) to *n* fuzzy values.

10. Define a resolution model for the Swiss safety regulations, similar to the one for the Dutch safety regulation in Figure VI.13.

CHAPTER VII

UNCERTAINTY AND NORMATIVE CHOICE

1. The Structural Model for Decision Making Under Uncertainty

1.1 Conceptual Aspects

In Chapter V we discussed value functions to rank alternatives under certainty. The purpose of this chapter is to discuss an extension of the concepts of value theory to a normative approach for decision making under uncertainty: **utility theory**. In the presence of certainty, the decision maker's dilemma refers to the tradeoffs between different criteria. For example, imagine a contest where the winner can choose one prize out of the set of all prizes, the second ranked person can then choose from the remaining prizes, and so on. If we assume that all consequences of the prices (e.g., monetary value, satisfaction, etc.) are known for sure, the persons choosing from the set of prizes are facing a typical choice situation under certainty. Examples of such prices are a TV set, a stereo, a bicycle, etc., with multidimensional value-tags attached to them. The problem is then simply to trade off preferences for different criteria, such as gains, satisfaction, and quality.

Imagine now that instead of offering these types of prizes, the winners are offered prizes which involve uncertainty. For example, a beach vacation at a not yet determined place and time (which can be great if the place and time of year is chosen right, but which might be miserable otherwise), a ticket to a not yet determined soccer game (which can be great if it is for an international competition game, but which is worthless if it is for a friendly game of the local team), or even a lottery ticket for which the winner is asked to pay a small amount but which offers him/her the chance to win a fortune (which is great in case of win, but bad if the investment is lost). Such alternatives involving uncertainty concerning their outcomes are called **lotteries**. A pure or real lottery involves at least two possible outcomes (states or scenarios), each with a probability associated to it, where the sum of the probabilities is 1.

A 1-dimensional lottery (one criterion) is thus defined as:

$$L := [(p_1,a_1),...,(p_k,a_k),...,(p_v,a_v)],$$

where a_k is the outcome of the lottery for the k-th scenario, s_k, which occurs with probability p_k, with overall v scenarios. An m-dimensional lottery is correspondingly defined as:

$$L := [(p_1,(e_1^1,...,e_m^1)),...,(p_k,(e_1^k,...,e_m^k)),...,(p_v,(e_1^v,...,e_m^v))].$$

With this definition of a lottery, we see that value theory is a special case of utility theory. For decision making under certainty (value theory, Chapter V), an alternative can be seen as a **degenerated lottery** or **certainty outcome** with one scenario occurring with probability 1. For a **binary lottery** we introduce the following notation:

$$L := [(p_1,a_1),(1\text{-}p_1,a_2)] \equiv [a_1,p_1,a_2], \text{ and } L := [(p_1,u_1),(1\text{-}p_1,u_2)] \equiv [u_1,p_1,u_2],$$

where u_1 is the utility value (subjective preference value) of alternative a_1.

A degenerated lottery for alternative a_j can be written as $[a_j,1.0,\text{-}]$. The utility function has a lot of similarities with the value function discussed in Chapter V. However, the main difference lies in the elicitation of these functions. The elicitation of utility functions is done by asking the decision maker preferences for uncertain outcomes, while the elicitation of value functions is done by comparing alternatives whose consequences are know. Lotteries are often represented graphically as shown in Figure VII.1.

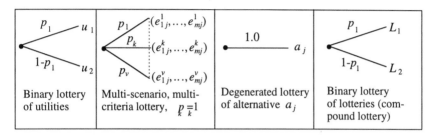

| Binary lottery of utilities | Multi-scenario, multi-criteria lottery, $\underset{k}{p}=1$ | Degenerated lottery of alternative a_j | Binary lottery of lotteries (compound lottery) |

Figure VII.1: Examples of lotteries.

The issue to be addressed now is how to compare different lotteries; that is, how to order lotteries according to the decision maker's preference. The mid-value splitting technique (Chapter V) would not be appropriate because it does not consider uncertainty. What we need is a preference elicitation method that produces higher expected utilities for more preferred lotteries. Analogous to value theory, we will discuss a set of axioms from which we can conclude that a rational decision maker, who complies with these axioms, seeks to maximize the expected utility.

The principle of computing the overall utility for an alternative is illustrated in Figure VII.2 (compare it to Figure V.2). For each criterion, c_i $(i=1,...,m)$, a component utility function is assessed with a method that will be discussed in Section 2.3. With these component utility functions, the evaluation values e_{ij}^k are transformed into utility values u_{ij}^k. Then, the utilities are aggregated over all criteria to a utility vector with component u_j^k for each alternative. Finally, the overall utility of an alternative a_j $(j=1,...,n)$ is computed by aggregating these values over all scenarios, s_k $(k=1,...,v)$, by computing the expected utility: $u(a_j) = \underset{k}{E}[u(e_j^k)]$.

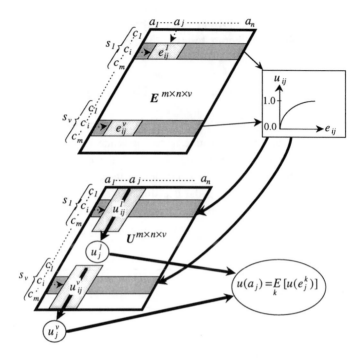

Figure VII.2: Principle of preference aggregation for expected utilities, $E[u_j]$.

We mentioned in Chapter V that there is quite some controversy in the literature concerning value theory due to its normative character. When we add uncertainty to the problem, the controversy about utility theory becomes even greater. If we imagine a theory that transforms a multidimensional evaluation under uncertainty (with m criteria and v scenarios, we have $m \times v$ evaluations) into one single value, we might expect some criticism if we assign a normative character to this outcome. In other words, if one were to conclude that one lottery ought to be preferred to another lottery because of the larger expected utility value, then the criticisms are justified. With such an interpretation, however, utility theory has been misunderstood or misused. Utility theory does not aim at imposing decisions but at describing preferences. Thus, the fact that the expected utility is higher is a *consequence* of the decision maker's preference over the lotteries, not a *cause*.

Note: In Chapter V we mentioned that economists use the term utility theory for what we called value theory. Utility theory, as we discuss it in this chapter, is referred to by economists as **expected utility theory**.

1.2 The Structural Model

The structural model of decision making under uncertainty is shown in Figure VII.3. The alternatives, a_j ($j=1,...,n$), are evaluated with respect to all criteria, c_i ($i=1,...,m$), and scenarios, s_k ($k=1,...,v$). Component utility functions are assessed by the decision maker, with which the evaluation values, e_{ij}^k, can be transformed into utility values u_{ij}^k. The aggregation across criteria leads to u_j^k. The preference ranking of the alternatives is based on the expected utility: $u_j \equiv E[u_j] = \Sigma_k p_j^k u_j^k$.

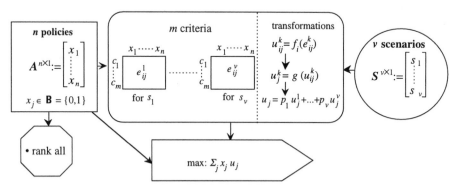

Figure VII.3: Structural model of utility assessment.

2. The Formal Model of Utility Theory

2.1 Motivation of Utility Theory

Utility theory is rooted in the theory of games as introduced by Von Neumann and Morgenstern in the 1940s. In fact, issues in conflict analysis and gambling motivated the development of expected utility theory. To see why utility theory goes beyond the concepts of value theory, we'll take a simple example. Assume your value function corresponds to the monetary outcome ($) of the alternatives. If you get offered the choice between the following two lotteries, $L_1 := [100,0.5,50]$ and $L_2 := [75,1.0,-]$, which of the two do you prefer: a 50% chance of winning $100 or $50, or $75 for certain. Before we proceed, let's introduce the **expected value**, $E[L]$, of a lottery $L := [(p_1,e_1), ..., (p_k,e_k), ..., (p_v,e_v)]$:

$$E[L] = \sum_{k=1}^{v} p_k e_k \, .$$

If u_k is the utility of an alternative a_k, and $L := [(p_1,u_1),...,(p_k,u_k),...,(p_v,u_v)]$ a lottery, then

$$E[u(L)] = \sum_{k=1}^{v} p_k u_k$$

is the **expected utility** of the lottery L.

Coming back to our choice problem, you probably have no preference between the two lotteries. The reason for your indifference is that the two lotteries have the same **expected monetary value** (EMV): $E[L_1]=E[L_2]=\$75$. However, two lotteries: $L_1:= [100000,0.1,10]$ and $L_2:=[10009,1.0,-]$ also have the same EMV: $E[L_1]=E[L_2]= \$10009$; your choice is between a low probability (0.1) of a very high win (\$100,000) and a certainty value of \$10,009, also a high win. The certainty of \$10,009 might sound good enough for you to take that amount and renounce to the lottery. This example shows, that the EMV is not necessarily a good rule to choose between lotteries.

The expected value of a lottery is called the lottery's **certainty equivalent**. A person who prefers the certainty equivalent over the lottery is considered **risk averse**. One who is indifferent between the lottery and its certainty equivalent, is considered **risk neutral**. And finally, one who prefers the lottery over the certainty equivalent is considered **risk prone**.

Before we address the axioms of utility theory, let's look at an example of a utility function that captures preferences correctly even in the face of seemingly contradictory evidence offered by the EMVs. For the two lotteries (gains in \$), $L_1:=[100000,0.5,-50000]$ and $L_2:=[100,0.5,-50]$, we have with the EMV, $E[L_1]=25000 > E[L_2]=25$. However, it can certainly be assumed that most people would prefer L_2 over L_1.

Thus, let's define the following utility function (where positive outcomes represent wins and negative losses): $u(100000)=1$, $u(-50000)=0$, $u(100)=0.7$, and $u(-50)=0.6$. Then, we get $E[u(L_1)]=0.5<E[u(L_2)]=0.65$, which agrees with the assumption of $L_2 \succ L_1$.

If $u(L)$ and $w(L)$ are two utility functions describing the same preferences, then there is a positive linear (affine) transformation between the two: $u(L)=\alpha u(L)+\beta$, $\alpha>0$. That is, utility functions are invariant under a positive linear transformation.

We can see this for binary lotteries; let $L_1:=[u_1,p,u_2] \succ L_2:=[u_3,q,u_4]$, and $w=\alpha u+\beta$ ($\alpha>0$). Thus, we get $E[L_1] > E[L_2]$, which is $pu_1+(1-p)u_2 > qu_3+(1-q)u_4$. Transforming these two expected utilities gives: $E[L_1]=p(\alpha u_1+\beta)+(1-p)(\alpha u_2+\beta)$, and $E[L_2]=q(\alpha u_3+\beta)+(1-q)(\alpha a u_4+\beta)$. These can be rewritten as $\alpha[pu_1+(1-p)u_2]+\beta[p+(1-p)]=\alpha E[L_1]+\beta$, and $\alpha[qu_3+(1-q)u_4]+\beta[q+(1-q)]=\alpha E[L_2]+\beta$. Since $\alpha E[L_1]+\beta > \alpha E[L_2]+\beta$, we see indeed that an interval scale is appropriate for expected utilities. Two utility functions that are related to each other by such a positive linear transformation are called **strategically equivalent**.

2.2 Axioms of Utility Theory

In Chapter V, Section 2.1, we introduced a set of axioms for rational behavior when no uncertainty is involved as the basis for value theory. For decision making under risk, Von Neumann and Morgenstern [1947] have established a set of **axioms** that implies the expected utility model. However, instead of comparing individual alternatives, lotteries are addressed. These axioms are, with $\{L_i\}$ being a set of lotteries:

- <u>Completeness</u>: For any two lotteries, L_j and L_k, either $L_j \succ L_k$, or $L_k \succ L_j$, or $L_j \sim L_k$.

- <u>Transitivity</u>: If $L_i \succsim L_j$, and $L_j \succsim L_k$, then $L_i \succsim L_k$.

- <u>Continuity</u>: Given the lotteries L_i, L_j, and L_k, such that $L_i \succ L_j \succ L_k$, then there exists a probability $p \in (0,1)$ such that $L_j \sim pL_i + (1-p)L_k$.

- <u>Independence</u>: If $L_i \succ L_j$, then for any $p \in (0,1]$ and any L_k: $pL_i + (1-p)L_k \succ pL_j + (1-p)L_k$.

These four axioms of utility theory imply that a **utility function** $u(\cdot)$ exists (can be determined) such that for any two lotteries, $L_1 := [(p_{11}, a_{11}), ..., (p_{1v}, a_{1v})]$ and $L_2 := [(p_{21}, a_{21}), ..., (p_{2w}, a_{2w})]$, the magnitude order of the expected utilities of these lotteries, $E[u(L_1)] = p_{11} \times u_{11} + ... + p_{1v} \times u_{1v}$ and $E[u(L_2)] = p_{21} \times u_{21} + ... + p_{2w} \times u_{2w}$, reflects (not imposes) the preference order of the two lotteries, where $u_{ij} \equiv u(a_{ij})$:

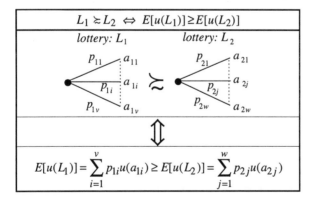

$$L_1 \succsim L_2 \iff E[u(L_1)] \geq E[u(L_2)]$$

$$E[u(L_1)] = \sum_{i=1}^{v} p_{1i} u(a_{1i}) \geq E[u(L_2)] = \sum_{j=1}^{w} p_{2j} u(a_{2j})$$

If a decision maker's preferences over lotteries are consistent with these four axioms, then his/her preferences for uncertain outcomes can be represented by an expected utility model. A decision maker who complies with the axioms of expected utility theory is called **rational**.

It should be noted that the arrow in the definition above goes primarily from preference order ($\succsim$) to magnitude order ($\geq$). What does it mean if one computes the expected utilities for two lotteries? Can the resulting order relation lead to conclusions about the preference relation? The point is that a utility function is personal (if one exists). Thus, it reflects a decision maker's personal preference structure. For this specific decision maker, we would therefore assume that the ordinal relation corresponds to the preference relation. However, if for any two lotteries the magnitude order and the preference order are different, then the utility function must be reassessed. If this does not remedy the situation, one must check if the decision maker still complies with the axioms of expected utility theory.

In the vast literature about expected utility theory, many other systems of axioms have been proposed. Some of the not yet explicitly emphasized aspects are:

- **Substitution**: A degenerated lottery, $[L_i, 1.0, -]$, can be replaced by any equivalent non-degenerated lottery. Note that every branch in a lottery is a degenerated lottery:

- **Compound lotteries**: A decision maker is indifferent between a compound lottery and a simple lottery with the corresponding evaluation and probability values.

e.g.,

- **Unequal probabilities**: For two lotteries with the same outcomes, $L_1=[e_1,p_1,e_2]$ and $L_2=[e_1,p_2,e_2]$, where $e_1 \succ e_2$, the decision maker prefers the lottery with higher probability on e_1:

for $P_1 \succ P_2$

This statement is easily evaluated by computing the expected values of the two lotteries.

2.3 Assessing Component Utility Functions

The **utility function** is a subjective preference function over lotteries. It is assessed considering the ranges defined by the best and worst outcomes with respect to each criterion.

A multidimensional utility function is often expressed in terms of **component utility functions**, one for each criterion. A component utility function is in fact a **conditional utility function**, because it may depend on the outcomes with respect to the other criteria. However, if the conditional utility function does not depend on the other criteria we have **utility independence** (see Section 2.5). In this case, the component utility functions are assessed as **marginal utility functions**, that is, independent of the other criteria.

Farquhar [1984] discusses 24 techniques to elicit utility functions. Some of these methods are given in Table VII.1, where the underlined element has to be assessed by the decision maker, where $\Re$ is the preference or indifference relation.

	Standard Gamble Methods	**Paired Gamble Methods**
preference comparison	$[e_1,p_1,e_2] \; \underline{\Re} \; e_3$	$[e_1,p_1,e_2] \; \underline{\Re} \; [e_3,p_3,e_4]$
probability comparison	$[e_1,\underline{p_1},e_2] \sim e_3$ $(e_1 \succ e_3 \succ e_2)$	$[e_1,\underline{p_1},e_2] \sim [e_3,p_3,e_4]$ $(e_1 \succ (e_3,e_4) \succ e_2)$
value equivalence	$[\underline{e_1},p_1,e_2] \sim e_3$	$[\underline{e_1},p_1,e_2] \sim [e_3,p_3,e_4]$
certainty equivalence	$[e_1,p_1,e_2] \sim \underline{e_3}$	-

Table VII.1: Methods for assessing utility functions.

As an example of how to assess a utility function, we will discuss briefly the **probability comparison method** for a standard gamble. Given below is a set of alternatives, $A=\{a_j\}$, where each alternative is measured with respect to only one criterion. For example, let's assume a decision maker considers five construction projects which are evaluated in terms of their construction costs for two scenarios, s_1 and s_2 (e.g., low and high economic growth).

costs: 10^6	a_1	a_2	a_3	a_4	a_5
s_1	10	8	12	5	15
s_2	11	7	9	7	13

Table VII.2: Evaluation values for one criterion (costs).

As the first step, analogous to the value function (Chapter V), we determine the range of the evaluation values and assign them the following utility values: $u(e_{best})=1$, $u(e_{worst})=0$. For our example we have: $u(5)=1$, and $u(15)=0$.

To assess the utility value of any other outcome, e_j, we ask the decision maker for the probability, p_j, for which s/he is indifferent between the certainty equivalent e_j and the lottery with the probabilities p_j for the best outcome and $(1-p_j)$ for the worst outcome:

$$[e_{best}, p_j, e_{worst}] \sim e_j.$$

The interactive question-answer dialog for the outcome $e_j = 10$, could be as follows:

Analyst: If you can choose between a project that costs 10^6 ($10m) and the lottery [$5m,0.5,$15m], which of the two would you take.

Decision maker: I would take the $10m project, because the risk to end up with a $15m project is too high.

Analyst: Well, then let's increase the attractiveness of the lottery by increasing the probability of the lottery to 0.9. Which of the two options looks now better to you: $10m for sure or the lottery [$5m,0.9,$15m]?

Decision maker: Now I would prefer the lottery.

Analyst: Okay, then this increase was too high. Now tell me for which probability (p_j) between 0.5 and 0.9 would you be indifferent between costs of $10m and the lottery [$5m,p_j,$15m].

Decision Maker: (thinks for a while). I think if the probability were 0.8, I would be indifferent between the two options.

We know now that the decision maker has specified the following indifference between the lottery and the certain outcome: [$5m,0.8,$15m] $\sim$ $10m. We can write this in terms of utilities as follows:

Because the certainty equivalent is indifferent to the lottery, we can write down the equation, with $u(e_{worst})=0$ and $u(e_{best})=1$:

$$\underline{u(\$10m)}=0.8u(\$5m)+0.2u(\$15m)= 0.8\times1.0+0.2\times0.0=\underline{0.8}.$$

Now we know the utilities of three alternatives, and we proceed in the same way to assess the utilities of all the other alternatives. If we continue with e_{worst} and e_{best}, we get in general:

$$\underline{u(e_j)} = p_j \times u(e_{best}) + (1-p_j) \times u(e_{worst}) = p_j \times 1.0 + (1-p_j) \times 0.0 = \underline{p_j}.$$

However, we do not always have to use the lottery with the best and worst outcomes: $[e_{best}, p_j, e_{worst}] \sim e_j$. Once we have determined several other utilities, we can use any two outcomes whose utilities are already determined (e_i, e_k) to asses new utilities:

$$\underline{u(e_j)} = p_j \times u(e_i) + (1-p_j) \times u(e_k).$$

This relation will also be used to make checks, by asking the decision maker if s/he is in fact indifferent between the two options: $[e_i, p_j, e_k] \sim e_j$, where the utilities of all three outcomes have been already assessed. If s/he disagrees, we must continue the assessment process until the inconsistencies are resolved. This **direct assessment approach**, based on a question-answer dialog, is appropriate if the number of alternatives is up to 50 [Keeney and Raiffa, 1993]. Otherwise, a curve (or function) should be estimated and corrected by doing some checks.

It should be noted that when we talk about outcomes we do not necessarily refer to the decision options as defined by the decision maker. Rather, we refer to any outcome between e_{worst} and e_{best}. We could also determine the utility of the outcome, say \$6m, although it does not belong to any of the alternatives considered in Table VII.2. In practical situations it is often sufficient to assess the utilities of a few outcomes and then to approximate a function through these utilities (e.g., the function used in Chapter V, Figure V.13). A possible result of our utility assessments for the outcomes in Table VII.2 is given in Figure VII.4.

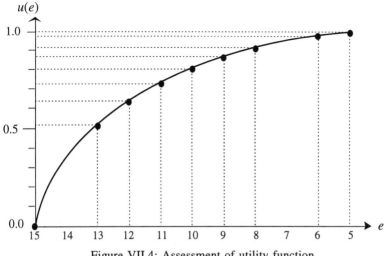

Figure VII.4: Assessment of utility function.

The utility values for the evaluation values in Table VII.2 are given in Table VII.3.

10^6	a_1	a_2	a_3	a_4	a_5
$s_1: p_1=0.6$	$u(10)=0.80$	$u(8)=0.91$	$u(12)=0.65$	$u(5)=1.00$	$u(15)=0.00$
$s_2: p_2=0.4$	$u(11)=0.72$	$u(12)=0.65$	$u(9)=0.87$	$u(13)=0.52$	$u(6)=0.98$
$E[u]$	0.77	0.81	0.74	0.81	0.39

Table VII.3: Utility values for corresponding evaluation values.

With these utilities we compute the expected utility for each alternative. A rational decision maker would rank the alternatives according to the magnitude of the expected utility. It should be noted that all alternatives are Pareto optimal. Based on the expected utility we get the following preference order: $(a_2 \sim a_4) \succ a_1 \succ a_3 \succ a_5$.

A major benefit of utility theory is that any complex decision situation involving lotteries can be reduced to a decision situation between two lotteries with both having only the worst and best evaluation values, from which we can compute the certainty equivalents.

For example, assume that a decision maker must decide between two construction projects, and there is uncertainty about the economic situation and the material used. The costs for the different scenarios and their utilities are shown in Figure VII.5. The three representations are equivalent; however, the middle (binary lotteries with best and worst outcome) and the right (certainty equivalents) representation make it easier to assess the preference between the two projects.

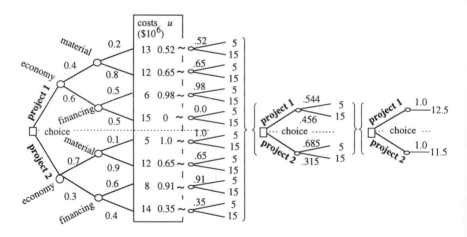

Figure VII.5: Reducing complexity in a decision situation under uncertainty.

The first step in reducing this complex decision situation is to replace the certainty outcomes (gains) in the left representation with the equivalent lottery with best (5) and worst (15) outcomes. The utilities are given in Table VII.3. The utility $u(13)$ is equivalent to the lottery $[u(5),0.52,u(15)]$. Each utility $u(x)=p$ is replaced by the lottery $[u(5),p,u(15)]$. Then, the left representation can be reduced to the middle representation. The middle representation can be reduced into the certainty equivalent representation by looking up in Figure VII.4 the evaluation values which correspond to 0.544 (which is about 12.5) and 0.685 (which is about 11.5), respectively. Thus, project 2 (11.5) seems to be preferred to project 1 (12.5).

2.4 Interpretation of the Utility Function and Risk Attitudes

Analogous to the value functions discussed in Chapter V (Figure V.9), we can distinguish three different shapes of the utility graphs, where one utility function might take on several of these shapes simultaneously (Figure VII.6).

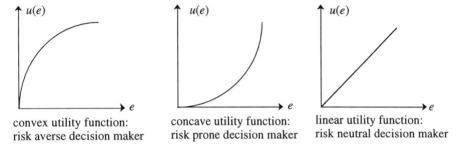

| convex utility function: | concave utility function: | linear utility function: |
| risk averse decision maker | risk prone decision maker | risk neutral decision maker |

Figure VII.6: Risk attitudes.

The **convex** utility graph ($d^2u/d(e)^2 < 0$) depicts a decision maker who is **risk averse**: one who prefers the certainty equivalent over the lottery. A **concave** utility graph ($d^2u/d(e)^2 > 0$) depicts a **risk prone** decision maker: one who prefers the lottery over the certainty equivalent. Finally, a **linear** utility graph ($d^2u/d(e)^2=0$, $du/de=$constant) depicts a **risk neutral** decision maker: one who is indifferent between the lottery and the certainty equivalent.

Note that a risk neutral decision maker's utility function is defined as follows (for $u \in [0,1]$): $u(e_j)=(e_j-e_{worst})/(e_{best}-e_{worst})$, which is a straight line from $(e_{worst},0)$ to $(e_{best},1)$. We also see that for $u \in [0,1]$, the probability p_j of the lottery $[e_{best},p_j,e_{worst}]$ $\sim e_j$ corresponds to the utility of the outcome e_j. Thus, risk aversion means that the utility (and thus the probability) is larger (i.e., closer to the maximum utility value) than the certainty equivalent. We can thus generalize this finding for any other ranges as follows: If the utility is larger than the certainty equivalent, the decision maker is risk averse, if it is smaller, s/he is risk prone, and if it is the same, s/he is risk neutral. Figure VII.7 shows different utility functions for different ranges, and the corresponding decision maker's attitude.

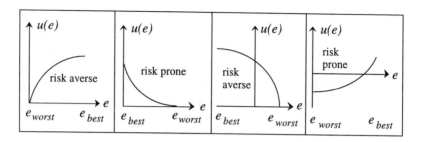

Figure VII.7: Different utility functions with varying ranges.

A measure for the degree of **risk aversion**, $r(e)$, is:

$$r(e) \equiv -\frac{d^2u / d(e)^2}{du / de} \equiv -\frac{u''}{u} = -\frac{d}{de}[\log u'(e)].$$

In words, the measure of risk aversion is the negative quotient of 'rate of changing slope' and 'slope.' For a risk neutral decision maker the utility function is linear which results in zero risk aversion because the slope does not change. For a risk averse decision maker the utility function is convex which results in a positive risk aversion because the slope is decreasing. For a risk prone decision maker the utility function is concave which results in a negative risk aversion because the slope is increasing.

Let's assume that we have the following utility function: $u(e)=\exp(ke)$, with $k>0$. Differentiating, we get $u'=k\exp(ke)$, and $u''=k^2\exp(ke)$. Thus, $r(e)=-k$, which means that the decision maker has a constant risk aversion.

In Section 2.1 we discussed that utility functions (u, w) are unique up to linear transformations $(w=\alpha u+\beta, \alpha>0)$. This means that u and w agree on the preference order between any lotteries and are therefore called **strategically equivalent**. Another important characteristic of utility functions is that two utility functions u and w are strategically equivalent $(w=\alpha u+\beta,, \alpha>0)$ if and only if they have the same risk aversion function [Keeney and Raiffa, 1993, p. 160]. For w we get $w'=\alpha u'$ and $w''=\alpha u''$; inserting these relations into the definition for risk aversion gives: $r(e)=-u''/u'$, which shows that u and w have the same risk aversion if they are strategically equivalent.

2.5 Utility Independence

In Chapter V, we introduced the concept of preferential independence, a necessary condition for a value function to be additive. In the following we will introduce the concept of **utility independence** which is a necessary condition for a multidimensional utility function to be multiplicative or additive.

Preferential independence was defined as follows: if the alternatives are evaluated with two criteria, c_1 and c_2, then c_1 is said to be preferentially independent of c_2 if the preference relation between alternatives, a_j: (e_{1j}, e_{2j}), with varying levels of c_1 (x) and a common, fixed level of c_2 (α) is independent of that fixed level of c_2: $(x,\alpha) \succsim (y,\alpha) \Leftrightarrow (x,\beta) \succsim (y,\beta)$. If c_2 is also preferentially independent of c_1, then the two criteria are mutually preferentially independent.

Preferential independence can be seen as utility independence for degenerated lotteries. Thus, utility independence must be defined in terms of lotteries. For example, four lotteries are given, where the outcomes are evaluated with two criteria c_1 (costs in \$$10^6$) and c_2 (landloss in km^2), and where $\Re$ stands for a preference relation:

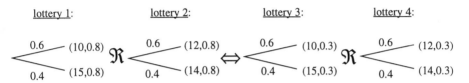

| lottery 1: | lottery 2: | lottery 3: | lottery 4: |

We see that all lotteries have the same probability distribution over the outcomes. Moreover, the landloss for the first two lotteries (0.8) and for the third and fourth lotteries (0.3) is the same. The costs of the first and third lotteries (10 and 15) and of the second and fourth lotteries (12 and 14) are the same. Thus, we would expect that the preference relation, $\Re$ (e.g., $\succsim$) must be the same between the first and second and third and fourth lottery.

From this example, we derive the definition of utility independence. It should be noted that if we change the probability of 0.6 to 1.0, and 0.4 to 0.0, then we get the definition of preferential independence. Thus, if utility independence holds, we can conclude that preferential independence also must hold. In other words, the definition of utility independence is equivalent to the definition of preferential independence if we replace the deterministic outcomes with lotteries. Or, the definition of preferential independence corresponds to the definition of utility independence with degenerated lotteries.

If the alternatives are evaluated with two criteria, c_1 and c_2, then c_1 is said to be **utility independent** of c_2 if the preference relation between lotteries, with varying levels of c_1 (x,y) and a common, fixed level of c_2 (α) is independent of that fixed level of c_2:

$$[(x,\alpha),p,(y,\alpha)] \succsim [(r,\alpha),p,(s,\alpha)] \Leftrightarrow [(x,\beta),p,(y,\beta)] \succsim [(r,\beta),p,(s,\beta)].$$

The criterion c_2 is said to be utility independent of c_1 if the preference relation between lotteries, with varying levels of c_2 (x,y) and a common, fixed level of c_1 (α) is independent of that fixed level of c_1.

$$[(\alpha,x),p,(\alpha,y)] \succsim [(\alpha,r),p,(\alpha,s)] \Leftrightarrow [(\beta,x),p,(\beta,y)] \succsim [(\beta,r),p,(\beta,s)].$$

If the two criteria are each utility independent of each other, they are **mutually utility independent**.

For more than two criteria, the definition of (mutual) utility independence is analogous to the one for two criteria, and especially to the definition of preferential independence. Let $C=\{c_1,...,c_m\}$ be the set of all criteria. Then, $C_j \subset C$ is a decomposition of C, $C_j \neq \varnothing$, $C_j \neq C$. The complement of the decomposition is defined as $\overline{C}_j = C \setminus C_j$. Thus, we have $C_j \cup \overline{C}_j = C$, and $C_j \cap \overline{C}_j = \varnothing$. Let $\overline{C}_{j:\alpha}$ be the complement of the decomposition C_j fixed at level α. Then, the decomposition C_j is utility independent of its complement $\overline{C}_j$ if the following relations hold, $\Re$ ($\succsim$):

$$
\begin{array}{c}
p \diagup (C_j, \overline{C}_{j:\,\alpha}) \\[-2pt]
\diagdown \\[-6pt]
1\text{-}p \diagdown (C_k, \overline{C}_{k:\,\alpha})
\end{array}
\Re
\begin{array}{c}
p \diagup (C_r, \overline{C}_{r:\,\alpha}) \\[-2pt]
\diagdown \\[-6pt]
1\text{-}p \diagdown (C_s, \overline{C}_{s:\,\alpha})
\end{array}
\Leftrightarrow
\begin{array}{c}
p \diagup (C_j, \overline{C}_{j:\,\beta}) \\[-2pt]
\diagdown \\[-6pt]
1\text{-}p \diagdown (C_k, \overline{C}_{k:\,\beta})
\end{array}
\Re
\begin{array}{c}
p \diagup (C_r, \overline{C}_{r:\,\beta}) \\[-2pt]
\diagdown \\[-6pt]
1\text{-}p \diagdown (C_s, \overline{C}_{s:\,\beta})
\end{array}
$$

A compact and quick method for the practical assessment of utility independence will be discussed as part of the resolution model in Section 3.2.

2.6 Multidimensional Utility Functions

A) <u>Multilinear Utility Functions</u>

If we can show that all criteria are utility independent of their complements, but full mutually utility independence does not hold, then the multidimensional utility function is **multilinear**. For a 3-dimensional utility function this means [Keeney and Raiffa, 1993, p. 293]:

$$
\begin{aligned}
u(e_1,e_2,e_3) = {}& k_1 u_1(e_1) + k_2 u_2(e_2) + k_3 u_3(e_3) \\
& + k_{12} u_1(e_1)u_2(e_2) + k_{13} u_1(e_1)u_3(e_3) + k_{23} u_2(e_2)u_3(e_3) \\
& + k_{123} u_1(e_1)u_2(e_2)u_3(e_3)
\end{aligned}
$$

where:

1. $u(e_1,e_2,e_3)$ is normalized by $u(e_{1w},e_{2w},e_{3w})=0$ and $u(e_{1b},e_{2b},e_{3b})=1$, where e_{ib} is the best outcome and e_{iw} the worst outcome for criterion c_i.
2. $u_i(e_i)$ is a conditional utility function on e_i normalized by $u(e_{iw})=0$ and $u(e_{ib})=1$ ($i=1,2,3$).
3. $k_i = u(e_{ib}, \overline{e}_{iw})$.
4. $k_{ij} = u(e_{ib},e_{jb},\overline{e}_{ij:w}) - k_i - k_j$ ($i \neq j$, $i,j=1,2,3$).
5. $k_{123} = 1 - k_{12} - k_{13} - k_{23} - k_1 - k_2 - k_3$.

These equations, which must hold simultaneously, allow us to determine a 3-dimensional utility function as follows. First, we determine the component utility functions for c_1, c_2, and c_3 as discussed in the previous section. Then, the three first-order scaling constants (k_1,k_2,k_3) are determined. For example, k_1 is determined by asking the decision maker for which k_1 s/he is indifferent between the deterministic outcome (e_{1b},e_{2w},e_{3w}) and the lottery $[(e_{1b},e_{2b},e_{3b}), k_1, (e_{1w},e_{2w},e_{3w})]$. This gives us the following equation for the utilities: $u(e_{1b},e_{2w},e_{3w})=k_1$. With the first-order scaling constants, the second and third order scaling constants can be determined. It should be noted that the second-order and third-order scaling constants could be negative. If the number of criteria (m) is large, the assessment of a general multilinear utility function is very time consuming.

B) Multiplicative Utility Functions

If the criteria are mutually utility independent, then the general form of an m-dimensional utility function has a **multiplicative** form [Keeney and Raiffa, 1993, p. 288]:

$$1+ku(e_1,...,e_m) = \prod_{i=1}^{m}[1+kk_iu_i(e_i)],$$

where:

1. $u(e_1,...,e_m)$ is normalized by $u(e_{1w},...,e_{mw})=0$ and $u(e_{1b},...,e_{mb})=1$.
2. $u_i(e_i)$ is a conditional utility function on e_i normalized by $u(e_{iw})=0$ and $u(e_{ib})=1$.
3. $k_i = u(e_{ib}, \bar{e}_{iw})$.
4. $1+k = \prod_{i=1}^{m}(1+kk_i)$.

For example, for a 3-dimensional utility function, where the three criteria are mutually utility independent, the multiplicative utility function has the following form [Keeney and Raiffa, 1993, p. 289]:

$$
\begin{aligned}
u(e_1,e_2,e_3) &= k_1u_1(e_1) + k_2u_2(e_2) + k_3u_3(e_3) \\
&+ kk_1k_2u_1(e_1)u_2(e_2) + kk_1k_3u_1(e_1)u_3(e_3) + kk_2k_3u_2(e_2)u_3(e_3) \\
&+ k^2k_1k_2k_3u_1(e_1)u_2(e_2)u_3(e_3),
\end{aligned}
$$

where:

1. $u(e_1,e_2,e_3)$ is normalized by $u(e_{1w},e_{2w},e_{3w})=0$ and $u(e_{1b},e_{2b},e_{3b})=1$.
2. $u_i(e_i)$ is a conditional utility function on e_i normalized by $u(e_{iw})=0$ and $u(e_{ib})=1$, $(i=1,2,3)$.
3. $k_i = u(e_{ib}, \bar{e}_{iw})$, $(i=1,2,3)$.
4. $1+k=(1+kk_1)(1+kk_2)(1+kk_3)$.

The difference to the multilinear model is the scaling constant k. But what about this scaling constant k in the multiplicative model? Obviously, if $k=0$, then we have the additive model (see next section). But if the additive model cannot be applied, that is, $\Sigma_i k_i \neq 1$, we have to compute the root(s) of the polynomial that implicitly defines k. Keeney and Raiffa [1993, p 347] show that for $\Sigma_i k_i > 1$ there is exactly one root, if the search for k is confined to the open interval $(-1,0)$. For $\Sigma_i k_i < 1$ there is exactly one root, if the search for k is confined to the open interval $(0, \infty)$.

C) Additive Utility Functions

A stronger concept than utility independence is **additive independence**. If for two alternatives a_i: (e_{1i},e_{2i}) and a_j: (e_{1j},e_{2j}), the following lotteries are indifferent (where $(e_{1i},e_{2i}) \neq (e_{1j},e_{2j})$):

$$[(e_{1i},e_{2i}),0.5,(e_{1j},e_{2j})] \sim [(e_{1i},e_{2j}),0.5,(e_{1j},e_{2i})],$$

then c_1 and c_2 are said to be additively independent. Additive independence is important because it implies an additive multidimensional utility function [Keeney and Raiffa, 1993, p. 295]. Additive independence means that the preferences over lotteries depend only on their marginal probability distributions and not on the joint probability distributions of the multidimensional outcomes.

Additive independence implies $\Sigma_i k_i = 1$ in the multiplicative model discussed under B. This means that given mutual utility independence holds, we determine the scaling constants. If they add up to one, we use the following **additive** form:

$$u(e_1,...,e_m) = \sum_{i=1}^{m} k_i u_i(e_i),$$

where:

1. $u(e_1,...,e_m)$ is normalized by $u(e_{1w},...,e_{mw})=0$ and $u(e_{1b},...,e_{mb})=1$.
2. $u_i(e_i)$ is a conditional utility function on e_i normalized by $u(e_{iw})=0$ and $u(e_{ib})=1$.
3. $k_i = u(e_{ib},\overline{e}_{iw})$.

For three additively independent criteria we have:

$$u(e_1,e_2,e_3) = k_1 u_1(e_1) + k_2 u_2(e_2) + k_3 u_3(e_3),$$

where:

1. $u(e_1,e_2,e_3)$ is normalized by $u(e_{1w},e_{2w},e_{3w})=0$ and $u(e_{1b},e_{2b},e_{3b})=1$.
2. $u_i(e_i)$ is a conditional utility function on e_i normalized by $u(e_{iw})=0$ and $u(e_{ib})=1$, $(i=1,2,3)$.
3. $k_i = u(e_{ib},\overline{e}_{iw})$, $(i=1,2,3)$.

The additive utility function is certainly more appealing because it does not have higher order terms. As a concluding remark we might add that the additive model is a special case of the multiplicative model, which in turn is a special case of the multilinear model. For practical purposes, however, we would hope to be able to use at least the multiplicative model, if not even the additive model.

3. The Resolution Model

3.1 The General Approach

The general procedure for assessing multidimensional utility functions is shown in Figure VII.8; for a detailed description about the art of assessing multidimensional utility functions see [Keeney, 1977].

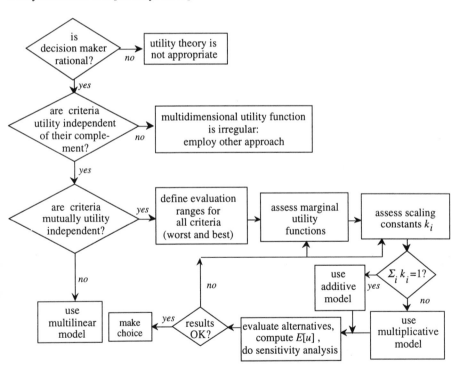

Figure VII.8: General process to determine a multidimensional utility function.

this is not the case, then the multidimensional utility function is irregular, and a procedure other than the ones discussed here must be employed. If the criteria are mutually utility independent, then the multidimensional utility function is multiplicative. If the scaling constants add up to unity, then we can use the additive model.

Once the multidimensional utility function is determined, we can evaluate the alternatives and compute the expected utilities over all scenarios. Then, sensitivity analysis is performed. If the results are not stable, the component utility functions and the scaling constants might have to be reassessed.

In the following example, we will use the simplified Swiss *AlpTransit* project to discuss in greater detail the steps of assessing a multidimensional utility function. Switzerland is investigating the options for building a new rail tunnel through the Alps to increase transport capacity. The project is estimated to cost up $10 billions, and its realization will take up to ten years. The Swiss government is currently investigating five alternatives: Lötschberg/Simplon (L/S); Gotthard (GOT); Ypsilon (YPS), which is an extended version of GOT; Splügen 1 (SP1); and the extended version of SP1, Splügen 2 (SP2). Multicriteria evaluations have been performed with more than twelve criteria. In this example, we will use only five criteria with altered evaluation values. The evaluation matrix for these five alternatives and five criteria is shown in Table VII.4.

alternatives:	L/S		GOT		YPS		SP1		SP2	
scenarios:	L	H	L	H	L	H	L	H	L	H
probabilities:	0.3	0.7	0.3	0.7	0.3	0.7	0.3	0.7	0.3	0.7
economic benefit (c_1) [%] $k_1 = 0.4$	1 0.64	4.4 0.98	2.6 0.86	5.1^b 1.00	1.6 0.74	3.5 0.93	0.6 0.55	3.0 0.89	-1^w 0.00	0.8 0.60
time savings (c_2) 10^6 [h/y] $k_2 = 0.1$	2.2^w 0.0	2.3 0.00	10.0 0.44	11.1 0.60	11.5 0.68	12.9^b 1.00	2.2^w 0.00	2.4 0.00	2.5 0.01	2.7 0.01
energy use (c_3) 10^7 [kWh/y] $k_3 = 0.15$	57 1.00	103 0.81	88 0.89	145 0.43	88 0.89	170^w 0.00	54^b 1.00	135 0.55	79 0.92	145 0.43
landloss (c_4) [ha] $k_4 = 0.05$	125^b 1.00	125^b 1.00	249 0.92	249 0.92	540^w 0.00	540^w 0.00	415 0.62	415 0.62	434 0.55	434 0.55
risks (c_5) 10^{-6} [1/y] $k_5 = 0.3$	6 0.65	7 0.53	2 0.95	3 0.90	1^b 1.00	4 0.83	8 0.38	10^w 0.00	7 0.53	9 0.21
u_j:	0.701	0.723	0.853	0.841	0.798	0.721	0.515	0.470	0.326	0.396
$E[u_j]$:	0.701		0.845		0.744		0.484		0.375	
final ranking:	3		1		2		4		5	

Table VII.4: Score card of simplified *AlpTransit* project.

The evaluations have been done for two scenarios, to account for the possibilities of low (L) and a high (H) economic growth over the next 25 years. The economic benefit (c_1) reflects the benefit that Switzerland expects to reap by building the new *AlpTransit* rail tunnel. As a result of this project, savings in travel time (c_2) are expected. The increase in rail transport results in additional energy use (c_3). Moreover, new construction will lead to loss of land (c_4). Finally, the new system will create new transportation risks (c_5).

3.2 Assessment of Mutual Utility Independence

The first step in deriving a multidimensional utility function is to identify the worst and the best outcomes with respect to all criteria. The best values have a super script 'b' and the worst values a super script 'w.'

Because of the rather large number of criteria (5), we would like to use either the additive or the multiplicative model for the multidimensional utility function. Thus, we must show that the criteria are mutually utility independent. Just as we did for the assessment of preferential independence of m criteria (Chapter V), here, we also would have to do 2^m-m assessments to check that the m criteria are mutually utility independent. We then explained why only m-1 assessments are necessary to check for preferential independence. Fortunately, Keeney and Raiffa [1993, p. 292] prove the following theorem for the assessment of utility independence:

Let A be a set of alternatives which are described with m (≥ 3) criteria $\{c_i\}$. Then, the following statements are equivalent:

• the m criteria $\{c_i\}$ are mutually utility independent,
• c_k is utility independent of its complement $\bar{c}_k$, and
• $\{c_k, c_j\}$ is preferentially independent of its complement $\{\bar{c}_k, \bar{c}_j\}$, $j=1,...,m; j \neq k$.

This theorem is very helpful because it reduces the original 2^m-2 assessments of utility independence to m assessments. First, the analyst chooses one criterion, c_k, and builds pairs with all of the other criteria. S/he then must check if each of the $(m$-1) pairs $\{c_k, c_j\}$ is preferentially independent of its complement (for all $j \neq k$). Second, s/he must check if criterion c_k is utility independent of its complement $\bar{c}_k$. If this is the case, it can be concluded that all criteria are mutually utility independence.

For example, let's consider c_1 (economy) and c_5 (risks) and generate some numeric pairs; for example, (5,1), (4,2), (3,8), etc. The preferences for these pairs must be independent of the level of the other three criteria. We would, for example, ask the decision maker to compare $(5, \alpha, \beta, \gamma, 1)$ with $(4, \alpha, \beta, \gamma, 2)$. If the decision maker is indifferent between such pairs (for any α, β, and γ), we would conclude that

---

preferential independence between (c_1,c_5) and its complement holds. Then, we would check this for all other pairs (c_1,c_2), (c_1,c_3), and (c_1,c_4).

Finally, we have to check if c_1 is utility independent of its complement. This is done by checking if for two ordered lotteries the order is independent of the varying level of the other criteria. For example, we can take as the two levels of the other criteria the best and the worst outcomes. This is shown in the figure below.

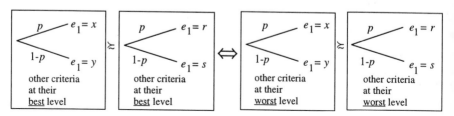

Assuming that utility independence could be shown for this criterion, we can conclude that the decision maker's multidimensional utility function must be either additive or multiplicative.

To decide if the additive model can be used, we must assess the scaling constants k_i's and see if they add up to one. If they do, we know that the additive model is appropriate. If the sum of the scaling constants is not equal to one we would have to use the multiplicative model. The next task is therefore to assess the scaling constants.

3.3 Assessment of Scaling Constants and Selection of Model

The first step in determining the scaling constants is to rank them in descending order. To find this ranking, we refer to the score table and assume a hypothetical alternative $a:=(w,...,w)$. To identify the largest scaling constant, we ask the decision maker to exchange one w for b. The chosen criterion thus has the largest scaling constant. To find the second largest scaling constant, we ask the decision maker to choose the next criterion where s/he wants to exchange the w for b. This procedure goes on until we have $a:=(b,...,b)$.

It is obvious that this ordering does not necessarily reflect the preferences of the decision maker for the different criteria. Rather, it depends on the actual and case specific ranges of the criteria; that is, on how much we can gain if we can move from worst to best level.

For our example we might get the following ordering of the scaling constants:

$$k_1 \succ k_5 \succ k_3 \succ k_2 \succ k_4.$$

The next step is to determine the numerical values of the scaling constants, k_i's. This is done by asking the decision maker for the probability p_i such that the following two lotteries are indifferent:

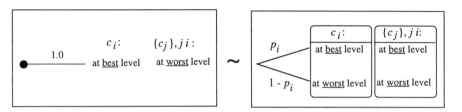

This means that we get with m criteria (where w_k means that e_k=worst, and b_k means that e_k=best):

$$u(w_1,...,w_{i-1},b_i,w_{i+1},...,w_m) = p_i \times u(all\ at\ best) + (1-p_i) \times u(all\ at\ worst).$$

The left part of the equality is equal to k_i, while the right part is $p_i \times 1 + (1-p_i) \times 0 = p_i$. Thus we get: $k_i = p_i$.

By proceeding this way with all scaling constants, we also see that the numeric values for the k_i's depend on the value ranges defined by the worst and the best outcomes. For example, the range for the economic benefit is 6.1%, from -1% (worst) to 5.1% (best). With these values, together with the best and worst values of the criterion, the decision maker might determine k_1=0.4. Let's now change the best economic value to, let's say, 100%, and leave all other values the same. Then, considering that the worst values for all criteria are not too bad, the decision maker might lean more toward the certainty outcome. Thus, s/he would choose the lottery only if the chance of having all values at the best level is higher than before. Thus, for example, k_1=0.7. Consequently, the larger the range between worst and best value, the greater the scaling constant. Von Nitzsch and Weber [1993] investigated experimentally the effect of criteria ranges on the scaling constants and concluded that decision makers do not adjust their judgments properly when the range is altered.

When all scaling constants are assessed, we can use the additive model if the k_i's add up to unity. If this is not the case, we must use the more general multiplicative model. For the multiplicative model, we have to compute the value k. The scaling constant k, however, is defined implicitly. For our example we assume that we found the following scaling constants: k_1=0.4, k_5=0.3, k_3=0.15, k_2=0.1, k_4=0.05. As we see, the scaling constants add up to one which means that we can use the additive model.

3.4 Assessment of Component Utility Functions and Evaluation of Alternatives

The component utility functions should be assessed as illustrated in Section 2.3, using one of the many utility elicitation methods mentioned in Table VII.1. However, to simplify the process, we take the exponential function proposed in Chapter V, Section 3.4, and replace v_{ij} with u_{ij}:

$$u_{ij} = \frac{e^{\rho e_{ij}} - e^{\rho e_{i,worst}}}{e^{\rho e_{i,best}} - e^{\rho e_{i,worst}}}.$$

The only utility that we assess is $u[(e_{best}-e_{worst})/2]$, by asking the decision maker for which p s/he is indifferent between the lottery $[e_{best},p,e_{worst}]$ and the certainty outcome $(e_{best}-e_{worst})/2$. As we discussed in Section 2.3, we get $u[(e_{best}-e_{worst})/2]=p$, since $u(e_{best})=1$ and $u(e_{worst})=0$. With these three utilities we compute ρ, such that the function defined by u_{ij}, goes through these three points. Below are the five component utility functions for our example with the assessed p values.

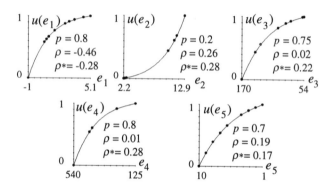

Obviously, the risk aversion function, $r(u)$, for a monotonically increasing utility function (higher outcomes are preferred to lower ones, e.g., gains) is $r=-u''/u'=-\rho$. For a monotonically decreasing utility function (lower values are preferred to higher ones, e.g., costs), the risk aversion function is $r=\rho$. A risk neutral decision maker has $\rho=0$. The reader should confirm this. In order to compare the risk aversions of the different criteria, we have to normalize the risk aversion values. We will do this to match the graph in Figure V.13 (Chapter V): $\rho^*=\rho \times abs(e_{best}-e_{worst})/10$. With the ρ^* values in the graphs above we see that the decision maker is most risk averse for the economic benefit (c_1) and the loss of land (c_4), while s/he is risk prone only for the time savings (c_2).

In case we have a non-monotonic component utility function, such as for the annual precipitation in an agricultural area, we would proceed as was suggested in Chapter V for value functions. The outcomes above the maximum utility are mapped into indifferent outcomes below the maximum utility. The utility function is then assessed for the monotonically increasing part between the worst and best outcome.

The utility values are entered below the evaluation values in Table VII.4. They are derived from the component utility functions. With these utility values and the scaling constants, the additive model is used to compute the overall utilities for each alternative and the different scenarios. Finally, we compute the expected utility. The ranking of the alternatives is based on the expected utilities. We see that we get the following preference order: GOT $\succ$ YPS $\succ$ L/S $\succ$ SP1 $\succ$ SP2.

3.5 Comments about Utility Theory

Utility theory has been criticized both from an operational point of view [McCord and de Neufville, 1983], as well as from behavioral and psychological perspectives [Kahneman et al., 1982]. In several laboratory experiments, behavioral scientists have identified biases in human judgment which contradict the axioms of utility theory [Kahneman et al., 1982]. They claim that subjective probabilities are not an appropriate measure to express a decision maker's uncertainties and that utilities not are not suited to reflect his/her preferences [Beach, 1997].

Consequently, alternative approaches have been proposed, such as **prospect theory** [Kahneman and Tversky, 1979], where utilities are replaced by prospects and probabilities by weights. The prospects are evaluated in terms of gains and losses with respect to some reference point. Thereby, the prospect function can reflect different attitudes toward risk, such as risk aversion for gains and risk proneness for losses [Weber and Miliman, 1997]. For example, if grades for a class range from 0 to 10, with 6 being the lowest sufficient grade, then a student who wants a sufficient grade might be risk seeking in the range between 0 and 6 (losses) and risk averse in the rage between 6 and 10 (gains).

Some of the aspects responsible for behavioral biases in assessments are briefly mentioned. **Conjunction fallacy** refers to judging probabilities by their plausibility, which causes humans to overemphasize details which results in overestimation of probabilities. **Optimism** refers to the overestimation of event probabilities due to a subjective positive attitude. **Representativeness** means that people assign an event to a set based on how familiar they are with the set instead of the event (e.g., a decision maker assumes a person to have a certain profession (or not) based on the decision maker's familiarity with the profession, rather than the person). **Availability** says that probability assessments are based on past experiences (e.g., overestimation of probability of death due to smoking because of familiarity with such a case). **Anchoring and adjustment** is the process used to estimate uncertain values; however, adjustments typically are insufficient, so that different starting points (anchors) lead to different results. Finally, the **certainty effect** says that people tend to overweight gains that are certain.

The certainty effect is best illustrated with the well-known **paradox of Allais** [1953]. Assume you compare two situations, S_1 and S_2, each offering a choice between two lotteries, as illustrated in Figure VII.9.

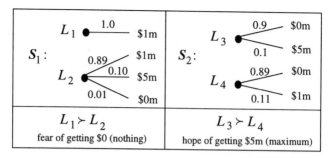

Figure VII.9: Allais' paradox.

The question is now, which lottery is preferred in both situations? In S_1, most people might prefer L_1 over L_2 because the small chance of winning $5m is accompanied by the possibility of getting nothing. In S_2, most people might prefer L_3 over L_4 because the chance of $5m is almost the same as the chance for $1m. If we look at the utilities, then we would define $u(\$5m)=1$, and $u(\$0m)=0$. According to expected utility theory, we conclude:

- for S_1: $u(\$1m) > 0.89{\times}u(\$1m) + 0.10{\times}1.0$, thus: $u(\$1m) > 0.1/0.11$, and

- for S_2: $0.1{\times}1.0 > 0.11{\times}u(\$1m)$, thus: $u(\$1m) < 0.1/0.11$.

Thus, we see that the preferences over the lotteries lead to a contradiction of expected utility theory. Howard [1992] refers to this contradiction as a phenomenon and not a paradox. The reason for this is that most people are fearful in S_1 of getting nothing (chance is 0.01), while in S_2 they are greedy although the chance of not getting more is also 0.01. However, expected utility requires one to be *consistent* in his/her attitudes toward uncertainty and thus either to be greedy or fearful in both situations.

4. Summary and Further Readings

When the evaluation of the alternatives is uncertain, an ordinal utility function that reflects the preferences over uncertain outcomes (called lotteries) is sought. The ordinal utility function should for any two lotteries map the preference order into the same magnitude order of expected utilities. Value theory is a special case of utility theory with only degenerated lotteries.

The axioms of expected utility (completeness, transitivity, continuity, and independence) imply the existence of a utility function with the above characteristics. A decision maker who accepts the axioms of expected utility is called rational. Based on the axioms and the assessment by the decision maker, a normative but subjective utility function over all criteria can be determined.

This multidimensional utility function is derived from the component utility functions which are normalized by $u(e_{worst})=0$ and $u(e_{best})=1$. Several methods to assess the component utility functions are known. One of the most frequently used methods is the probability comparison method. The decision maker is asked to identify the probability of the binary lottery with best and worst possible outcomes, such that s/he is indifferent between this lottery and the certainty equivalent for which the utility is to be determined.

If the component utility function is convex, the decision maker is said to be risk averse. If it is concave, s/he is said to be risk prone, and if it is linear, s/he is said to be risk neutral. The measure of risk aversion, which is defined as the negative of the second derivative divided by the first derivative, is invariant under positive linear transformations.

If each criterion is utility independent of its complement, then the multidimensional utility function is at least multilinear. If the criteria are also mutually utility independent, then the multidimensional utility function is at least multiplicative. If, in addition, the scaling constants add up to unity, the multidimensional utility function is additive. If none of the above conditions hold, then the multidimensional utility function is irregular.

Basically all books on decision analysis discuss some aspects of utility theory. An in-depth treatment of utility theory can be found in Keeney and Raiffa [1993]. A discussion of behavioral and psychological aspects of decision making with normative models can be found in [Kahneman et al., 1982], [Beach, 1997], Bazerman [1998] and [Payne et al., 1993]. A more recent review of the state-of-the-art in utility theory can be found in Edwards [1992]. Recent results concerning the relations between utility functions, $u(e)$, and value functions, $v(e)$, can be found in [Smidts, 1997] and [Weber and Miliman, 1997].

Utility theory is the basis for multi-actor decision making under uncertainty (game theory) which will be addressed in Chapter IX. Thus, most books about game theory address the concepts of utility theory, such as Von Neumann and Morgenstern [1947], and Luce and Raiffa [1985].

5. Problems

1. Identify decision situations where you are risk averse, risk neutral, and risk prone.

2. R. Howard [1992] discusses the following decision problem. Assume you are offered the choice between the following two lotteries, L_1 and L_2. A fair die is rolled and for the two lotteries you get the following gains, depending on the die number:

Die #:	1	2	3	4	5	6
p:	1/6	1/6	1/6	1/6	1/6	1/6
L_1: $	500	600	700	800	900	1000
L_2: $	600	700	800	900	1000	500

- Which of the two lotteries do you prefer (do not read the Hint after Problem 10)?
- Read the Hint after Problem 10; how would you decide now?

3. A decision maker has the following preference relation, where the utility values, u_1 and u_2, are derived from a utility function that has been assessed for him/her:

Comment on his/her attitudes toward uncertain outcomes.

4. A decision maker identified the following preferences for different lotteries of financial gains: $[\$80,0.8,\$20] \sim [\$60,1.0,-]$, $[\$80,0.3,\$20] \sim [\$30,1.0,-]$, $[\$60,3/5,30] \sim [\$50,1.0,-]$.

 If we know that the decision maker assessed $u(\$80)=1.0$ and $u(\$20)=0.0$, determine the utilities of $60, $50, and $30, as well as the decision maker's risk attitude.

5. An analyst has identified for a decision maker the following utility values for four outcomes:

 case 1: $u(100) = 1.0, u(75) = 0.80, u(50) = 0.6, u(25) = 0.35, u(0) = 0$
 case 2: $u(100) = 1.0, u(75) = 0.60, u(50) = 0.4, u(25) = 0.20, u(0) = 0$
 case 3: $u(100) = 1.0, u(75) = 0.75, u(50) = 0.5, u(25) = 0.25, u(0) = 0$
 case 4: $u(100) = 1.0, u(75) = 0.80, u(50) = 0.5, u(25) = 0.15, u(0) = 0$

Comment on the decision maker's risk attitude in the four cases.

Chapter VII: Uncertainty and Normative Choice

6. The three scaling constants of a 3-dimensional utility function, where mutual utility independence holds, have been assessed for three decision makers (d_i). d_1: $k_1=0.40$, $k_2=0.35$, $k_3=0.25$; d_2: $k_1=0.30$, $k_2=0.25$, $k_3=0.20$; and d_3: $k_1=0.60$, $k_2=0.50$, $k_3=0.40$.

Determine which multidimensional utility model is appropriate for the three decision makers and compute the scaling constant k.

7. Assume there are two actors involved in the *AlpTransit* project with extreme opposite attitudes. One actor is the Swiss trucking association which opposes the rail project because it fears that rail will be promoted over road transport. The other actor is the local environmental and citizen group, which strongly favors the project because it reduces traffic emissions. Guess both actors' component utility functions and scaling constants. Then, determine for both of them a 5-dimensional utility function and rank the five alternatives according to the expected utility. Discuss the results. What does it mean to assess a component utility function for the criterion *risk* (c_5)?

8. Compare and contrast the utility functions discussed in this chapter with the value functions discussed in Chapter V.

9. Discuss the behavioral biases in estimating probabilities and events that are mentioned in Section 3.5. Give examples from your own experience. Discuss strengths and limitations of the expected utility model.

10. A famous paradox (or rather dilemma) is the Monty Hall paradox [Vos Savant, 1996]. Assume you are at a game show and you are given the choice of three doors, where behind one door a car is placed. If you choose the door with the car, the car is yours, otherwise you get nothing. Assume now the game starts and you chose a door; before you announce your choice, the host opens one of the other two doors, which of course does not have the car. Would this lead you to reconsider your choice?

• Hint for Problem 2: You have probably figured out that the two lotteries have the same expected value. However, if you compare the monetary gains for each number, you see that in five out of the six outcomes your gain is bigger if you choose L_2. Does this make L_2 more attractive than L_1?

CHAPTER VIII

SEQUENTIAL DECISION MAKING

1. The Structure of Sequential Decisions

1.1 Concept of Probabilistic Influence Diagrams

In this chapter we will focus on sequential decision problems that must be made in the presence of informed uncertainty; that is, uncertainty that can be quantified with probability distributions (also called decision making under risk). These sequential decisions will result in a sequence of actions, also referred to as policies.

In terms of the structural model we will use a compact form of what we have used so far. First of all, we assume that the preferences of the actions are expressed in terms of utilities (which might have been derived from a multicriteria evaluation as discussed in Chapter VII). Second, we assume that the content goal is always to maximize the expected utility. Third, we assume that the decision maker must choose exactly one action from each decision node. Because content and structural goals are obvious, we will omit them from the structural model. An example of an extended (left) and corresponding compact (right) structural model for sequential decision making is given in Figure VIII.1.

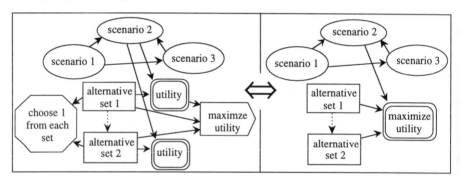

Figure VIII.1: Example of extended (left) and corresponding compact (right) structural model for sequential decision making.

In both structural models in Figure VIII.1 we have a dashed arrow (**informational arc**) from decision node A_1 to decision node A_2. This is to illustrate that at the moment decision A_2 is to be made, decision A_1 is already known; that is, decision A_1 serves as information for decision A_2.

An example of a structural model for a sequential decision problem under uncertainty is shown in Figure VIII.2 (left). It is about a European nation that has to decide on a traffic policy, in the presence of uncertainty. Obviously, the uncertain outcome of the EU traffic policy (stringent or relaxed), has an influence on the

National policy of whether to promote LPG or clean engines. This decision influences the air pollution which can be high or acceptable. Based on the air pollution and the national policy, the nation must decide on a speed limit. The utility depends on the two decisions and on the level of air pollution.

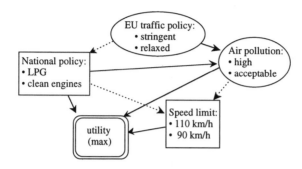

Figure VIII.2: Structural model.

A structural model with sequential decision options under uncertainty is called a **probabilistic influence diagram** [Schachter, 1986]. We thus have three classes of nodes (Figure VIII.3): **decision** nodes (rectangles; i.e., alternatives), **chance** nodes (circles; partitions of the uncertainty space), and **utility** nodes (double rounded rectangles; utilities). The literature knows a fourth type of node: the deterministic node. But because we will not use it in the following discussions it is omitted without loss of generality. We recognize two types of arcs: **conditional** arcs (full line): pointing to evaluation and uncertainty nodes, and **informational** arcs (dashed line): pointing to decision (alternative) nodes. If node *A* points to node *B* then node *A* is a **predecessor** of node B, and node *B* is a **successor** of node *A*.

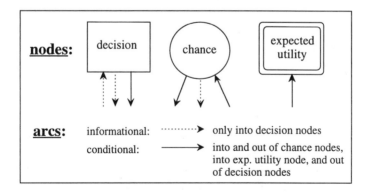

Figure VIII.3: Nodes and arcs in probabilistic influence diagrams.

1.2 The Meaning of Influences in Probabilistic Influence Diagrams

Influence diagrams are a tool to solve and analyze dependencies in decision problems [Bodily, 1985]. As mentioned by Howard [1989], the influence diagram technique can be used to summarize, display, communicate, and analyze relationships between objects, decision variables, and variables representing the information and knowledge that influence key decisions. Influence diagrams have been used in many different fields, sometimes for situations other than decision problems [Beroggi and Aebi, 1996]. For example, system dynamics models also illustrate relationships between variables and objects but they are formalized with stochastic differential equations. Causal models also depict the relationships between variables but they are formalized with linear regression equations. Probabilistic influence diagrams depict probabilistic dependencies (conditional arcs) and available information (informational arcs).

An **informational arc** means that the state of the preceding node is known at the time the decision must be made. Thus, the optimal decision for the national policy will be determined conditioned on the EU traffic policy (Figure VIII.3). The optimal decision for the speed limit will be determined conditioned on the air pollution and the national policy. A possible solution is shown in the following diagram:

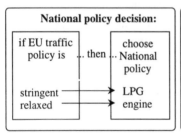

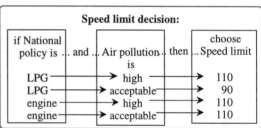

The diagram reads as follows: if the EU policy is *stringent*, then choose national policy to promote *LPG*; with the national policy being *engine*, and the air pollution being *high*, then choose speed limit *110*; with the national policy being *engine*, and the air pollution being *acceptable*, then choose speed limit *90*. On the other hand, if the EU traffic policy is *relaxed*, then choose as national policy to promote *engine*; with the national policy being *engine*, choose speed limit *110*, regardless of the air pollution.

Because previous decisions provide information for subsequent decisions, all informational nodes (which are direct predecessors) on previous decisions are also information for subsequent decisions. We can illustrate this explicitly by adding all **no-forgetting** arcs. The only no-forgetting arc for the model in Figure VIII.2 is the one from the EU traffic policy to the speed limit. **Conditional arcs** require that a node be defined in terms of the preceding nodes. Thus, a conditional arc pointing to a chance node means that we must define the chance node as a **conditional probability distribution** (as we did in Chapter VI). Chance nodes with no preceding nodes are defined in terms of a **marginal probability distribution**. It is important to emphasize that because conditional arcs might be reversed with Bayes' theorem, they do not necessarily depict **causal relations**. For example, the fact that the moon might cause a dog to bark can be illustrated by drawing an arc from the node *moon* to the node *dog*. However, the reversal of this causal relation does not make any sense - no matter how long a dog barks, the moon will not appear as a result of the barking. On the other hand, if we estimate the probability of a dog barking when the moon is out (conditional probability, $p(bark|moon)$), the probability of the moon being out (marginal probability, $p(moon)$, and the marginal probability of the a dog barking, $p(bark)$, then we can reverse the probabilistic relation with Bayes' theorem; that is, we can compute $p(moon|bark)=p(bark|moon)\times p(moon)/p(bark)$.

The nodes pointing to the utility node make up the preference structure. For example, the EU policy has no influence on the decision maker's utility, but the air pollution does. Moreover, we expect that all decision nodes influence the utility.

The decision options are defined in the customary way: by listing the decision alternatives. An informational arc does not mean that the decisions must be defined in a conditional form: rather, it specifies how we would like to have the results - conditioned on the incoming nodes. It is therefore obvious that all decision nodes must be connected in a direct path that leads into the utility node.

There are several requirements for a probabilistic influence diagram. If the following three requirements are satisfied, then we have a **regular** probabilistic influence diagram [Schachter, 1986]:

- there are no cycles in the diagram,

- there is at most one utility node with no successors, and

- there is a directed path that contains all of the decision nodes.

Probabilistic influence diagrams have some important characteristics:

- If only decision nodes point to the utility node, the solution is simply the combination of alternatives with highest utility; that is, the problem can be solved purely deterministically. In such cases, informational arcs and chance nodes have no meaning. This can be verified by omitting the arc between air pollution and utility in the diagram in Figure VIII.2.

- The partitions of the uncertainty space define a **joint** probability distribution over the resulting scenarios. In the example above, we have two partitions (EU traffic policy and air pollution), each with two outcomes. where the air pollution depends on EU traffic policy. Thus, the probability distribution for EU traffic policy is a **marginal** probability distribution, while the distribution for the air pollution is a **conditional** probability distribution.

- Chance nodes with no successors (called sinks or **barren nodes**) have no influence on the decision problem and can therefore be omitted.

- Every probabilistic influence diagram can be represented as a symmetric decision tree. The sequence of the elements (nodes) in the decision tree shows the order in which the chance variables are observed and decisions are made. The decision tree for the example in Figure VIII.2 is (see Problem 6):

Before we address the formal model and the procedures to resolve a probabilistic influence diagram, let's introduce a simplified version of the often cited oil wildcatter problem [Raiffa, 1968]. A person holds an option for a piece of land which might have some oil in the ground. The amount of oil is not known. Seismic testing can be done to get some information about the structure of the ground and the likelihood of the presence of oil. These results could then be used to decide whether to drill for oil or to abandon the plot. Testing and drilling cost money but if there were a lot of oil, then drilling would be profitable.

Given this somewhat vaguely described problem, let's look at some influence diagrams and their meanings. We assume that there are the two decision options: *test* (test or don't test) and *drill* (drill or don't drill); and the uncertainties are the geological *structure* (no structure, closed structure, or open structure), the *amount of oil* (dry, wet, or soaking), and the *monetary outcomes* (utility node). Figure VIII.4 shows on the left an influence diagram, saying that first a decision must be made whether to test or not, followed by the decision to drill or not. The amount of oil depends on the geological structure. The utility is determined by the amount of oil and the two decisions.

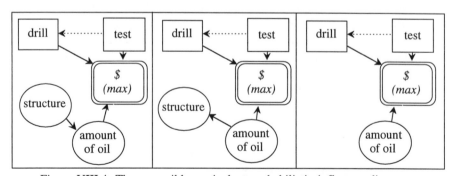

Figure VIII.4: Three possibly equivalent probabilistic influence diagrams.

We discussed in Chapter VI that an arc between two chance nodes can be reversed using Bayes' theorem. The diagram in Figure VIII.4 in the middle is thus equivalent to the one on the left. From this new diagram we see that the structure has no relevance anymore (barren node) and it can be removed from the diagram, resulting in the diagram in Figure VIII.4 on the right. Arc reversal and barren node removal are two of the four basic resolution steps for solving a dynamic decision problem under uncertainty. The other resolution steps and the resolution model (algorithm) will be discussed in Section 3 of this chapter.

Because there is no uncertainty node between the two decision stages in Figure VIII.4, we could simply merge the two decision nodes into one decision node by considering all four combinations: test and drill, test and don't drill, don't test and drill, and don't test and don't drill. The one with highest expected utility would then be implemented.

Let's take a look at a numerical example for this problem. We assume that the decision maker's utility function is: $u(\$)=\$/1000$ (i.e., the decision maker is risk neutral). The seismic test costs \$10,000 and drilling costs \$70,000. If the ground is dry, the gain is \$0, if it is wet \$120,000, and if it is soaking \$270,000. To get the net gains, we must thus subtract \$70,000 for drilling and \$10,000 if we decide to test. With the probabilities for the amount of oil being $p(\text{dry})=0.5$, $p(\text{wet})=0.3$, and $p(\text{soaking})=0.2$, we get the results of Figure VIII.5. The best policy is not to test but to go right ahead with drilling which results in an expected utility of 20.

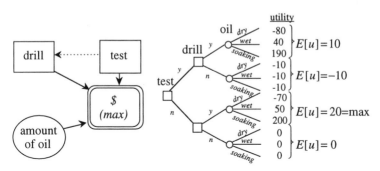

Figure VIII.5: Influence diagram and corresponding decision tree.

Starting at the very right of the decision tree, we compute the expected utilities. This gives four values, two for drilling (10 and 20) and two for not drilling (-10 and 0). Because 20 is the largest value and it is obtained if we do not test, we conclude that the best decision sequence is not to test and to drill. We have thus performed two operations, known in **decision tree** analysis as **averaging out** (computing the expected utility) and **folding back** (cutting off the not drill branches because they have smallest expected utility).

What would be the best policy if the decision maker could know in advance the amount of oil, and how much would this knowledge be worth? To answer this question, we need to know the expected utilities conditioned on the amount of oil.

This can be achieved by adding an informational arc from the amount of oil to the drill decision, as shown in Figure VIII.6 (left).

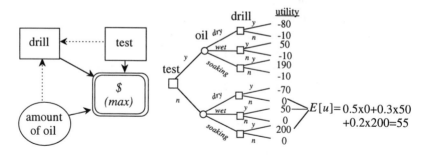

Figure VIII.6: Influence diagram and corresponding decision tree.

From the computations given in Figure VIII.5 we know that the best policy is not to test. With this decision in mind, we determine according to Figure VIII.6 the best policy for drilling, given that we have the information about the amount of oil. If it is dry we would choose not to drill, if it is wet we would decide to drill (net win is 50), and if it is soaking we would decide to drill (net win is 200). The expected utility of adopting this policy is 0.5×0+0.3×50+0.2×200=55. Without the knowledge of the amount of oil, we had an expected utility of 20. Thus, the **value of information** is 55-20=35; that is, the (risk neutral) person holding the plot should be willing to pay $35,000 for this additional information about the distribution of the amount of oil in the ground. More on the value of information will be discussed in Section 4.2.

The definition of the oil exploration problem as given in Figures VIII.4-6 does not really reflect what we are looking for. Performing a test should provide some results that help in the consecutive decision whether to drill or not. The test, if we decide to perform it, will give us information on the structure; that is, it will tell us the probability distribution of the structure. To reflect this gain of information, different structural models are possible, as shown in Figure VIII.7.

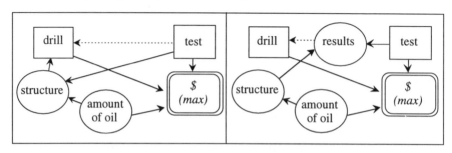

Figure VIII.7: Possibly equivalent structural models reflecting that testing provides information whether or not to drill.

One possibility is to define the structure node conditioned on the test decision and the amount of oil, and have it influence the drill decision (Figure VIII.7, left). Another possibility is to introduce a results node (conditioned on the structure) that provides some information after the test, if a test is conducted (Figure VIII.7, right).

Both structural models in Figure VIII.7 provide us with conditional results telling whether to drill or not, depending on the test results. The decision whether to test or not will based on what provides us with the highest expected utility. To assure that the two structural models really result in the same decision, we must formalize the two models appropriately.

2. The Formal Model

2.1 Defining Probabilistic Relations

The formalization of probabilistic influence diagrams uses techniques that we have encountered in previous chapters. Chance nodes reflect partitions of the uncertainty space (in terms of states) and are formalized with marginal and conditional probability distributions (Chapter VI). Preferences are expressed in terms of utilities, and the optimal decision strategy is based on the expected utility (Chapter VII).

To discuss the formalization of probabilistic influence diagrams, we continue with the oil explorer problem and use data as provided by Raiffa [1968]. If the seismic test is done, the information that results from that test is given in Table VIII.1. If the seismic test is not done, all that is known about the uncertainty is the marginal probability distribution of the amount oil (as given in Table VIII.1). The costs are the same as used in the previous section.

| p(oil,structure) | structure | | | |
oil	no (NS)	open (OS)	closed (CS)	p(oil)
dry (D)	0.30	0.15	0.05	0.5
wet (W)	0.09	0.12	0.09	0.3
soaking (S)	0.02	0.08	0.10	0.2
p(structure)	0.41	0.35	0.24	**1.0**

Table VIII.1: Uncertainty values for the oil exploration problem.

2.2 Decision Trees

With the **decision tree** approach the problem can be represented as shown in Figure VIII.8. The decision tree of the oil exploration problem shows nicely the asymmetry of this decision problem which is caused by the definition of the test node. If there is no test, the structure has no influence on the drill decision. However, if there is a test, the test results are given as shown in the decision tree. The optimal decision is not to test but to start drilling right away, which gives us an expected utility of 20. If we do not test and abandon the plot, we have an expected utility of 0. If we test and not drill, we have an expected utility of -10, and if we test and drill, we have and expected utility of 10.

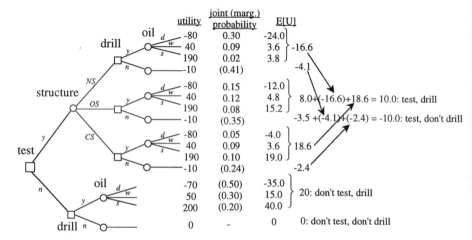

Figure VIII.8: Decision tree (asymmetric) for oil exploration problem.

It is possible to transform this asymmetric decision tree (or decision problem) into a symmetric one. To do this, we just neglect the information about the structure, and assume a uniform distribution. Figure VIII.9 shows how to expand the don't test branch to get a symmetric decision tree.

To make the asymmetric decision tree of Figure VIII.8 symmetric, we must decide for each of the three *structure* branches whether or not to drill. For this we add the complete don't-test branch to the decision tree (Figure VIII.9). It should be noted that for the don't-test branch, the marginal distribution of *structure* is uniform; $p(NS)=1/3$, $p(OS)=1/3$, and $p(CS)=1/3$.

If we take at every drill decision point the option with highest expected utility, we get for the don't-test branch the following value (Figure VIII.9): $20/3+20/3+20/3=20$. Doing the same for the test branch we get (Figure VIII.8): $-4.1+8.0+18.6=22.5$. Comparing the two values, we see that if we can have foreknowledge of the structure, we must decide to test (because $22.5 > 20$). This expected utility is obtained if we decide as follows, given that we do test: if the structure is NS then we don't drill (-4.1 > -16.6), if it is OS then we drill (8.0 > -3.5), and if it is CS then we also drill (18.6 > -2.4).

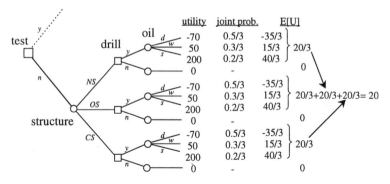

Figure VIII.9: The don't-test branch of the oil exploration problem.

2.3 Probabilistic Influence Diagrams

As with the decision tree technique, the **probabilistic influence diagram** technique also provides different structural models that lead to the same result, saying which decision combination (policy) should be chosen based on the highest expected utility (Figure VIII.7). However, the diagrams must be formalized appropriately. A possible influence diagram for the symmetric decision tree in Figure VIII.9 is given in Figure VIII.7, left.

It should be remembered that even if a decision node has a predecessor node, we must define the decision node in marginal terms. The arc is therefore only an informational arc, saying that at the moment this decision must be made, the outcome of the predecessor node is known. Thus, we would define the test decision node as follows (binary decision variables):

- test,
- don't test.

The drill decision node is defined as follows (binary decision variables):

- drill,
- don't drill.

The amount of oil is defined as the marginal distribution, given in Table VIII.1:

- $p(\text{dry})$ = 0.5,
- $p(\text{wet})$ = 0.3,
- $p(\text{soaking})$ = 0.2.

The conditional distribution, p(structure|oil) is given in the table below. It should be noted that the don't-test probability distribution is uniform:

		amount of oil		
	structure	dry	wet	soaking
test	no	0.6	0.3	0.1
	open	0.3	0.4	0.4
	closed	0.1	0.3	0.5
don't test	no	1/3	1/3	1/3
	open	1/3	1/3	1/3
	closed	1/3	1/3	1/3

The utility node is defined in terms of the two decision options and the amount of oil:

		amount of oil		
	utility	dry	wet	soaking
test	drill	-80	40	190
	don't drill	-10	-10	-10
don't test	drill	-70	50	200
	don't drill	0	0	0

The formalization of the influence diagram in Figure VIII.7, right, is as follows. The two decision nodes, as well as the amount of oil, are defined as before. The structure is defined as a conditional distribution with values derived from Table VIII.1:

amount of oil	structure		
	no	open	closed
dry	0.6	0.3	0.1
wet	0.3	0.4	0.3
soaking	0.1	0.4	0.5

The test results are defined as follows:

		structure		
	result	dry	wet	soaking
test	dry	1.0	0	0
	wet	0	1.0	0
	soaking	0	0	1.0
don't test	dry	1/3	1/3	1/3
	wet	1/3	1/3	1/3
	soaking	1/3	1/3	1/3

The utility node is also defined as before. The optimal decision for both diagrams in Figure VIII.7 is the same as for the decision tree solution: to test; and if the structure is NS then not to drill, and if the structure is CS or OS then to drill. The corresponding expected utility of this policy is 22.5.

To see why the two diagrams in Figure VIII.7 with the discussed formalizations lead to the same decisions, we must first introduce the resolution model for probabilistic influence diagrams.

3. The Resolution Model

3.1 Resolution Steps

Several approaches to resolution modeling for influence diagrams have been discussed in the literature and implemented with computerized decision support systems. In this section we will discuss the **node elimination algorithm** for probabilistic influence diagrams as introduced by Schachter [1986].

We assume that the decision maker wants to choose the alternative with highest expected utility (e.g., measured in monetary units). The resolution of a probabilistic influence diagram is based on four basic transformations:

- barren node removal
- chance node removal
- decision node removal
- arc reversal

These transformations have the following characteristics:

- after any of these transformations, the modified diagram is still a probabilistic influence diagram (however, only in terms of information preservation; the causal model might be lost), and

- all of these transformations preserve the expected utility of the decision options; that is, the new influence diagrams result in the same optimal decision strategy and associated expected utility.

For the following discussions, we will introduce the following notation for sets and nodes:

- u: utility (effectiveness) node,
- C_u: set of conditional predecessors of the utility node u,
- D: set of all decision nodes (actions),
- C: set of all chance nodes,
- C_i: set of conditional predecessor nodes of node i,
- S_i: set of direct successor nodes of node i, and
- I_i: set of conditional predecessor nodes of decision node i.

With this notation, the four basic transformations, on which the resolution model is based (Figure VIII.20), are summarized in Figure VIII.10.

Transformation	Description
A) Barren node removal 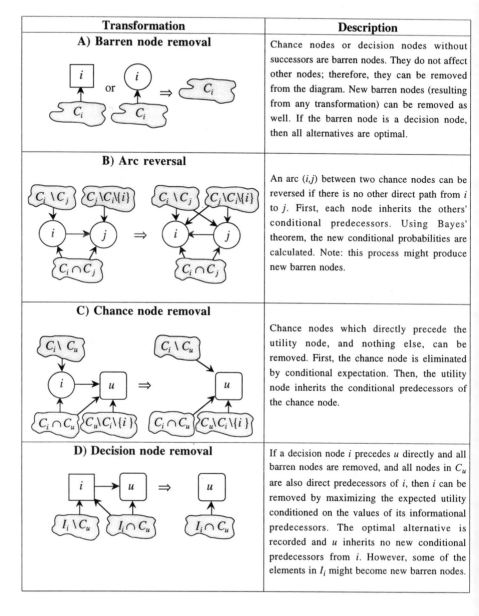	Chance nodes or decision nodes without successors are barren nodes. They do not affect other nodes; therefore, they can be removed from the diagram. New barren nodes (resulting from any transformation) can be removed as well. If the barren node is a decision node, then all alternatives are optimal.
B) Arc reversal	An arc (i,j) between two chance nodes can be reversed if there is no other direct path from i to j. First, each node inherits the others' conditional predecessors. Using Bayes' theorem, the new conditional probabilities are calculated. Note: this process might produce new barren nodes.
C) Chance node removal	Chance nodes which directly precede the utility node, and nothing else, can be removed. First, the chance node is eliminated by conditional expectation. Then, the utility node inherits the conditional predecessors of the chance node.
D) Decision node removal	If a decision node i precedes u directly and all barren nodes are removed, and all nodes in C_u are also direct predecessors of i, then i can be removed by maximizing the expected utility conditioned on the values of its informational predecessors. The optimal alternative is recorded and u inherits no new conditional predecessors from i. However, some of the elements in I_i might become new barren nodes.

Figure VIII.10. The four basic transformations for the evaluation (resolution) of probabilistic influence diagrams [Schachter, 1986].

A) <u>Barren Node Removal</u>

A barren node is a chance or decision node that has no successors, but can have multiple predecessors (including decisions). Thus, a barren node has no impact on the optimal policy and can be removed from the diagram, with all of its incoming arcs deleted.

Barren nodes can be defined in the influence diagram as part of the conceptual model. However, appropriate arc reversals can transfer a chance node into a barren node that can then be omitted from the structural model. Moreover, barren nodes can also occur when a decision node gets removed. If a decision node is a barren node, then it can also be omitted. In such a case, all of the decision options are optimal.

B) <u>Arc Reversal</u>

Arc reversal might be necessary to transfer the structural model into a model for which one of the three node removal operations can be performed (barren node, chance node, or decision node removal). Arc reversal is possible under two conditions:

- the arc is between two chance nodes, and

- reversing an arc does not create cycles in the structural model.

Arc reversal in a probabilistic influence diagram amounts to applying Bayes' theorem; that is, reversing the conditional probabilities without affecting the joint probability distribution. In Chapter VI we discussed arc reversal for a model with only chance nodes. Here, we will replace the chance node C of Figure VI.4 with a decision node C, also with two outcomes and the same marginal distribution for A and conditional distribution for B. The structural model and the arc reversal between A and B are shown in Figure VIII.11.

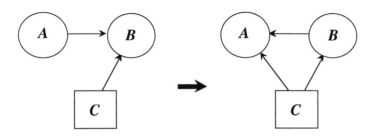

Figure VIII.11: Arc reversal with a decision node (C) as part of the influence diagram.

The original values are as follows:

A: • $p(a_1) = 0.2$, $p(a_2) = 0.8$

B: • $p(b_1|a_1,c_1) = 0.3$, $p(b_2|a_1,c_1) = 0.7$
    • $p(b_1|a_2,c_1) = 0.6$, $p(b_2|a_2,c_1) = 0.4$
    • $p(b_1|a_1,c_2) = 0.8$, $p(b_2|a_1,c_2) = 0.2$
    • $p(b_1|a_2,c_2) = 0.1$, $p(b_2|a_2,c_2) = 0.9$

C: • alternatives: c_1 and c_2

Because of the decision node, we do not get a three-dimensional joint distribution (as in Figure VI.4) but two 2-dimensional joint distributions, as shown in Figure VIII.12.

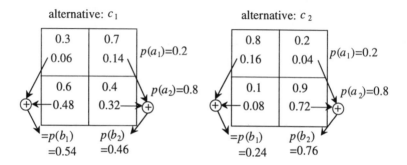

Figure VIII.12: Joint probability distributions, one for each of the two alternatives.

The corresponding computations are the following:

for alternative c_1:
• $p(a_1,b_1,c_1) = p(b_1|a_1,c_1) \times p(a_1)$ $= 0.3 \times 0.2$ $= 0.06$,
• $p(a_2,b_1,c_1) = p(b_1|a_2,c_1) \times p(a_2)$ $= 0.6 \times 0.8$ $= 0.48$,
• $p(a_1,b_2,c_1) = p(b_2|a_1,c_1) \times p(a_1)$ $= 0.7 \times 0.2$ $= 0.14$,
• $p(a_2,b_2,c_1) = p(b_2|a_2,c_1) \times p(a_2)$ $= 0.4 \times 0.8$ $= 0.32$,
                                                $\Sigma = 1.00$

for alternative c_2:
• $p(a_1,b_1,c_2) = p(b_1|a_1,c_2) \times p(a_1)$ $= 0.8 \times 0.2$ $= 0.16$,
• $p(a_2,b_1,c_2) = p(b_1|a_2,c_2) \times p(a_2)$ $= 0.1 \times 0.8$ $= 0.08$,
• $p(a_1,b_2,c_2) = p(b_2|a_1,c_2) \times p(a_1)$ $= 0.2 \times 0.2$ $= 0.04$,
• $p(a_2,b_2,c_2) = p(b_2|a_2,c_2) \times p(a_2)$ $= 0.9 \times 0.8$ $= 0.72$,
                                                $\Sigma = 1.00$

Reversing the arc between *A* and *B* gives:

A:
- $p(a_1|b_1,c_1) = p(a_1,b_1,c_1)/p(b_1|c_1) = (0.06)/0.54 = 0.111$
- $p(a_2|b_1,c_1) = p(a_2,b_1,c_1)/p(b_1|c_1) = (0.48)/0.54 = 0.889$
- $p(a_1|b_2,c_1) = p(a_1,b_2,c_1)/p(b_2|c_1) = (0.14)/0.46 = 0.304$
- $p(a_2|b_2,c_1) = p(a_2,b_2,c_1)/p(b_2|c_1) = (0.32)/0.46 = 0.696$

- $p(a_1|b_1,c_2) = p(a_1,b_1,c_2)/p(b_1|c_2) = (0.16)/0.24 = 0.667$
- $p(a_2|b_1,c_2) = p(a_2,b_1,c_2)/p(b_1|c_2) = (0.08)/0.24 = 0.333$
- $p(a_1|b_2,c_2) = p(a_1,b_2,c_2)/p(b_2|c_2) = (0.04)/0.76 = 0.053$
- $p(a_2|b_2,c_2) = p(a_2,b_2,c_2)/p(b_2|c_2) = (0.72)/0.76 = 0.947$

B:
- $p(b_1|c_1) = p(b_1,c_1)/p(c_1) = (0.024+0.192)/0.4 = 0.54$
- $p(b_2|c_1) = p(b_2,c_1)/p(c_1) = (0.056+0.128)/0.4 = 0.46$
- $p(b_1|c_2) = p(b_1,c_2)/p(c_2) = (0.096+0.048)/0.6 = 0.24$
- $p(b_2|c_2) = p(b_2,c_2)/p(c_2) = (0.024+0.432)/0.6 = 0.76$

C:
- alternatives: c_1 and c_2

If we compare these results with the ones obtained when the node *C* was a chance node instead of a decision node (see Chapter VI, Section 2.2, D), we see that the new conditional distributions are the same. Using the same reasoning from Chapter VI, we can conclude two characteristics of an admissible arc reversal that we already mentioned in Chapter VI.

- the joint probability distribution remains unchanged, and

- the two nodes, between which the arc has been reversed, inherit each others' predecessor nodes.

The new diagram after the arc reversal is completed describes the same joint probability distribution as the original one. However, if we reverse the arc back to its original direction, the arcs that were added in the first transformation are not removed. This could only be done if the original diagram had been saved prior to arc reversal.

C) Chance Node Removal

To discuss the removal of chance and decision nodes, we introduce a practical example. A community has installed a velocity monitoring system along its busiest road. The monitoring system is supposed to detect speeding cars and trucks, and to prepare automatically speeding tickets in case of violations. About 80% of all vehicles violate the speed limit and the monitoring system is quite reliable in detecting speeding cars (90%). However, there is a rather high chance of a vehicle being identified by the monitoring system of speeding even though it was not speeding (30%). The community's problem is to decide whether or not to fine the driver, based on the findings by the monitoring system.

The primary goal of the community is not to fine drivers but rather to reduce the incidence of speeding. The highest utility (on a scale from 0 to 100) is therefore assigned to those situations where the community decides not to issue a ticket for a vehicle that was not speeding (80). The least preferred situation is when the community decides not to issue a speeding ticket even though an offense has been committed (0). Issuing a ticket in case of speeding is valued also high (60), while issuing a ticket when no offense has been committed is rather low (20).

Figure VIII.13 shows on the left the original structural model for this decision problem. The abbreviations used in the structural model have the following meaning:

Velocity: • speed: the velocity is above the speed limit
                • slow: the velocity is below the speed limit

Monitor: • fine: the monitoring system detects a speeding car
                • no fine: the monitoring system does not detect a speeding car

Issue ticket?: • yes: the community issues a ticket
                • no: the community does not issue a ticket

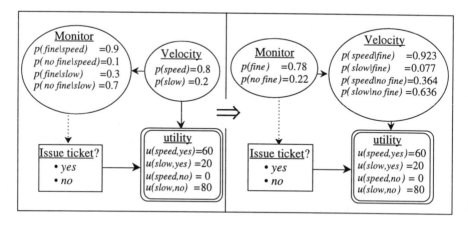

Figure VIII.13: Velocity monitoring system (left) with arc reversal (right).

The original structural model of the speed monitor problem of Figure VIII.13 (left) has been transformed into the one on the right hand side, by reversing the arc between *velocity* and *monitor*. This new structural model is conceptually different but formally the same as the original diagram; that is, the causal relation is lost but the optimal decision and the corresponding utility are still the same, and the joint probability distribution is preserved. The advantage of having the structural model with reversed arc is that the chance node *velocity* can now be removed, because it is a chance node which directly precedes the utility node and nothing else. First, the chance node is eliminated by **conditional expectation**. Then, the utility node inherits the conditional predecessors of the chance node. The result of this transformation is illustrated in Figure VIII.14. On the left is the structural model with reversed arc (which corresponds to the one in Figure VIII.13 on the right), and on the right is the new structural model with removed chance node *velocity*.

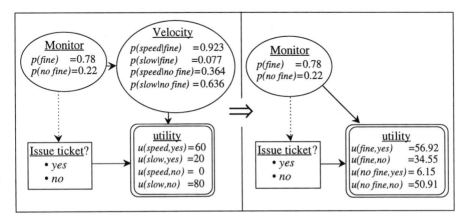

Figure VIII.14: Removal of chance node *velocity* by conditional expectation.

To compute the conditional expectations for the decision tree, we **average out** the chance node *velocity*. This procedure is shown in the decision tree of Figure VIII.15.

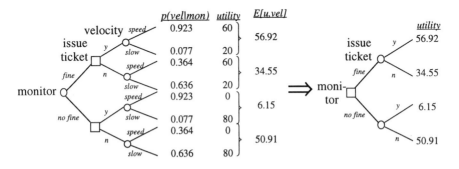

Figure VIII.15: Removal of chance node *velocity* by averaging out in a tree.

D) Decision Node Removal

The right structural model in Figure VIII.14 complies with the requirements for decision node removal, as defined in Figure VIII.10, because the decision node precedes the utility node directly and all barren nodes have been removed. Moreover, all direct predecessors of the utility node are also direct predecessors of this decision node. Therefore, the decision node can be removed by maximizing the expected utility. The removal of the decision node is illustrated in Figure VIII.16.

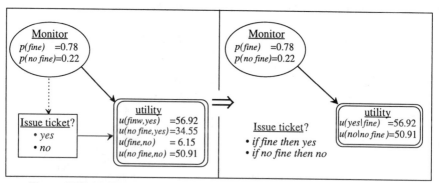

Figure VIII.16: Removal of decision node by maximizing expected utility.

As we see from Figure VIII.16, the removal of a decision node does not involve any computation but corresponds to the **folding back** procedure for a decision tree. This means that the optimal decision option (in this case conditioned on the monitor outcome) is determined. The solution is recorded: if *fine* then *yes*, if *no fine* then *no*. At this point the optimal decision is known.

The last step in resolving the decision problem is to compute the corresponding expected utility by removing the last chance node *monitor* (Figure VIII.17).

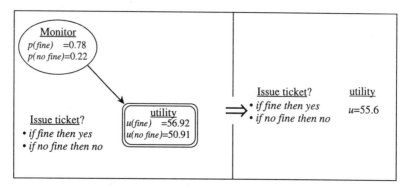

Figure VIII.17: Removal of chance node *monitor*.

In terms of the decision tree, we compute the last chance node elimination step as follows:

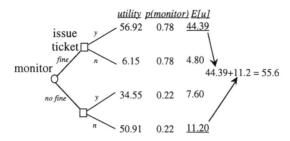

Figure VIII.18: Computation of utility when following the optimal policy.

3.2 Node Elimination Algorithm

The node elimination steps we employed to resolve the speed monitoring decision problems are summarized in Figure VIII.19. First we reversed the arc between *velocity* (V) and *monitor* (M). Then, we eliminated the chance node *velocity* (V). The next step was to eliminate the decision node *issue ticket* (I). With the removal of the decision node, we computed the optimal decision policy. The last step was to remove the chance node *monitor* (M) which gave us the expected utility (55.6) for employing this policy.

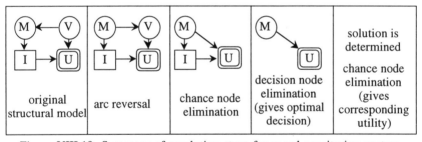

Figure VIII.19: Summary of resolution steps for speed monitoring system.

Schachter [1986] proposes a **node elimination algorithm** which is based on these four transformations. Two restrictions are imposed on the use of this algorithm:

- the decision nodes must be chronologically ordered; thus, they state an unambiguous representation of an individual decision maker; and

- the influence diagram contains the no-forgetting arcs; that is, each decision node and its direct predecessors directly influence all successor decision nodes.

These restrictions represent the decision maker's ability to recall past decisions. The resolution model (algorithm) to resolve the decision problem is summarized in Figure VIII.20.

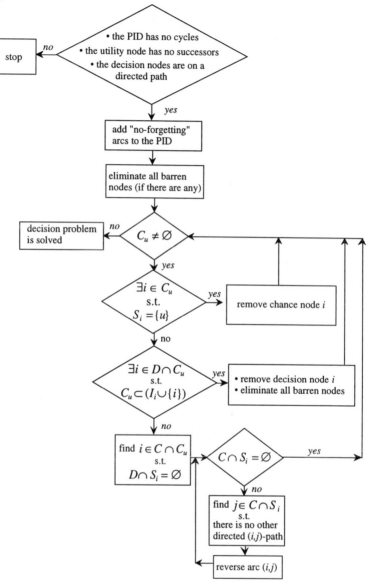

Figure VIII.20: Flow diagram of the resolution algorithm to evaluate probabilistic influence diagrams.

First, the algorithm checks if the probabilistic influence diagram (PID) has no cycles, only one utility node (u) with no successors, and if all decisions are on a directed path. Then, it adds the no-forgetting arcs and removes all barren nodes, if there are any. The recursion goes as long as the set of predecessor nodes to the utility node, C_u, is not empty. In the loop, the algorithm tries first to remove a chance node, and if this is done or not possible it tries to remove a decision node. If no chance or decision nodes can be removed, it makes appropriate arc reversals, which allows then chance nodes to be removed.

To illustrate the resolution algorithm, we take the oil exploration problem that was discussed earlier. The steps from the original structural model to the complete resolution of the decision problem are given in Figure VIII.21.

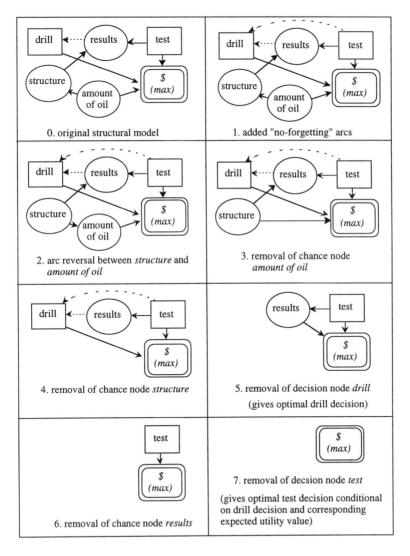

Figure VIII.21: Resolution of oil exploration problem.

4. Sensitivity Analysis

4.1 Probabilistic Sensitivity

For all marginal chance nodes, that is, chance nodes with no predecessor nodes, the probabilities can be altered to study the resulting impact on the solution. This procedure is called **probabilistic sensitivity**. Coming back to the velocity monitoring problem, the only marginal chance node is the velocity. Figure VIII.22 shows the probabilistic sensitivity for *speed*.

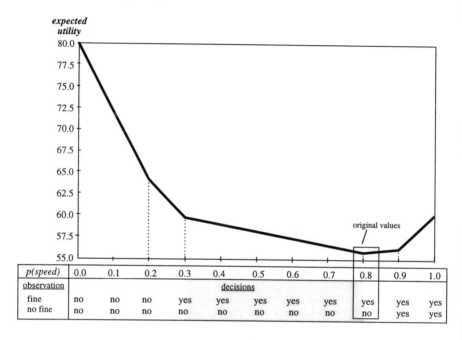

Figure VIII.22: Probabilistic sensitivity for *speed*.

The impact on the results refers both to the optimal decision alternatives and to the corresponding utility. Because the optimal decision was determined conditioned on the monitoring system (*fine* or *no fine*), the probabilistic sensitivity is also stated conditionally. From Figure VIII.22 we see that for the original probability, $p(speed)=0.8$, the optimal solution was to issue a ticket (*yes*) if a speeding vehicle was observed, and not to issue a ticket otherwise. The decision to issue a ticket remains the same (*yes*) for $p(speed)$ shrinking down to 0.3, below this value, a ticket should not be issued. The range between $p(speed)=0.3$ and $p(speed)=0.8$ defines the range of "reasonable solutions:" to issue a ticket only if the monitoring system detects a violation. For $p(speed)<0.3$ the community should never issue a ticket, even if the monitoring system detects a violation, while for $p(speed) > 0.8$, the community should always issue a ticket, even if the monitoring system does not detect a violation.

4.2 Value of Information

The **value of information** can be computed for every chance node with respect to any decision. It is the difference between perfect and expected information. Thus, there are two types of chance nodes for which we want to compute the value of information: the ones that point to the decision node, and the ones that do not point to the decision node. This means that the value of information refers to:

- **gains**, by knowing the outcome of a chance node which does not influence the decision prior to making a decision, and

- **losses**, by not knowing the outcome of a chance node which does influence the decision.

Figure VIII.23 illustrates the concept of how to compute the value of information for conditional and not conditional chance nodes. Note that in the original diagram, M is a conditional arc for the decision I, while V is not a conditional chance node for the decision I. It should further be noted that the computation of the value of information is done simply by changing the structural model and not the formal model. This means that arcs are added or removed, but no numeric values are altered. If we could not observe the monitoring outcome (*fine* or *no fine*), the loss would be 3.6 units. If it were known whether the velocity was *speed* or *slow*, the gains would be 8.4 units.

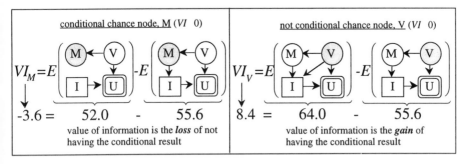

Figure VIII.23: Value of information (VI) for conditional (loss) and not conditional (gain) chance nodes.

5. Summary and Further Readings

A probabilistic influence diagram is a structural model for sequential decision problems. It consists of multiple chance and decision nodes, and a utility node. Arcs into decision nodes are informational, while those into chance and utility nodes are conditional.

A regular probabilistic influence diagram has no cycles and exactly one utility node. In addition, the decision nodes are on a directed path. Consequently, subsequent decisions are determined in terms of all the preceding decisions.

A probabilistic influence diagram reflects the dependencies between chance nodes, and the availability of information for decision nodes. In general, however, it does not necessarily capture causal relations. Dependencies are formalized with marginal and conditional probabilities, and the utility is assessed in terms of the predecessor of the utility node.

There are four admissible transformations for a probabilistic influence diagram: barren node removal, chance node removal, decision node removal, and arc reversal. These transformations preserve the formalization of the model and the results, but not the causal meaning of the original structural model.

A node removal algorithm eliminates nodes until only the utility node remains. It first tests if the diagram has no cycles, if the utility node has no successors, and if the decision nodes are on a directed path. Then it adds the no-forgetting arcs and it eliminates all barren nodes. The recursion then begins in the following order: chance node elimination, decision node elimination including removal of new barren nodes, and arc reversal.

For chance nodes with no predecessors (marginal chance nodes) a probabilistic sensitivity analysis can be performed. The probability of each possible outcome is varied from zero to one, and the resulting decisions and utilities are computed.

For each chance node the value of information can be computed with respect to any decision node. If the chance node is a predecessor of the decision node, the value of information is the loss of not being able to observe this chance node prior to making the decision. If the chance node is not a predecessor of the decision node, the value of information is the gain of being able to observe the chance node prior to making the decision. The value of information is computed by altering only the structural and not the formal model.

More theoretical backgrounds on probabilistic influence diagrams can be found in the papers by Schachter [1986] and Howard [1989]. An introduction to the concepts of influence diagrams can be found in Clemen [1996], and a collection of articles on influence diagrams is given in Oliver and Smith [1990].

6. Problems

1. What are the differences between *causal models, influence diagrams* (in the broader sense), and *probabilistic influence diagrams?*

2. Why is it not possible for a probabilistic influence diagram to have cycles? Discuss it with an example.

3. Discuss for the probabilistic influence diagrams in Figure VIII.24 which are admissible and which are not, which nodes do not contribute to the decision, which chance nodes are probabilistically independent, etc.

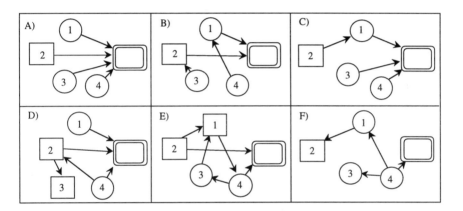

Figure VIII.24: Different types of structural models.

4. For the probabilistic influence diagrams in Figure VIII.24, discuss the number of marginal and conditional assessments for each node if every chance and decision node has two outcomes.

5. Given is a probability table for two partitions $A=\{a_1,a_2\}$ and $B=\{b_1,b_2\}$, each with two states (see Section 2.3). How can you tell if the probability values in the table stand for the joint probability $p(a_i,b_j)$, the conditional probability $p(a_i|b_j)$, or the conditional probability $p(b_i|a_j)$? How can you tell from the values in the table if A and B are probabilistically independent? Use some numerical examples to discuss these different cases.

6. Use both the node elimination algorithm and the decision tree approach to solve the problem given in Figure VIII.2. Assume the following values for the probabilities and utilities:

- p(EU traffic policy is stringent) = 0.3
- p(EU traffic policy is relaxed) = 0.7

p(air pollution):

	Air pollution	EU traffic policy	
		stringent	relaxed
LPG	high	0.2	0.4
	acceptable	0.8	0.6
engine	high	0.1	0.35
	acceptable	0.9	0.65

National policy	LPG	LPG	LPG	LPG	engine	engine	engine	engine
Speed limit	110	110	90	90	110	110	90	90
Air pollution	high	accept.	high	accept.	high	accept.	high	accept.
utilities	20	5	10	50	80	30	40	10

7. For the probabilistic influence diagrams in Figure VIII.24, discuss for which chance nodes you could assess probabilistic sensitivity. Further, discuss how you would compute the value of information for each chance node.

8. Solve the probabilistic influence diagrams discussed in Section 2.3 (Figure VIII.7) to confirm that the solution is the same as the one obtained with the decision tree in Section 2.2. Then, look at the diagram below which is very similar to the first diagram discussed in Section 2.3 (Figure VIII.7, left). Moreover, the numeric values have been derived from the same evaluation table. However, the solution to this problem definition is different: first of all, it should *not be tested* (while the other solution was to *do the test*). Given that we *don't test*, we should *not drill* if the structure is NS, and *drill* if the structure is OS or CS (this part of the solution is the same as for the model in Figure VIII.7, left, discussed in Section 2.3). The corresponding expected value of this policy is 40.1 (while the other was 22.5). This new diagram and its values are:

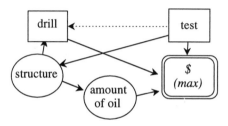

The conditional distributions for *oil* and *structure* are given in the tables below. It should be noted that the probability distribution of *structure* for when the decision is made *not to test*, is uniform:

• Amount of oil: p(oil|structure)

	structure		
oil	no	open	closed
dry	.30/.41	.15/.35	.05/.24
wet	.09/.41	.12/.35	.09/.24
soaking	.02/.41	.08/.35	.10/.24

• Seismic structure: p(structure|test decision)

	test decision	
structure	test	don't test
no	0.41	1/3
open	0.35	1/3
closed	0.24	1/3

The difference between this diagram and the first diagram in Section 2.3 (Figure VIII.7, left) is only the reversed arc between *structure* and *oil*. However, if a decision influences a chance node, then the decision problem may be asymmetric. Thus, if this asymmetry holds for the whole problem, then the decision node also should influence the subsequent chance nodes (in this case the *oil* node). If we add an arc from *test* to (amount of) *oil*, and add the corresponding values, we will get the same decision as in Section 2.3 (Figure VIII.7, left). The revised diagram is then the following:

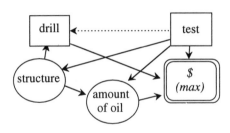

The values of the *oil* node are then:

	oil	structure		
		no	open	closed
test	dry	.30/.41	.15/.35	.05/.24
	wet	.09/.41	.12/.35	.09/.24
	soaking	.02/.41	.08/.35	.10/.24
don't test	dry	0.5	0.5	0.5
	wet	0.3	0.3	0.3
	soaking	0.2	0.2	0.2

Discuss this situation.

9. If any of the admissible transformations is performed on a probabilistic influence diagram, discuss what happens to:

- the causal meaning of the original diagram,
- the formal relations of the diagram, and
- the solution to the decision problem.

10. A politician at the ministry of transportation is thinking about the alternatives of promoting rail over road transport for shipping hazardous materials. Rail is usually safer than road but if there is an accident, the consequences of a rail accident are more severe than those of a road accident. The decision which of the two transportation modes should be promoted depends on the expected economic growth. Economists have identified several factors which determine this growth. Among these are the political stability of the country, the development of the European market, and the developments in the world economy. Of course, the world economy influences the European economy. The decision which of the two transportation modes should be promoted directly influences the environment (e.g., air pollution) and the situation on the national highways. These two outcomes affect the investment policy for road construction. The ultimate policy in this transportation problem depends on the environmental impact and the investment policy for road construction.

- Identify chance, decision, and utility nodes for this simplified decision problem. Add other meaningful nodes, and draw an influence diagram.

- Discuss how many outcomes each node has, and how you would formalize the decision problem.

• Discuss the types of conditional solutions that you will receive with your model. Finally, discuss the steps that the resolution model would go through in order to solve the problem and how you would perform a sensitivity analysis.

CHAPTER IX

MULTI-ACTOR DECISION MAKING

1. Structural Models in Multi-Actor Settings

1.1 Conceptual Aspects

A **decision maker** has been defined in Chapter I as a person who participates in the assessment of the decision options and in the choice process. People who are not decision makers but are relevant to the problem solving process are called **stakeholders**. **Actor** is a generic term that refers to both decision makers and stakeholders. When multiple actors participate in the decision making process we have a multiactor setting. Being able to make a decision means being able to assess the alternatives under consideration and to make a choice. Thus, we would expect that multiple decision makers might disagree in their assessments as well as in their choices.

Multiactor settings can be characterized in different ways. The most relevant distinction is between multiactor settings which must make a single choice as a group, and those where each decision maker can make a different personal choice. In both cases, the decision makers might communicate with each other, hoping either to reach a consensus that will suit all of the decision makers as much as possible or just to consult on individual choices. Multiactor settings which have to come up with a single choice are engaged in **group decision making**, while those where each decision maker can make a personal choice are engaged in **conflict resolution** and **negotiation**.

1.2 Finding the Best Alternative in Group Decision Making

The objective of group decision making is to find a solution that suits the group as a whole. Hopefully, this will also be a satisfying compromise solution for each group member. We assume that all group members want to optimize the performance of the group as a whole rather than pursuing their individual objectives. Differences among group members can therefore be seen as different views on how to achieve the best for the group, rather than on how to maximize personal benefit. Figure IX.1 shows the structural model of group decision making where each decision maker assesses the preferences for the decision alternatives. The assessments of the preferences by each decision makers could be done with respect to multiple criteria. In such cases, we would have to address two types of preference aggregations, one across the criteria (for each decision maker) and one across the decision makers. There are two ways to aggregate these preferences, first across the criteria and then across the decision makers, or first across the decision makers for each criterion and then across the criteria.

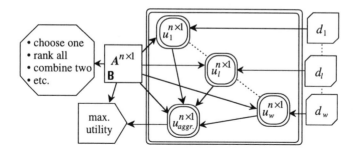

Figure IX.1: Group decision making with utility aggregation.

The issues in group decision making refer to the assessments of the alternatives, the prioritization of the decision makers, and the aggregation of preferences. If the group members are of varying importance, like the shareholders of a company, their assessments have also different weights. If the group constitutes a society, then we are dealing with **democratic theory** and address the issue of **social choice**. Finally, if the group is a business unit, we might have a 'dictatorial' choice process or preference aggregations according to preestablished rules.

1.3 Finding the Best Alternative for each Individual Decision Maker

Conflict resolution and negotiation are elements of **game theory** [Luce and Raiffa, 1985]. Each decision maker can choose the solution which suits him/her best. In this chapter, we address a few selected aspects of game theory. Elements which characterize conflict situations include **uncertainty** of the outcomes (complete vs. incomplete information) and **dynamics** of the conflict situation (static vs. dynamic decision making). However, we will confine our discussion to static conflict situations under certainty.

The number of **decision makers** in conflict with each other must be at least two. Two-person conflict situations are the most thoroughly studied conflict situations. Decision makers do not have to be individuals; they may represent interest groups, companies, or nations. Each decision maker may choose from a predefined set of alternatives, also called strategies. If a decision maker chooses exactly one alternative, we say s/he adopts a **pure strategy**; if the choice of the alternative is uncertain (described by a probability distribution over the alternatives), we say that s/he adopts a **mixed (randomized) strategy**.

A two-actor conflict situation where one decision maker wins what the other loses (i.e., the preferences of the two decision makers are in exact opposition) is called a **strictly competitive** game, or a **zero-sum** conflict situation. It is obvious that even strict competitors, such as nations at war, have a common interest in avoiding unnecessary losses. Therefore, they are willing to exchange at least some information. **Cooperative** conflict situations, as opposed to **non-cooperative** conflict situations, are based on cooperation principles such as strength, compensation, and fairness.

A two-actor non-cooperative conflict situation can be addressed from one actor's point of view. Recall the two-actor conflict situation discussed in Problem 7-10 in Chapter II. Two students, Sue and Joe must each decide for one of two classes (math. and English). The structural model for Sue's point of view is given in Figure IX.2. It should be noted that Joe's choice is unknown to Sue. Sue's objective is to choose the class which maximizes her utility. However, her overall utility depends on which class Joe will choose. Her decision will therefore depend on her estimation of Joe's probability for choosing the two classes (see text in Section 3.2,B). The problem can be modeled from Joe's point of view in the same way.

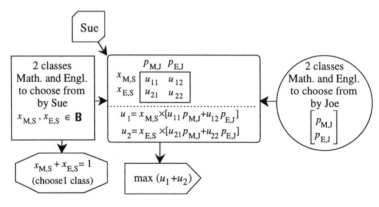

Figure IX.2: Sue's conflict decision situation with Joe's uncertain behavior.

2. Formal and Resolution Models in Group Decision Making

2.1 Aggregation of Ordinal Assessments

A) Voting Principles

Decision situations where the decision makers have to select one alternative or rank all alternatives are called **voting procedures**. Fair voting procedures have been subject of debate and research for more than 200 years. Let's assume that six decision makers $(d_1,...,d_6)$ provide the following rankings for four decision options $(a_1,...,a_4)$:

ranks	d_1	d_2	d_3	d_4	d_5	d_6	Σ
a_1	1	4	1	4	3	2	15
a_2	4	1	4	3	1	1	14
a_3	3	2	2	1	2	3	13
a_4	2	3	3	2	4	4	18

If we do not allow ties, then with four alternatives there are 24 ($4\times3\times2\times1$) different possible ranks per decision maker, and the sum of all ranks provided by the six decision makers is 60 (6×10). In 1770, the French mathematician Jean-Charles de Borda proposed to the *Académie des Sciences de Paris* a rule to aggregate n rankings into a single ranking. The rule, called the **Borda count**, has become a standard procedure for aggregating ranks. The rule is intuitive; it starts by adding up all ranks for each alternative. The alternative with the lowest sum is the best one - in the example above, it is alternative a_3.

If we eliminate the worst alternative, a_4, and adjust the ranks so they now range from 1 to 3, we get the following result:

ranks	d_1	d_2	d_3	d_4	d_5	d_6	Σ
a_1	1	3	1	3	3	2	13
a_2	3	1	3	2	1	1	11
a_3	2	2	2	1	2	3	12

The best alternative is now a_2 instead of a_3. Deleting an alternative can thus lead to **rank reversal** - a phenomenon we already encountered in Chapter IV for paired comparisons. The Borda count might therefore lead to strategic voting by assigning to the most feared competitor the lowest rank although s/he would not deserve it. A fairer approach would be to have the decision makers vote again between the two most preferred candidates.

In 1785, the Marquis de Condorcet criticized the Borda count and proposed an alternative procedure, called the **Condorcet's method** or **majority rule**. This rule says to rank a_i over a_j ($a_i \succ a_j$) if the number of voters preferring a_i over a_j is larger than the number of voters preferring a_j over a_i. This relation, however, is not necessarily transitive, and can lead to circular ranking, called the **Paradox of Condorcet**.

To illustrate the Paradox of Condorcet, assume that three decision makers, d_1, d_2, and d_3, provides the following preferences for three alternatives a_1, a_2, and a_3:

$$d_1: a_1 \succ a_2 \succ a_3,$$
$$d_2: a_2 \succ a_3 \succ a_1,$$
$$d_3: a_3 \succ a_1 \succ a_2.$$

The majority rule says then: $a_1 \succ a_2$, $a_3 \succ a_1$, and $a_2 \succ a_3$ which contradicts transitivity. Condorcet consequently concludes that because of this paradox, the majority rule is not a true aggregation method. The majority rule can in fact be viewed as an early version of the outranking procedures discussed in Chapter IV.

These kinds of shortcomings in group preference aggregation, or social choices, led K.J. Arrow [1951] to investigate the question of whether there exists a group preference aggregation method for **ordinal** preferences which complies with the following four **social choice axioms**:

- Transitivity: the aggregation procedure must produce a transitive group preference order for the alternatives being considered.

- Pareto optimality: If all decision makers prefer one alternative over the others, then the aggregated preferences must do the same.

- Binary relevance: The aggregated preference between two alternatives depends only on the decision makers' assessment of the preferences between these two alternatives and not on the other alternatives. This assumption of independence of irrelevant alternatives assures that rank reversal through deletion of an alternative is not possible.

- No dictatorship: There is no decision maker whose assessment becomes the overall group assessment.

Arrow proved that if two or more decision makers have assessed three or more alternatives on an ordinal scale, there exists no aggregation procedure which simultaneously satisfies these four social choice axioms. This result is known as **Arrow's impossibility theorem**, which has been extensively discussed in the literature. Some additional conditions should be assumed. For example, all decision makers' assessments must be considered (nobody gets neglected) and any set of preference assessments by the decision makers must result in a group preference order. The Borda count complies with all of these conditions except the condition for binary relevance (rank reversal). Condorcet's method does not comply with transitivity.

B) Statistical Approaches

Instead of aggregating the rank orders of multiple decision makers into one group rank order, we might also look at the statistical relevance of the rank orders. In other words, we are interested in whether the rank orders provided by w decision makers are significantly different. Several non-parametric statistical methods have been proposed to study preference orders. One way to compare two rankings is to compute a correlation coefficient, called **Kendall's Tau**, τ_K, named after Sir Maurice Kendall [Kendall and Gibbons, 1990], which is 1.0 if the ranks are identical and -1.0 if there is complete disagreement. Let's assume we want to compare the two ranks of seven alternatives $(a_1,...,a_7)$ provided by two decision makers d_1 and d_2.

$P = 15: +$	4	3	3	3	2	0	
$Q = 6: -$	2	2	1	0	0	1	
ranks	a_1	a_2	a_3	a_4	a_5	a_6	a_7
$d_1:$	3	6	1	2	7	5	4
$d_2:$	1	4	3	2	7	5	6

Kendall's Tau is defined as:

$$\tau_K = \frac{2S}{n(n-1)}, \text{ where } S = P - Q,$$

with P and Q being the sum of positive and negative scores, respectively. To determine the scores, we compare all alternatives pairwise and assign a value of +1 (concordance) if the preference relation of two alternatives is the same for the two decision makers; if not we assign a value of -1 (discordance). Thus, we get a score of +1 for (a_1,a_2) (because both decision makers agree on $a_1 \succ a_2$), -1 for (a_1,a_3); -1 for (a_1,a_4); +1 for (a_1,a_5); +1 for (a_1,a_6); +1 for (a_1,a_7). Then, we continue with +1 for (a_2,a_3); +1 for (a_2,a_4); etc., until the last, -1 for (a_6,a_7). We then add up all the +1's which gives us P, and all the -1's which gives us Q (see top two rows in the table above). With n alternatives, we have $(n^2-n)/2$ comparisons. For this example we have $P+Q=(n^2-n)/2=15$, and $S=P-Q=15-6=9$. Kendall's Tau is: $\tau_K =2(15-6)/42=0.43$.

Note that $N=(n^2-n)/2$ is the total number of paired comparisons. Because Kendall's Tau is defined as $(\tau_k=S/N=(P-Q)/N)$ the proportion of concordant pairs (P/N) minus the proportion of discordant pairs (Q/N), it is also referred to as a **relative measure of concordance** between the two sets of n rankings. Here again, we have a strong analogy to the concepts discussed in Chapter IV.

The two ranks provided by the two decision makers are significantly different if S takes on an extreme (non-random) value. For $n=7$, S can take on the values $\{1,2,3,4,...,19,20,21\}$, where the negative values are given by symmetry. Perfect agreement between the two decision makers in their ranks gives $S=21$, while complete disagreement gives $S=-21$. Because $n=7$ is odd, we cannot get $S=0$; the smallest absolute value is $|S|=1$. If the two ranks are determined randomly, most of the values of S will cluster around small absolute values. Therefore, the two rankings are not independent if S is extreme, that is $|S|$ is large.

From Table 1 in [Kendall and Gibbons, 1990] we get $p(|S|\geq9)=2\times0.119$ $=0.238$. This probability is rather large (i.e., S is not enough extreme), and therefore we would not reject the null hypothesis of independence of the two rankings. For large $|S|$ (which results in a small probability), we would conclude that the assessments by the two decision makers are not independent and that the two ranks are not significantly different.

If there are more than two decision makers, we could compare the rankings pairwise. For w decision makers we would have to compute $(w^2-w)/2$ correlation coefficients (τ_K). However, we would prefer an overall measure that tells us how similar all w rankings are. **Kendall's coefficient of concordance**, W, is such an overall measure. Let's look at another example with six decision makers and four alternatives.

ranks	d_1	d_2	d_3	d_4	d_5	d_6	$\Sigma = R_i$
a_1	1	4	1	4	3	2	15
a_2	4	1	4	3	1	1	14
a_3	3	2	2	1	2	3	13
a_4	2	3	3	2	4	4	18

Kendall's coefficient of concordance, W, is based on the sum of squares of deviations of the rank sums (R_i) around their mean (15): $S=(15-15)^2+(14-15)^2+(13-15)^2+(18-15)^2=14$. Random ranks would tend to have $S=0$, and complete agreement by the decision makers (i.e., identical ranks) would yield the maximum value of S (S_{max}). For this example we get $S_{max}=(6-15)^2+(12-15)^2+(18-15)^2+(24-15)^2=180$. Note that to compute S_{max} we assumed the following rank order: $a_1 \succ a_2 \succ a_3 \succ a_4$. However, any permutation would give the same value of S_{max}.

The measure of relative agreement is defined as: $W=S/S_{max}=14/180 = 0.08$. With n alternatives and w decision makers, the sum of ranks is $wn(n+1)/2$. Thus, the average rank value for the alternatives is $w(n+1)/2$. With the actual rank sums (R_i), we can now compute the sum of squares of deviations as:

$$S = \sum_{i=1}^{n}\left[R_i - \frac{w(n+1)}{2}\right]^2 = \sum_{i=1}^{n} R_i^2 - \frac{nw^2(n+1)^2}{4}.$$

The maximum value of S (S_{max}) is obtained if the w rankings are in complete agreement. In this case, the rank sums would be some permutation of $1w$, $2w$, ..., nw. Thus:

$$S_{max} = \sum_{i=1}^{n}\left[iw - \frac{w(n+1)}{2}\right]^2 = \frac{w^2(n^3-1)}{12}.$$

Consequently, with n alternatives being ranked by w decision makers, we get as the Kendall coefficient of concordance

$$W = \frac{S}{S_{max}} = \frac{12S}{w^2(n^3-n)}, \text{ with } S = \sum_{i=1}^{n} R_i^2 - \frac{nw^2(n+1)^2}{4}.$$

If all decision makers come up with the same ranking, then $W=1$; we have $W=0$ if there is complete disagreement. For our example we get:

$$S=[13^2+14^2+15^2+18^2]-4\times36\times25/4= 914-900=14,$$

which confirms the results obtained earlier. Finally, we get $W=0.078$ which means that the six decision makers disagree quite a bit.

From Table 6 in [Kendall and Gibbons, 1990] we get the significance points of S. With $m=6$ and $n=4$ we read from the table: $S=75.7$ for a level of significance of 0.05, and $S=99.5$ for a level of significance of 0.01. Because our computed $S=14$ is smaller than the S value at both levels of significance we cannot conclude that the ranks are extreme and thus not independent. If we have had, for example, $S=80$, we could have concluded that the probability of obtaining a value of 80 or greater is less than 0.05. This would be reasonable to conclude that the ranks are extreme and thus not independent.

If any ranks are tied, Kendall and Gibbons [1990] propose the following definition for Kendall's coefficient of concordance:

$$W = \frac{S}{\frac{1}{12}w^2(n^3 - n) - w\sum U'}, \text{ with } U' = \frac{1}{12}\sum(u^3 - u).$$

The summation, $\Sigma U'$, is over all sets of tied ranks. As an example of tied ranks, let's adapt the previously used table. The utilities which were assigned by the six decision makers to the four actions are given in parenthesis in the following table:

rank (u)	d_1	d_2	d_3	d_4	d_5	d_6	$\Sigma=R_i$
a_1	1.5 (65)	3 (50)	1 (56)	2.5 (52)	3 (56)	2 (57)	13
a_2	3.5 (40)	1 (60)	4 (48)	2.5 (52)	1 (64)	1 (58)	13
a_3	3.5 (40)	3 (50)	2.5 (51)	2.5 (52)	2 (60)	3 (52)	16.5
a_4	1.5 (65)	3 (50)	2.5 (51)	2.5 (52)	4 (50)	4 (49)	17.5
type of ties	2 pairs	1 triple	1 pair	1 quadruple	no ties	no ties	

With $\Sigma R_i^2=916.5$, we compute $S=916.5-900=16.5$ from the above formula. The U' are computed as follows:

$$\begin{aligned}
d_1 &: 2(2^3 - 2)/12 = 1 \\
d_2 &: (3^3 - 3)/12 = 2 \\
d_3 &: (2^3 - 2)/12 = 1/2 \\
d_4 &: (4^3 - 4)/12 = 5 \\
d_5 &: (0^3 - 0)/12 = 0 \\
d_6 &: (0^3 - 0)/12 = \underline{0} \\
\Sigma U' &= 8.5
\end{aligned}$$

If we have had five alternatives with one pair and one triple, we get $U'=(2^3-2)/12$ (for the pair) + $(3^3-3)/12$ (for the triple) = 3. Kendall's coefficient of concordance for this example is thus:

$$W = \frac{16.5}{\frac{1}{12}6^2 \times (4^3 - 4) - 6 \times 8.5} = 0.123.$$

The test of significance of W with ties is somewhat more complicated if the extend or number of ties is large (see [Kendall and Gibbons, 1990]).

Let's assume that we have found that W is significant, that is, the ranks are not independent, the question is how to aggregate these ranks across the decision makers to one group rank. Kendall and Gibbons [1990] suggest the Borda count, that is, to rank the elements according to the sums of ranks assigned by the decision makers.

2.2 Aggregation of Cardinal Assessments

A) <u>Aggregation Process</u>

We saw in previous chapters that cardinal values are used to calculate probabilities, to assess objective evaluations, and to elicit subjective preferences. After all assessments have been done, the multidimensional evaluations are aggregated to one preference value per alternative (e.g., expected utility value). The question is when to aggregate which cardinal values. For example, should the multi-actor evaluations be aggregated to a group assessment before the values are normalized or after? Should probabilities and utilities be aggregated before or after the computation of the expected utility? Should any aggregation be done before or after the computation of each actor's ranking of the alternatives? Figure IX.3 gives an overview of when aggregations could be done.

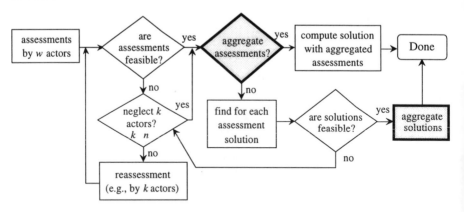

Figure IX.3: Procedure for aggregation of assessments and choices in multi-actor setting.

The procedure for aggregating assessments in a multi-actor setting, as shown in Figure IX.3, begins by asking each of the w decision makers to do an assessment. The w assessments are then checked for levels of difference, which could be done with Kendall's coefficient of concordance, W. If the assessments are not too different, then the w assessments are deemed feasible. If there are substantial differences, some of the assessments may be discarded. If this is undesirable, some (or all) of the decision makers could be asked to reconsider their assessments, possibly by taking into account the assessments done by the other decision makers. Then, either feasible assessments could be aggregated to one group assessment, or a solution could be determined for each decision maker. The group aggregated assessment can be used to find a solution to the decision problem. If for each decision maker a solution is computed, we need to check whether the multiple solutions are feasible (i.e., whether they are not too different from one another). If they are not feasible, we could consider neglecting some of the assessments. When the multiple assessments produce feasible solutions, we aggregate these to obtain a group solution.

B) Aggregation of Normative Evaluations

So far we have addressed *when* to aggregate assessments – now we examine *how*. There are different ways to proceed. Since we are dealing with cardinal scales, we would like the aggregation method to comply with the social choice axioms, as defined by Arrow [1951] and discussed in Section 2.1, A.

To get an idea of the problems involved in aggregation of multiple values let's assume that two decision makers (d_1, d_2) have assessed two alternatives (a_1, a_2) for two scenarios (s_1, s_2) with their corresponding probabilities (p_1, p_2):

	d_1			d_2			*aggregated*		
	s_1	s_2	$E[u]$	s_1	s_2	$E[u]$	s_1	s_2	$E[u]$
a_1	0.6	0.3	0.45	0.7	0.1	0.58	0.65	0.20	0.4925
a_2	0.2	0.8	0.50	0.8	0.1	0.66	0.50	0.45	0.4825
p	0.5	0.8		0.8	0.2		0.65	0.35	

The decision makers' assessments, both the utilities and the probabilities, are aggregated simply by taking the arithmetic averages. The two decision makers both prefer a_2 over a_1 but the aggregated values suggest the opposite. As we see, the aggregation of probabilities requires special attention.

An approach proposed in the literature to aggregate probabilities is to use one decision maker's assessment as information for another decision maker's assessment and upgrade the probability using Bayes' theorem [Morris, 1983], [Brodley, 1982]. Although more complex approaches might seem more appropriate, the **arithmetic average** is the most pragmatic approach. Examples of processes to elicit group probabilities are the **Delphi method** [Dalkey, 1970] and the **nominal group technique** [Gustafson et al., 1973].

The problem of aggregating subjective values and utilities to a group assessment is that interpersonal comparisons must be made. Although the value functions have been assessed using cardinal measures we do not know how to compare these values across decision makers. French [1988] shows that with given (interval) value functions for each decision maker, an ordinal group value function can be derived that complies with Arrow's axioms of social choice. However, joint value functions are difficult to determine. Often, monetary values are used to aggregate cardinal assessments. But here too, changes in project costs are perceived differently, for example by a rich and a poor community which have to decide on an infrastructure project.

Let's assume that each of the w decision makers ($w \geq 3$) has provided a value function v_i ($i=1,...,w$). Keeney and Raiffa [1993] show that an additive group value function $v_{Group}(x)$ (with v^* *being* a positive monotonic transformation of $v_i(x)$ reflecting his/her interpersonal comparison of the individuals' preferences)

$$v_{Group}(x) = \sum_{i=1}^{w} v^*[v_i(x)]$$

exists if and only if the following assumptions hold:

1) Preferential independence: If w-2 decision makers are indifferent between a pair of consequences, then the group preference between these two consequences should depend only on the remaining two decision makers' (d_i, d_j) individual preferences (u_i, u_j), and should be independent of the levels of preference of the w-2 decision makers.

2) Ordinal positive association: For two alternatives (a_i, a_j) that are equally preferred by all decision makers, if a_i is altered to a_i^* such that only one decision maker prefers it over a_j, while all other decision makers are still indifferent, then the group as a whole should also prefer a_i^* over a_j.

For decision making under uncertainty, Harsanyi [1955], and Keeney and Raiffa [1993] postulate sets of necessary and sufficient conditions for when a group utility function, $u_{Group}(x)$, must be additive (with $u_i(x)$ being the decision makers' individual utility functions):

$$u_{Group}(x) = \sum_{i=1}^{w} \lambda_i u_i(x).$$

Although the theoretical existence of additive group utility functions has been proved, the practical assessment is difficult. Keeney and Raiffa [1993] propose that the group identify a **supra decision maker** who verifies the assumptions and determines the scaling constants. However, in practical situations it might be difficult to find such a benevolent decision maker who behaves in the common interest by taking into consideration every group member's individual preferences. Moreover, although the decision makers must eventually agree on one alternative, the decision makers' positions might be quite firm, requiring negotiation processes within the group [Mumpower, 1991]. Resolving these situations in a purely democratic manner may be impossible. As French [1988] contemplates: "Democracy cannot exist in the form that many naively assume possible."

C) Aggregation of Descriptive Evaluations

Descriptive evaluations are based on paired assessments of the alternatives. Such assessments, provided by a group of decision makers, can be aggregated with different methods to a group assessment, from which a preference ranking of the alternatives under consideration can be derived. Ramanathan and Ganesh [1994] evaluate two approaches commonly used to aggregate multi-actor assessments in the analytic hierarchy process (AHP, see Chapters III and IV): the **geometric mean method** (GMM) and the **weighted arithmetic mean method** (WAMM). These two methods are evaluated in the context of five social choice axioms which are similar to the ones introduced by Arrow [1951]. These axioms are:

- Universal domain (A1): The preference aggregation method should be able to process all logically possible individual preferences.

- Pareto optimality (A2): Same as defined by Arrow.

- Independence of irrelevant alternatives (A3): Identical to Binary Relevance axiom as defined by Arrow.

- Non-dictatorship (A4): Same as defined by Arrow.

- Recognition (A5): The assessments of all decision makers are used to determine the overall group assessment; that is, no assessment is ignored.

The **geometric mean** method is criticized for not complying with the Pareto optimality axiom (A2). The GMM asks each decision maker to do all the necessary assessments, as if s/he were the sole decision maker. Then, all k_{ij}^l ($l=1,...,w$) assessments of the w decision makers are aggregated to one group assessment by computing the geometric mean:

$$k_{ij}^{group} = (k_{ij}^1 \times ... \times k_{ij}^w)^{1/w}.$$

To show the failure to satisfy the Pareto optimality axiom, Ramanathan and Ganesh [1994] use an example with five criteria and three alternatives. The summary of the individual and group priorities is as follows (for computational details see Ramanathan and Ganesh [1994], Appendix A):

	d_1	d_2	d_3	group
a_1	0.2765	0.2765	0.3084	0.3217
a_2	0.3486	0.3486	0.2394	0.3672
a_3	0.3749	0.3749	0.4522	0.3111

The example shows that all decision makers favor the third alternative, but the aggregated group assessment favors the second alternative. However, according to the Pareto optimality axiom, the group should also prefer the third alternative. The authors conclude that nonlinearities in the aggregation might be the reason for this shortcoming. Because the Pareto optimality axiom is widely accepted as a social choice axiom, this shortcoming is quite serious.

The **weighted arithmetic mean** method is criticized by Ramanathan and Ganesh [1994] for not complying with the independence of irrelevant alternatives axiom (A3). Rank reversal was discussed in Chapter IV; it is not a problem for group preference aggregation either. The WAMM using the w preferences of the n alternatives (a_j), computes the overall group preference as follows (where w_i is the weight of decision maker d_i):

$$k_j^{Group} = \sum_{i=1}^{w} w_i k_j^i , j = 1,...,n.$$

The WAMM requires that weights reflecting the importance of the decision makers be determined. A straight forward procedure is to ask a **supra decision maker** to do this assessment with paired comparisons.

Several remedies have been proposed to overcome the shortcomings of GMM and WAMM. Saaty [1989] introduces a distance measure for the relative deviation of a decision maker from the whole group. This **consistency ratio** should not exceed the value 0.1, in order for a decision maker's assessment to remain in the group (see also Chapter IV, Section 3.1). Van den Honert and Lootsma [1996] propose a modified geometric mean aggregation procedure which complies with all five social choice axioms, and therefore also with the Pareto optimality axiom.

However, Ramanathan and Ganesh [1994] propose an **eigenvector** method to derive the decision makers' weights, circumventing the need for a supra decision maker. The decision makers are seen as criteria for which each decision maker must assess the importance. That is, each decision maker must assess pairwise the importance of all the decision makers. For decision maker d_i we then get the following assessment (assuming the group consists of three decision makers):

$$K_i = \begin{bmatrix} 1 & k_{12|i} & k_{13|i} \\ k_{21|i} & 1 & k_{23|i} \\ k_{31|i} & k_{32|i} & 1 \end{bmatrix}.$$

For each decision maker, we would then compute the priorities of all decision makers, w_i as follows:

$$K_i w_i = \lambda w_i = \lambda \begin{bmatrix} w_{i1} \\ w_{i2} \\ w_{i3} \end{bmatrix}.$$

The value w_{ij} (j=1,2,3) is the preference which decision maker d_i assigns to decision maker d_j. For each decision maker we get a preference vector reflecting the priorities for all decision makers. Let

$$M = \begin{bmatrix} w_{11} & w_{21} & w_{31} \\ w_{12} & w_{22} & w_{32} \\ w_{13} & w_{23} & w_{33} \end{bmatrix}$$

be the matrix, with columns being the w_i's.

Let $w = [w_1, w_2, w_3]^T$ be the true but unknown weight of the decision makers that we would like to derive from the three weights determined by each decision maker. We use the same analogy of dominances introduced in Chapter III, although the matrix M is not reciprocal symmetric. Thus, we compute w as the eigenvector corresponding to the largest eigenvalue:

$$Mw = \lambda_{max} w.$$

Because the matrix M is a positive column stochastic matrix (i.e., the entries in the columns are all positive and sum up to one), we get $\lambda_{max} = 1$.

The non-dictatorship axiom (A4) would be violated if one decision maker had the weight w_i=1 (and thus all other w_i=0). The recognition axiom (A5) would be violated if at least one decision maker had a weight of zero. The authors then discuss that in order for this aggregation method not to violate the non-dictatorship and the recognition axioms, the matrix M must be irreducible. A matrix is reducible if it can be transformed into a matrix with a value of 0 as top right corner entry, solely by permuting its rows and columns.

To illustrate this theoretical discussion, let's take a numerical example with three decision makers, where each of them makes the following assessments of each others' priorities:

$$K_1 = \begin{bmatrix} 1 & 2 & 4 \\ 1/2 & 1 & 1 \\ 1/4 & 1 & 1 \end{bmatrix} \rightarrow w_1 = \begin{bmatrix} .5842 \\ .2318 \\ .1840 \end{bmatrix}, \; K_2 = \begin{bmatrix} 1 & 2 & 4 \\ 1/2 & 1 & 2 \\ 1/4 & 1/2 & 1 \end{bmatrix} \rightarrow w_2 = \begin{bmatrix} .5714 \\ .2857 \\ .1429 \end{bmatrix},$$

$$K_3 = \begin{bmatrix} 1 & 3 & 5 \\ 1/3 & 1 & 3 \\ 1/5 & 1/3 & 1 \end{bmatrix} \rightarrow w_3 = \begin{bmatrix} .6370 \\ .2583 \\ .1047 \end{bmatrix}.$$

This gives: $M = \begin{bmatrix} .5842 & .5714 & .6370 \\ .2318 & .2857 & .2583 \\ .1840 & .1429 & .1047 \end{bmatrix} \rightarrow w = \begin{bmatrix} .5895 \\ .2495 \\ .1608 \end{bmatrix}.$

As solution we get that decision maker d_1 is more important than d_2, followed by d_3.

2.3 Aggregation of Linguistic Assessments

So far we have discussed the aggregation of numeric assessments, over different criteria and different decision makers. However, some experts have argued that human decision makers would rather use linguistic values for their assessments. Different methods have been proposed for aggregating linguistic values. These approaches are based on **fuzzy logic theory** and involve descriptive rules as well as numeric procedures. We will outline one of these approaches.

Yager [1993] proposes a non-numeric, multi-criteria, and multi-person decision making method. The linguistic linear scale to evaluate an alternative for a criterion is the following: $S = \{s_7$: perfect $(P) \succ s_6$: very high $(VH) \succ s_5$: high $(H) \succ s_4$: medium $(M) \succ s_3$: low $(L) \succ s_2$: very low $(VL) \succ s_1$: none $(N)\}$. The 'max' operator $(\vee)$ and the 'min' operator $(\wedge)$ are defined as:

- $\max(e_i, e_j) \equiv e_i \vee e_j = e_i$ for $e_i \succsim e_j$.

- $\min(e_i, e_j) \equiv e_i \wedge e_j = e_j$ for $e_i \succsim e_j$.

This scale is used not only to evaluate the alternatives but also to evaluate the weights or importances of the criteria: $e(c_i) \equiv w_i$.

Yager introduces the negation of an evaluation (its inverse) as: $Neg(s_i) = s_{7-i+1}$. For the given linguistic scale, we therefore get: $Neg(P)=N$, $Neg(VH)=VL$, $Neg(H)=L$, $Neg(M)=M$, $Neg(L)=H$, $Neg(VL)=VH$, and $Neg(N)=P$.

A) Aggregation across Criteria

To aggregate the assessments across all criteria, Yager suggests the following aggregation method (for one decision maker and one alternative), where $d_k(a_j)$ is the aggregated evaluation of the j-th alternative (a_j) by the k-th decision maker (d_k) over all criteria, and $e_{ij|k}$ is the evaluation of the j-th alternative (a_j) for the i-th criterion (c_i) by the k-th decision maker (d_k):

$$d_k(a_j) = \min_i [Neg(w_i) \vee e_{ij|k}] = \min_i [\max(Neg(w_i), e_{ij|k})].$$

This means that in order to determine the criteria-aggregated evaluation for a decision maker, we must first look at the pairs $(w_i, e_{ij|k})$ and take the maximum; among all these maximums, we take the minimum as the criteria-aggregated value. Let's assume that a decision maker (d_k) provides the following evaluations (where $Neg(w_i) \equiv w_i'$):

	w_i	w_i'	a_1	$w_i' \vee e_{i1|k}$	a_2	$w_i' \vee e_{i2|k}$	a_3	$w_i' \vee e_{i3|k}$	a_4	$w_i' \vee e_{i4|k}$
c_1	H	L	P	P	M	M	M	M	L	L
c_2	M	M	H	H	M	M	P	P	M	M
c_3	L	H	H	H	VH	VH	L	H	VH	VH
$min(w_i' \vee e_{i1|k})$:				H		M		M		L

We conclude that for this decision maker, alternative a_1 is the most preferred (H), followed by a_2 and a_3 (M), while a_4 is the least preferred (L).

From this example with four alternatives and three criteria we can derive some characteristics of this preference aggregation method across the criteria. If a criterion has little importance, it gets a high w_i' value. But because the aggregation across the criteria is done with the 'min' operator, we see that a low importance criterion has no effect on the overall rating. We must be aware that it is not the absolute score of the criteria that is relevant but the relative score. For example, let's change the weights of all the criteria to very high (VH). Then, all alternatives get the same overall rating. On the other hand, if all criteria get the lowest rating (N), the alternative rating is determined solely by the evaluation values.

Yager notes that this aggregation procedure satisfies the properties of Pareto optimality, independence of irrelevant alternatives, positive association of individual scores with overall score, non-dictatorship, and symmetry. However, here we are talking about criteria, not about decision makers. Thus, these properties, which we originally defined as social choice axioms for preference aggregation across multiple decision makers, are now applied to criteria. However, it has been mentioned several times that decision makers and criteria both refer to having different points of view and therefore have some similarities as far as the preference aggregation is concerned.

B) Aggregation across Decision Makers

So far we have determined that each decision maker has a linguistic preference intensity for each alternative. Now we have to aggregate these preferences over all decision makers. The **cardinality** (q) of a set is defined as the number of its elements. Thus, the cardinality of the scale S introduced above is $q=7$. Yager introduces a function $Q(k)$ which maps the number of decision makers that agree on an alternative to the evaluation scale. The maximum value of $Q(k)$ is P (perfect), which is obtained if at least r decision makers agree on an alternative. This minimum number (r) is determined by the **supra decision maker**. For example, the supra decision maker might decide that if four ($r=4$) decision makers agree, then the agreement is perfect (P). With the scale values having the same indices as above, the agreement function is defined as

$$Q(k) = S_b(k), \quad \text{where } b_k = Int[1 + k\frac{q-1}{r}],$$

where $Int[\cdot]$ means to take the integer value that is closest to the real number; that is, to round to the closest integer value.

As an example, let's assume that the supra decision maker (using the scale introduced above with $q=7$), requires that at least four ($r=4$) decision makers must agree on a policy in order for the policy to be perfectly accepted. Then, if only two agree, we get $b_k=Int[1+2(7-1)/4]=4$. Thus, $S_4=M$ (medium). However, if four decision makers agree, we get $b_k=Int(1+4(7-1)/4)=7$. Thus, $S_7=P$ (perfect).

It should be noted that the total number of decision makers is used only to normalize the agreement value. Thus, the quality of the assessment is determined by the supra decision maker's choice of the minimum number of decision makers that have to agree on an alternative. However, we could also state that $r=w$, the number of decision makers participating in the assessment.

With this arithmetic average link function (i.e., $r=w$), Yager suggests an **ordered weighted averaging** (OWA) procedure to aggregate alternative preferences over all decision makers. First, the alternative scores of the decision makers (as computed for one decision maker in the numerical example above) are ordered from highest to lowest. Then, we take for each of these w scores $Q(j) \wedge B_j$ (minimum of the two values), where B_j is the j-th highest score of all the decision makers' scores. Among these w minimum values, we then take the maximum as the actor-aggregated preference for an alternative.

For example, let's assume we have four ($w=4$) decision makers (d_i), and they provides the following assessments for an alternative: d_1, H; d_2, L; d_3, M; and d_4, VH. Then, we order these assessments in descending order and get: B_1, VH; B_2, H; B_3, M; and B_4, L. We see that B_j is the j-th best score, and $Q(j)$ can be seen as an indicator of how important the supra decision maker feels that at least j decision makers agree on this assessment. Thus, the term $Q(j) \wedge B_j$ can be seen as a weighting of an alternative's j-th best score, and the number of decision makers agreeing with that score.

Next we compute the associated values $Q(j)$. These are (with $q=7$, and $k=1,2,3,4$): $Q(1)=L$, $Q(2)=M$, $Q(3)=VH$, and $Q(4)=P$. Then, we compute $Q(j) \wedge B_j$, from which we choose the maximum. Thus, the actor-aggregated preference of this specific alternative is:

$$\max[\min(L,VH),\min(M,H),\min(VH,M),\min(P,L)]=\max[L,M,M,L]=M.$$

This aggregation method over all decision makers is applied to all alternatives. As a result we get the rank-order of the alternatives according to their preferences, aggregated over all criteria and decision makers.

3. Formal and Resolution Models in Conflict Settings

3.1 Two-Actor Strictly Competitive Settings

Conflict situations, also called **games**, are characterized by the fact that each decision maker has his/her own set of alternatives from which to choose. For example, two nations in conflict have the options to submit new negotiation proposals, enforce trade restrictions, declare war, improve exchange of know-how, and so on.

Conflict situations are characterized by a number of aspects. One aspect is the number of decision makers involved in the conflict situation. Situations with more than two decision makers are called n-person games. Other aspects include whether the conflict is purely competitive, static or dynamic, and under certainty or uncertainty. If a conflict involves more than one choice by each decision maker, the chronologically ordered set of choices is called a **play**.

The foundation of game theory was laid in the 1920s with papers by Von Neumann (in German) and Borel (in French). The most significant early contribution, however, dates back to 1944 with the book *Theory of Games and Economic Behavior* by Von Neumann and Morgenstern (which was revised in 1947).

In this first section we will address the basic concepts of two-actor strictly competitive conflict situations. **Strictly competitive** means that for any combination of decisions made by the two decision makers, the utility of one decision maker is the disutility of the other decision maker. For this reason, strictly competitive situations are also called **zero-sum** conflict situations. We assume that for each decision pair, a utility has been determined that complies with the axioms of utility theory. We can represent the decision options of the two decision makers and the utilities in a table. The entries in the table reflect the utilities of decision maker d_1 and the disutilities of

d_2. That is, the utilities of decision maker d_2 are the negative values of the entries for d_1.

Let u_{ij}^1 be the utility of decision maker d_1 and u_{ij}^2 the utility of decision maker d_2. Then, we have $u_{ij}^1 = -u_{ij}^2$, or $u_{ij}^1 + u_{ij}^2 = 0$ (zero-sum conflict situation). For example, if one wins \$20, the other loses \$20. To simplify the discussion, we will call decision maker d_1 Row (with actions R_i, $i=1,...,m$) and decision maker d_2 Col (with actions C_j, $j=1,...,n$), and assume that both want to maximize their utilities (that is, Row prefers larger values, while Col prefers smaller values). We will discuss three approaches to solving two-person zero-sum conflict situations.

A) Pure Strategy Sequential Elimination

Let's assume that Row has four options and Col has three options. For example, if Row chooses R_2 and Col C_3, Row gets 7 units and Col loses 7 units.

	C_1	C_2	C_3
R_1	5	4	7
R_2	4	5	7
R_3	4	3	2
R_4	6	5	6

To find the best pair (R_i, C_j); that is, the pair that satisfies both decision makers, we use the concepts of **dominance**, **inferiority**, and **Pareto optimality**. First we check to see that none of the two decision makers has a dominant alternative. Then, we see that R_3 is an inferior alternative (because it is dominated by R_1, R_2, and R_4) and we know that Row will never choose it. Thus, we can eliminate this alternative from the table and get the reduced table below.

	C_1	C_2	C_3
R_1	5	4	7
R_2	4	5	7
R_4	6	5	6

In this new table, we see that C_3 has become an inferior (dominated) alternative and we can eliminate it.

	C_1	C_2
R_1	5	4
R_2	4	5
R_4	6	5

We see now that R_4 is the dominating alternative for Row, and Row will therefore choose it. Knowing this, Col will choose C_2 to minimize the damage. Thus, the solution to the conflict situation is (R_4, C_2) with a utility value of 5 for *Row* and -5 for *Col*.

B) Pure Strategy MinMax Test

Unfortunately, not all zero-sum conflict situations can be solved by aspects of dominance, inferiority, and efficiency. For example, if we change the utility for (R_4, C_2) from 5 to 2, then all rows and all columns in the last table are Pareto optimal; that is, we cannot reduce the conflict situation beyond this point.

However, we can try to find an **equilibrium pair**, if it exists. An equilibrium pair is a pair where neither player can do better by unilaterally choosing another alternative (i.e., strategy). To find an equilibrium pair we can use the

> **MinMax equilibrium test** for zero-sum conflict situations: A necessary and sufficient condition for a pair of alternatives to be in equilibrium is that the utility is the minimum value of its row and the maximum value of its column.

For the example above we see that 5=min(6,5,6) for *Row* and 5=max(4,5,3,5) for *Col*. Thus, (R_4, C_2) is an equilibrium pair. For the conflict situation given in the table below, we see that all strategies are Pareto optimal, and we cannot reduce the conflict situation beyond this point. However, according to the MinMax equilibrium test, we have an equilibrium pair (R_2, C_2) with value 4.

	C_1	C_2	C_3
R_1	7	2	1
R_2	5	4	6
R_3	2	3	7

The equilibrium point is the optimal solution for the two decision makers because neither would do better by choosing unilaterally another strategy.

We can also determine the equilibrium pair by finding for *Row* and *Col* the **security levels**, which are the MaxMin strategies (Chapter VI, Section 1.2). If the two security levels coincide, we have an equilibrium. For *Row*, the security level is in (R_2, C_2) with MaxMin=4 (out of the minimums 1, 4, and 2). For *Col*, the security level is also in (R_2, C_2) with MaxMin=-4 (out of the maximum values -7, -4, and -7). Because the two refer to the same entry in the table $(R_2, C_2$ with utility 4), we have found an equilibrium, or a solution to the conflict situation. For zero-sum conflict situations it can be shown that if (R_i, C_j) and (R_v, C_w) are in equilibrium, then (R_i, C_w) and (R_v, C_j) are also in equilibrium; moreover, the utilities for all equilibrium pairs are the same [Resnik, 1987, p 131].

C) Mixed (Randomized) Strategy Solutions

Choosing an alternative a_i out of a set of n possible alternatives $A=\{a_1,...,a_n\}$, is identical to the following strategy: $[(p_1,a_1),...,(p_i,a_i),...,(p_n,a_n)]$, where $p_k=0$ for $k \neq i$, and with $\Sigma_i p_i=1$. If a decision maker chooses an alternative a_i with corresponding $p_i=1$, and with $\Sigma_i p_i=1$, we say s/he chooses a **pure strategy**. On the other hand, if $p_k \neq 1$, for all k, we say s/he chooses a **mixed strategy**.

Although not every zero-sum conflict situation necessarily has a pure strategy solution (equilibrium point: (R_i, C_j)), the following theorem is of fundamental importance:

> **MaxMin theorem**: Every zero-sum conflict situation has a mixed (if not pure) strategy solution. If there is more than one strategy, then their expected utilities are equal. *Row*'s (mixed or pure) strategy maximizes his/her minimum utility, and *Col*'s (mixed or pure) strategy minimizes his/her maximum disutility. For a proof see [Luce and Raiffa, 1985, p. 391].

The analytic computation of the mixed-strategy equilibrium when *Row* and *Col* each have two strategies to choose from is briefly discussed with the following table:

	C_1	C_2
	q	$1-q$
$R_1: p$	u_{11}	u_{12}
$R_2: 1-p$	u_{21}	u_{22}

The equilibrium point can be reached by each decision maker independently of the other. The expected utilities, $E[u]$, of the two are the same (except that for *Row* we have gains, while for *Col* we have losses). Therefore, $E[u_{Row}]$ for *Row* must be the same regardless whether *Col* chooses C_1 or C_2. The same holds for *Col*. Thus, we compute the probabilities of the equilibrium points with $E[u_{Row}]$ and $E[u_{Col}]$ as follows:

• *Row*'s $E[u_{Row}]$ is independent of *Col*'s choice: $pu_{11}+(1-p)u_{21}=pu_{12}+(1-p)u_{22}$;

$$\text{thus: } p = \frac{u_{22} - u_{21}}{(u_{11} + u_{22}) - (u_{12} + u_{21})}.$$

• *Col*'s $E[u_{Col}]$ is independent of *Row*'s choice: $qu_{11}+(1-q)u_{12}=qu_{21}+(1-q)u_{22}$;

$$\text{thus: } q = \frac{u_{22} - u_{12}}{(u_{11} + u_{22}) - (u_{12} + u_{21})}.$$

With these probabilities we can now compute the equilibrium point of the conflict situation by evaluating any of the four expected utilities used in the two equations to compute p and q. The following numerical example illustrates this.

	C_1	C_2
	q	$1-q$
$R_1: p$	4	17
$R_2: 1-p$	9	-6

The solution of the conflict situation shown in the table above is computed as follows:

- $p = (-6-9)/[(4-6) - (17+9)] = (-15)/(-28) = 15/28$
- $q = (-6-17)/(-28) = (-23)/(-28) = 23/28$

- The value of the solution is:

$E[u_{Row}]$: $(15/28)(4)+(13/28)(9) = (15/28)(17)+(13/28)(-6) = 6.32$

$E[u_{Col}]$: $(23/28)(4)+(5/28)(17) = (23/28)(9)+(5/28)(-6) = 6.32$

A mixed strategy could mean rolling a die to decide which pure strategy to choose. Another interpretation of the solution of this type of conflict situation is to see the probabilities as the uncertainties of the opponent's choice. *Row* is uncertain about *Col's* choice. If *Row* chooses R_1, *Row's* expected utility is $E[u_{R1}]=4q+17(1-q)=17-13q$. If *Row* chooses R_2, *Row's* expected utility is $E[u_{R2}] =9q-6(1-q)=15q-6$. *Row's* security level can be computed from $17-13q=15q-6$. Thus, $q=23/28$. The interpretation of this solution is that if *Row* estimates $q<23/28$, then *Row* chooses R_1 because $E[u_{R1}] > E[u_{R2}]$, and if *Row* estimates $q>23/28$, then *Row* chooses R_2. For example, for $q=1/2$ we have $E[u_{R1}]=10.5 > E[u_{R2}]=1.5$, thus *Row* would choose R_1.

The same concept can be applied to *Col*: $E[u_{C1}]=4p+9(1-p)=-5p+9$, and $E[u_{C2}]=17p-6(1-p)=23p-6$. Thus, from $-5p+9=23p-6$, we get $p=15/28$. The interpretation of this result is that *Col* chooses C_2 if s/he estimates $p<15/28$, and C_1 for $p>15/28$. This approach motivates the following definition:

Nash equilibrium: Let $p*=(p_1*,...,p_m*)$ be *Row's* and $q*=(q_1*,...,q_n*)$ *Col's* mixed strategy. Then, the mixed strategies $(p*,q*)$ form a Nash equilibrium if $E[u_{p*q*}]\geq E[u_{pq*}]$ for all probability distributions p, and $E[u_{p*q*}]\geq E[u_{p*q}]$, for all probability distributions q. This means that each decision maker's mixed strategy is a best response to the other decision maker's mixed strategy (i.e., neither of the two can improve his/her position unilaterally).

The Nash equilibrium for the previous numerical example is $p*=(15/28,13/28)$ and $q*=(23/28,5/28)$, which gives $E[u_{Row}]=E[u_{Col}]=6.32$.

To summarize, we can say the following: a zero-sum conflict situation has always a solution in terms of a security level, for which the expected utilities are the same for both decision makers (but with opposed meaning; i.e., what the one wins is

---

what the other loses). To compute the solution, we first try to find a pure strategy. If a pure strategy solution does not exist, we compute a mixed strategy as discussed above.

The subjective probabilities $p*$ and $q*$ can also be interpreted as the preferences for the actions from which the two decision makers can choose. Thus, these probabilities are the **decision variables** of our problem. The decision makers choose them in such a way that the expected utility is maximized. A systematic way of computing the security level for strictly competitive 2-actor conflict situations (zero-sum games) will be discussed in Chapter X, Section 2.3, B.

3.2 Conflict Resolution and Negotiation Support

Zero-sum conflict situations can be solved analytically, assuming that a mixed strategy is something a decision maker would consider implementing. The analysis of non-zero sum conflict situations is more complex and depends heavily on one's point of view. Aspects of cooperation, negotiation, bargaining, and dynamic aspects (e.g., repetition of conflict situation) become more relevant. We will restrict our discussion to two classical examples.

A) Non-Pareto Optimal Equilibrium

The first example is the prisoners' dilemma, invented by Merrill Flood and Melvin Dresher in the 1950s. This conflict situation has been studied extensively because it represents many real-life problems, like strategic arms races. Assume that two prisoners, *Row* and *Col*, have been arrested for a minor offense but are suspected of a major crime. The investigator interrogates them separately and threatens with the following sentences (years of lost freedom) for confessing and not confessing to the major charge, where the first number of the pairs refers to *Row's* and the second to *Col's* sentence (e.g., if both do not confess, they each get one year imprisonment, or minus one year of freedom):

	$C_{confess}$	$C_{do\ not}$
$R_{confess}$	(-2,-2)	(0,-5)
$R_{do\ not}$	(-5,0)	(-1,-1)

Clearly, confessing is the dominant strategy for both prisoners. Thus, the conflict situation has the equilibrium point $(R_{confess}, C_{confess})$ which yields the utility of -2 as the value of the conflict solution. This means that its outcome should be fully predictable. However, if neither confesses, both would be better off. Thus, the Nash equilibrium is **non-Pareto optimal**. Not knowing what the other chooses, there is quite some risk involved in deciding not to confess. Anyhow, the fact is that following an individual rational approach leads to results more inferior then the results obtained by looking at the situation as a whole. Obviously, if the two could communicate, they would decide not to confess. However, if one knows that the other does not confess, s/he would be tempted to cooperate with the police, because s/he would then get no imprisonment at all.

In case of non-communication, the decision maker's probabilities for choosing between the two options are independent, and the joint probabilities are the product of the two marginal probabilities. Thus, their expected utilities are computed as follows:

- $E[u_{Row}] = pqu_{R_{11}} + (1-p)qu_{R_{21}} + p(1-q)u_{R_{12}} + (1-p)(1-q)u_{R_{22}}$
- $E[u_{Col}] = pqu_{C_{11}} + (1-p)qu_{C_{21}} + p(1-q)u_{C_{12}} + (1-p)(1-q)u_{C_{22}}$

By varying the probabilities p and q from 0 to 1, we get the space of feasible solutions in Figure IX.4 (shaded area).

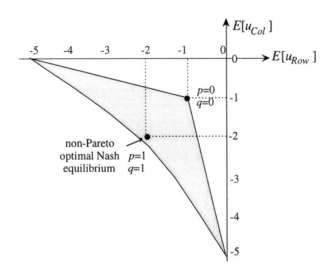

Figure IX.4: Solution space of non-cooperative conflict situation with non-Pareto Nash equilibrium in (-2,-2).

The non-Pareto Nash equilibrium strategy is $p=q=1$, which yields $E[u_{Row}]=-2$, and $E[u_{Col}]=-2$. However, the best possible compromise solution lies on the efficient frontier with $p=0$ and $q=0$, which yields $E[u_{Row}]=E[u_{Col}]=-1$.

The situation is somewhat different if the conflict situation is repeated multiple times and the decision makers can base their decisions on what the other decided in previous instances. An example of repetitive conflict situations is the annual price adjustment by manufacturers based on the competitors' prices. Of the many strategies proposed for repetitive conflict situations, the trivial **tit-for-tat** (TFT) **strategy** turns out to be very successful [Axelrod, 1984], because it has the highest average score. TFT simply instructs a decision maker to start out with cooperation and to choose in the next turn the opponent's choice for this turn. This means that if the opponent cooperates with the police, the decision maker should do so in the next turn, but if the opponent does not cooperate, then the decision maker should also not cooperate in the next turn.

B) Pure and Mixed Strategy Equilibrium

So far, we assumed that the decision makers cannot communicate. To see the added value of **communication**, let's look at the second classical conflict situation. Two people have plans for an evening. *Row* prefers to go to a sport event and *Col* to the theater. The conflict matrix is the following:

	C_{sport} (q)	$C_{theater}$ $(1-q)$
R_{sport} (p)	(2,1)	(-1,-1)
$R_{theater}$ $(1-p)$	(-1,-1)	(1,2)

The conflict table tells us that both decision makers are unhappy if they do not go out together (utility of -1), and that they are most happy if the other joins them to their favorite event. Figure IX.5, left, shows the solution space of this conflict situation. We see that there are two pure strategy Nash equilibria: $p=q=1$ (i.e., both going to the sport event) which yields $E[u_{Row}]=2$ and $E[u_{Col}]=1$, and $p=q=0$ (i.e., both going to the theater) which yields $E[u_{Row}]=1$ and $E[u_{Col}]=2$.

Assuming the decision makers do not communicate, then the uncertainties involved in such a conflict situation refer to the uncertainty about what the opponent is going to choose. If *Row* chooses sport, his/her expected utility is $E[u_{R,sport}]=2q-(1-q)=3q-1$. If *Row* chooses theater, his/her expected utility is $E[u_{R,theater}]=-q+(1-q)=1-2q$. Thus, from $3q-1=1-2q$, we get $q=2/5$. This means that if *Row* estimates $q<2/5$, Row chooses theater because $E[u_{R,theater}] > E[u_{R,sport}]$. The same concept applied to *Col* gives $p=3/5$, and *Col* chooses theater if s/he estimates $p<3/5$, because $E[C_{theater}] > E[C_{sport}]$. Thus, $p=3/5$ and $q=2/5$ is the mixed strategy non-Pareto Nash equilibrium (see Problem 7), with $E[u_{Row}]=E[u_{Col}]=1/5$.

C) Cooperation in Conflict Situations

If this conflict situation is repeated multiple times (e.g., every weekend), and the two decision makers can communicate with each other about what to do, the best strategy for the two is to alternate events each weekend. For example, go to the sport event on

---

the first weekend, go to the theater the next weekend, then sport, then theater, etc. This yields for both an average utility of 3/2. In other words, if they communicate, they would decide to toss a fair coin to decide whether to go together to the sport event or to the theater. This simply means that they correlate their mixed strategies.

Figure IX.5 explains the effect of correlating the mixed strategies. On the left we have the conflict region for all mixed strategies for the non-cooperative case. For every point (u_R, u_C) in the shaded solution space, there exists a strategy, expressed in terms of p and q. Because Row and Col do not communicate, p and q are independent, and the joint probabilities are the product of the marginal probabilities. For example, the probability that Row goes to the sport event and Col to the theater is $p(1-q)$. The expected utilities for $p=q=1/2$ are $E[u_{Row}]=E[u_{Col}]=1/4$.

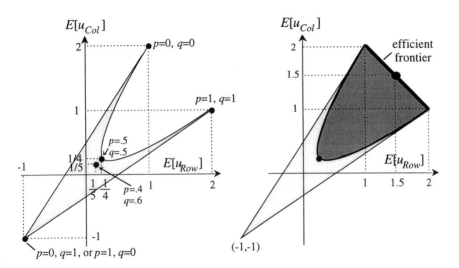

Figure IX.5: Solution space of non-cooperative (left) and cooperative (right) conflict.

The right part of Figure IX.5 shows the enlarged solution space in the cooperative case. **Perfect cooperation** means that we correlate p and q, such that the two decision makers go, for example, with a 50-50 chance either to the sport event or to the theater. Then, the expected utilities are

$$E[u_{Row}]=0.5\,u_{R_{11}}+0.5\,u_{R_{22}}=E[u_{Col}]=0.5\,u_{C_{11}}+0.5\,u_{C_{22}}=3/2.$$

For the **efficient frontier** (Figure IX.5) we have

$$E[u_{Row}]=k\,u_{R_{11}}+(1-k)\,u_{R_{22}} \text{ and } E[u_{Col}]=k\,u_{C_{11}}+(1-k)\,u_{C_{22}}, \text{ with } k\in[0,1].$$

Therefore, we see that cooperation can lead to better results than independent decisions.

The solution space for the cooperative conflict situation (Figure IX.5, right) is simply the **convexification** of the non-cooperative solution space. This means, that we enlarge the non-convex set of the non-cooperating situation with the minimal area to a convex set (polygon). The **efficient frontier** (Pareto optimal points) for this conflict situation is the line between (2,1) and (1,2): $u_C = -u_R + 3$.

D) Bargaining (Negotiation) Sets

Let's assume that the solution space looks like in Figure IX.6, and let the values of the point (u_{R_s}, u_{C_s}) be the non-Pareto **Nash equilibrium** (security level). The optimal utilities for *Row* and *Col* are $u_{R_{opt}}$ and $u_{C_{opt}}$, respectively. Although the decision makers might not reach them, we known that *Row* can make a non-cooperative decision and be sure of getting at least u_{R_s}, while *Col* can also make a non-cooperative decision and get at least u_{C_s}. Thus, the part of the efficient frontier which lies between the dotted lines is the **negotiation** or **bargaining set**.

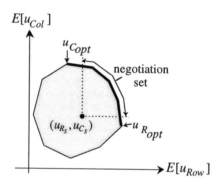

Figure IX.6: Negotiation (bargaining) set.

The question now is, given the Nash equilibrium, where in the negotiation set does the optimal solution point, called the **negotiation point**, lie? Nash suggests taking as the solution the point (u_R^*, u_C^*) such that

$$\max: (u_R^* - u_{R_s})(u_C^* - u_{C_s}).$$

This solution point is called the **Nash point**. For the prisoners' dilemma (Section 3.2, A), the Nash point was (-1,-1). The Nash point satisfies the following four conditions:

- <u>Pareto optimality</u>: the Nash point is unique and lies within the negotiation set (i.e., there is only one such point).

- **Invariance with respect to utility transformations**: if two bargaining situations are unique up to a linear transformation of the decision makers' utility scales, then the Nash points are also unique up to the same linear transformation.

- **Symmetry**: For a symmetric conflict situation (i.e., the solution space above the 45° slope corresponds to the solution space below the 45° slope, and the security level is on the 45° slope), the Nash point lies also on the 45° slope.

- **Independence of irrelevant alternatives**: If the negotiation space of one situation is contained in the space of another, and both situations have the same security level, then, if the Nash point of the larger space falls within the smaller, the two situations have the same Nash point.

Let's assume the solution space of a conflict situation between two decision makers the one given in Figure IX.7, where a coordinate transformation has been done to transfer the security level to the origin.

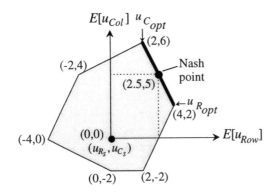

Figure IX.7: Example of conflict situation.

For this example, the negotiation set is identical to the Pareto optimal set - the line between the points (4,2) and (2,6): $u_C = -2u_R + 10$. To compute the Nash point, we maximize the function $(u_R - 0)(u_C - 0)$ on the negotiation set: max $f(u_R) = -2(u_R)^2 + 10u_R$. This is done by setting the first derivative equal to zero: $-4u_R + 10 = 0$. Thus, $u_R = 2.5$ and $u_C = 5$.

The next step in the discussion of conflict situations is to address conflict situations with more than two decision makers. This leads into the field of **coalition theory** and **multi-group negotiation processes**. For an elaborated treatment on theories of coalition formation see [Kahan and Rapoport, 1984]. We will terminate the discussion at this point and refer to the classical literature mentioned in the next section.

4. Summary and Further Readings

Multi-actor decision making can be divided into problems where the decision makers must come up with only one choice of alternative (group decision making) and where each decision maker can choose from his/her own set of alternatives (conflict resolution and negotiation).

Group decision making addresses the similarity of multiple assessments and the aggregation of multiple assessments. The similarity of multiple ordinal assessments can be measured with Kendall's coefficient of concordance. A more in-depth treatment of rank correlation methods can be found in Kendall and Gibbons [1990].

Preference aggregation in group decision making can be done for ordinal or cardinal values. Arrow [1951] formulated a set of axioms with which an aggregation method should comply; however, the Borda count does not comply with the binary relevance axiom, and the majority rule does not comply with the transitivity axiom.

Aggregation methods for cardinal values have been proposed for value and utility functions. The major problem is that interpersonal comparisons must be made. For a more in-depth discussion see Keeney and Raiffa [1993] and French [1988]. Aggregation procedures are also proposed for descriptive assessments by viewing the decision makers as criteria. The decision makers' weight is the maximum eigenvector of a positive column stochastic matrix. Aggregation of linguistic preferences has been proposed using a linguistic linear scale and an ordered weighted averaging procedure [Yager, 1993]. For an in-depth treatment of group decision making with multiple criteria see [Hwang and Lin, 1987].

Conflict situations are characterized by the number of decision makers, the possibility of cooperation and negotiation, the dynamics, and the uncertainty. Choosing only one action is called a pure strategy, while defining a probability distribution over all actions is called a mixed strategy. The uncertainty involved in mixed strategies is also seen as one decision maker's uncertainty about the other decision maker's choice of the different pure strategies. Two-person zero-sum conflict situations might be solved with a pure strategy by applying the dominance principle. If this approach fails, the security level might be determined. If this also fails, a mixed strategy solution always exists.

The solution space of two-person non-zero sum games is larger for cooperating than for non-cooperating decision makers. The efficient frontier (Pareto optimal set) is defined by the security level (Nash equilibrium). The negotiation or bargaining set is the subset of the efficient frontier that lies beyond the security level. The bargaining solution lies in the negotiation set and is called the Nash point. It is computed by maximizing the product of the coordinate differences between the Nash point and the security level. Game theory is a large area. For a more elaborate introduction to game theory see Luce and Raiffa [1985].

5. Problems

1. Make up an example based on the U.S. presidential election of 1996 with the three candidates Bill Clinton, Bob Dole, and Ross Perrot. Then, take five voters and discuss the concept of rank reversal in the Borda count.

2. a) Use the different approaches discussed for aggregating assessments across criteria and decision makers, to discuss the similarities/dissimilarities between criteria and decision makers. What is, for example, an interpretation of Kendall's Tau and Arrow's impossibility theorem in terms of criteria?

   b) Discuss the different concepts of concordance and discordance which have been discussed so far. These include the methods discussed in Chapter IV and Kendall's approach (Section 2.1, B). What analogies can you draw?

3. Assume that two decision makers have determined their preference order for four alternatives. Determine the probability distribution for S (difference between concordance and discordance) and discuss for which values of S you think that the two rankings are not independent.

4. Convince yourself that using the majority rule sequentially leads to inconsistent rankings. You can do this by comparing first a_1 to a_2 and then the better of the two to a_3. Then, compare first a_2 to a_3 and then the better of the two to a_1. Use the following rankings by three decision makers:

	d_1	d_2	d_3
a_1	1	2	3
a_2	2	3	1
a_3	3	1	2

5. Give a numerical example of a two-actor zero-sum conflict situation (where each decision maker has three decision options), such that no pure-strategy equilibrium point exists.

6. The mixed-strategy expected utilities in a 2-actor non-cooperative conflict situation where each actor has two decision options are linear in probabilities. Draw for the prisoner dilemma the solution space in terms of these probabilities at an increment of 0.05. Discuss how this **dynamic plot** could be used for **virtual negotiations** between two non-cooperative decision makers.

7. Use the concept of dynamic plots and virtual negotiation and the example in Section 3.2, B (Figure IX.5, left) to discuss why the non-Pareto mixed strategy Nash equilibrium might not be the solution to this conflict situation.

8. Solve the following zero-sum conflict situations. Discuss how *Row* and *Col* would decide based on estimates for q and p, respectively.

a)

	C_1	C_2
	q	$1-q$
$R_1: p$	3	2
$R_2: 1-p$	4	6

b)

	C_1	C_2
	q	$1-q$
$R_1: p$	3	5
$R_2: 1-p$	8	2

9. Assume that for the example of the two people deciding where to go out for an evening (sport or theater, Section 3.2, B), the two cooperate (Figure IX.5, right):

(a) find the Nash point geometrically, and

(b) compute the Nash point analytically.

10. Throughout this chapter, several sets of axioms of social choice were introduced, such as defined by Arrow (Section 2.1, A), Keeney and Raiffa (Section 2.2, B), and Ramanathan and Ganesh (Section 2.2, C).

a) Discuss these sets of axioms and compare them to one another. Discuss which axioms you think are more/less important.

b) Define different decision situations (environments) and appropriate axioms. For example, imagine an ad-hoc crisis management center for an off-shore oil spill, where several decision makers are involved (e.g., coast guard, port authority, cargo owner, local fishermen, ecologists). Which choice axioms would you choose/define?

CHAPTER X

CONSTRAINT-BASED POLICY OPTIMIZATION

1. Structural Model

1.1 Basic Concepts and Types of Decision Problems

The previous chapters addressed extensively decision problems where the **decision variables** could take on binary values and where the objective was to find the best alternative out of the set of feasible alternatives. In this chapter we will address decision problems with binary-, integer-, and real-valued decision variables. This means that we see the decision variable x_j as the intensity of employing alternative a_j. For example, imagine a project manager who must decide how many hours employees should spend on a certain project. One does not want to know which employees to assign to that project, but for how many hours each employee should be assigned to the project. The number of hours for each employee are the decision variables and the time-unit for each employee the basic alternative. A solution to this problem could be to have one employee work 5 hours ($x_1=5$), another 7 hours ($x_2=7$), and a third 8.5 hours ($x_3=8.5$).

These sort of problems also operate with an **evaluation matrix**, $E^{m \times n}$, where the entry in row i and column j, e_{ij}, is the **unit-effectiveness** value for engaging in action a_j with respect to criterion c_i. Thereby, it is assumed that engaging in an action (a_j) results in the unit effectiveness (e_{ij}), and not engaging in an action results in zero effectiveness (0). The question is, what is the effectiveness if we employ an action multiple times? For example, if the unit-gain is $15 for a specific employee working one hour on a project, how much is the gain if the same employee works, say, 7.5 hours.

The simplest and in many cases sufficient assumption is that the resulting gain is the product of unit-gain ($15) and intensity (7.5), which gives a resulting gain of $112.5=\$15 \times 7.5$. In other words, we assume that the effectiveness values are proportional to the intensities, x_j, where the proportionality factor is the unit-gain. This is the fundamental assumption of **proportionality of effectiveness** within actions: the effectiveness of an action which is employed with an intensity x_j is proportional to the unit-effectiveness e_{ij} given in the evaluation table; $e_{ij} x_j$. For example, if the cost of a person working on a project is $100, then the cost of that person working only 70% on that project is $70.

A second often made assumption refers to the overall effectiveness across all actions. For example, what is the overall project-gain of all employees together? We could assume that for three employees (actions, a_1, a_2, a_3), the total effectiveness is simply the sum of the individual effectiveness values ($e_i=e_{i1}+e_{i2}+e_{i3}$). This characteristic of **additivity of effectiveness** across actions is the second fundamental assumption. Decision problems that comply with these two assumptions are called **linear**. Thus, the two assumptions for a linear model are (where e_{ij} is the unit-effectiveness as given in the evaluation table, and x_j is the intensity with which an action is performed):

- proportionality of effectiveness within actions: $e_{ij} \times x_j$, for all i, j
- additivity of effectiveness across actions: $e_i = \sum_{j=1}^{n} e_{ij}$, for all i

These two characteristics make up what in the literature is known as **linear programming** (LP). A special case of linear programming is when the decision variables are restricted to integer values, called **integer programming** (IP). A special case of integer programming is when the decision variables are restricted to binary values, called **binary integer programming** (BIP). Decision models with all three types of decision variables are called **mixed-integer programming** (MIP). The differences among these three approaches, however, lie not at the formalization level but at the resolution level, as they call for different resolution methods. Thus, the two fundamental characteristics of proportionality and additivity of effectiveness values still hold for all three approaches.

In Chapter V we addressed non-linear decision problems and discussed how to solve them with the Lagrange multiplier method. Many problems are non-linear, especially when subjective preferences are to be taken into account. However, there is a large class of decision problems that can be modeled as linear systems. Moreover, many decision problems that are formalized in non-linear terms can be reformulated to become linear, if appropriate decision variables can be identified.

1.2 The Structural Model

The structural model of multi-action decision making under certainty is illustrated in Figure X.1. The unit-effectiveness e_{ij} of alternative a_j with respect to criterion c_i, is given in the evaluation matrix $E^{m \times n}$. The decision variable x_j is the intensity of alternative a_j. The content goals refer to aspirations (expressed as minimizations or maximizations) and constraints (expressed as inequalities '$\geq$' and '$>$,' or equalities '$=$').

The structural goals are also expressed as aspiration (e.g., minimize the number of people to work on a project) and as constraints (e.g., have at most 10 people work on a project). Often, auxiliary decision variables (y_k) are introduced to state logistic statements about the content goals (e.g., goal priorities).

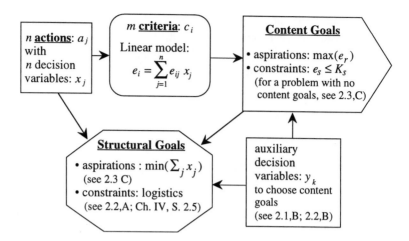

Figure X.1: Structural model of linear constraint-optimization decision problem.

The problems we are dealing with in this chapter involve optimizations of policies in the presence of constraints. These types of problems are called **constrained-optimization problems**, since the potential or feasible alternatives are not given explicitly but are defined implicitly through constraints. A major challenge is therefore not only to find the most preferred alternative but to find first feasible alternatives. In Chapter V, Section 3.6 we discussed already the concept of non-linear constrained-optimization problems and a resolution model based on the Lagrange multiplier method. In this chapter we discuss different resolution models in Section 3 for linear problems and different types of decision variables.

2. Formal Models

2.1 Content Goals: Constraints and Aspiration Levels

A) Single-Criterion Optimization

To discuss the concept of linear constraint-optimization of policies we introduce an example. A city government plans to allocate resources (e.g., expressed in dollars) for the next planning period to two different programs: the social program (a_1) gets assigned an amount x_1 of resource units and the transportation program (a_2) an amount x_2 of resource units. The allocation of resources is not necessarily an investment. The government could allocate negative resources to a program ($x_i<0$), forcing the program (a_i) to hand in parts of its resources to the government. Thus, the decision variables, x_1 and x_2, can be positive or negative real numbers. A positive value ($x_j>0$) indicates an allocation of resources to program a_j and a negative value ($x_j<0$) means that program a_j must donate resources to the government.

Six different criteria are used to evaluate the effectiveness of these two programs in terms of the resources allocated to them. The first criterion (c_1) refers to the labor time. For every resource unit (e.g., dollar) allocated to the social program, 4 units ($e_{11}=4$) of labor time are necessary, while for every resource unit allocated to the transportation program, only 1 unit ($e_{12}=1$) of labor time is necessary. The goal of the city government is to restrict the total labor time to 32 units ($4x_1+x_2\leq32$).

Investing in these two programs brings some short-term profit (c_2). For each resource unit allocated to the social program, the government can expect 4 units ($e_{21}=4$) of short-term profit, while for each resource unit allocated to the transportation program the resulting profit is expected to be 6 units ($e_{22}=6$). The government is willing to accept short-term losses but not more than 24 units ($4x_1+6x_2\geq-24$).

The third criterion (c_3) refers to external resources (e.g., work done by consulting companies) which is taken care of by a special budget. The social program needs 1 unit ($e_{31}=1$) of external resource per allocated resource unit, while the transport program requires 3 units ($e_{32}=3$). The city government wants to limit the total external resources to 23 units ($x_1+3x_2\leq23$).

Investing in these programs has some impact on public safety (c_4). The social program has an impact on the reduction of the crime rate (2 units per allocated resource unit, $e_{41}=2$), while the transportation program increases the public risk (i.e., reduces the safety) at the rate of 1 unit per allocated resource unit ($e_{42}=-1$). The goal is to keep public safety from falling below 10 safety units ($2x_1-x_2\geq-10$).

The city government has defined a measure of self-sufficiency (c_5). Transportation projects become very quickly self-sufficient, while social programs need continuous investments. Thus, the self-sufficiency rate of the social program is negative (factor 4, $e_{51}=-4$), and the transportation program is positive (factor 3, $e_{52}=3$). Public programs need investment, but the government wants to keep the total self-sufficiency degree from falling below 48 units ($-4x_1+3x_2\geq-48$).

The last criterion is the long-term profit (expressed as an index) of the allocations (c_6). The social program has a one-to-one relation between allocated resource unit and long-term profit ($e_{61}=1$), while the transportation program is more profitable, at a two-to-one ratio ($e_{62}=2$).

The criteria c_1 to c_5 are used to define constraints, while the criterion c_6 is used to define the aspiration level. The equation that describes the aspiration (maximize or minimize) is called the **goal function** (in the literature also referred to as objective function). The goal of this example is to maximize the long-term profit (c_6: x_1+2x_2). We assume a linear model. It should be noted that if both projects must hand in resources to the government ($x_1<0$, $x_2<0$), then the long-term profit is negative (loss).

In summary we have the following elements for our decision problem:

Decision variables: • x_1: amount of resources allocated to social program,
• x_2: amount of resources allocated to transportation program.

Content Goals:

$E^{6\times2}$:

		criteria	criteria values	total effectiveness	content goal
	c_1	labor time	$e_1 =$	$4x_1 + x_2$	≤ 32
	c_2	short term profit	$e_2 =$	$4x_1 + 6x_2$	≥ -24
	c_3	external resources	$e_3 =$	$x_1 + 3x_2$	≤ 23
	c_4	public safety	$e_4 =$	$2x_1 - x_2$	≥ -10
	c_5	self-sufficiency	$e_5 =$	$-4x_1 + 3x_2$	≥ -48
goal function:	c_6	long-term profit	$e_6 =$	$x_1 + 2x_2$	maximize

The above table can be rewritten by rearranging the values as shown in the table below. The values for the criteria, $\tilde{e}_1$ to $\tilde{e}_5$, are called **slack variables**.

		criteria	slack variables		
	c_1	labor time	$0 \leq \tilde{e}_1 =$	$-4x_1 - x_2$	$+ 32$
	c_2	short term profit	$0 \leq \tilde{e}_2 =$	$4x_1 + 6x_2$	$+ 24$
	c_3	external resources	$0 \leq \tilde{e}_3 =$	$-x_1 - 3x_2$	$+ 23$
	c_4	public safety	$0 \leq \tilde{e}_4 =$	$2x_1 - x_2$	$+ 10$
	c_5	self-sufficiency	$0 \leq \tilde{e}_5 =$	$-4x_1 + 3x_2$	$+ 48$
goal function:	c_6	long-term profit	max: $e_6 =$	$x_1 + 2x_2$	$+ 0$

The resource allocation problem has two real-valued decision variables, six criteria, and six goals, where five are constraints, and one is an aspiration (max e_6). We can write the goals also in matrix notation as follows:

constraints:

$$\begin{bmatrix} 0 \\ 0 \\ 0 \\ 0 \\ 0 \end{bmatrix} \leq \begin{bmatrix} \tilde{e}_1 \\ \tilde{e}_2 \\ \tilde{e}_3 \\ \tilde{e}_4 \\ \tilde{e}_5 \end{bmatrix} = \begin{bmatrix} -4 & -1 \\ 4 & 6 \\ -1 & -3 \\ 2 & -1 \\ -4 & 3 \end{bmatrix} \times \begin{bmatrix} x_1 \\ x_2 \end{bmatrix} + \begin{bmatrix} 32 \\ 24 \\ 23 \\ 10 \\ 48 \end{bmatrix}$$

aspiration:

$$\text{max: } e_6 = \begin{bmatrix} 1 & 2 \end{bmatrix} \times \begin{bmatrix} x_1 \\ x_2 \end{bmatrix}$$

For this formal system in matrix notation, we introduce the corresponding tableau:

		x_1	x_2	1
$0 \leq$	$\tilde{e}_1 =$	-4	-1	32
$0 \leq$	$\tilde{e}_2 =$	4	6	24
$0 \leq$	$\tilde{e}_3 =$	-1	-3	23
$0 \leq$	$\tilde{e}_4 =$	2	-1	10
$0 \leq$	$\tilde{e}_5 =$	-4	3	48
goal function: max	$e_6 =$	1	2	0

The variables at the top of the tableau (in this case the decision variables x_1 and x_2) are called **basic variables**. The variables in the second column (in this case the slack variables $\tilde{e}_1$ to $\tilde{e}_5$) are called **non-basic variables**. If we set the basic variables equal to zero, then the non-basic variables are equal to the right-most column. This policy is called a **basic feasible solution** because it satisfies: $0 \leq \tilde{e}_i$, $i=1,...,5$. If we set the total effectiveness values for all criteria equal to zero ($\tilde{e}_i=0$), the goals define lines in the x_1-x_2 plane, as shown in Figure X.2. The small arrows indicate the direction of maximal growth of the slack variables and the goal function.

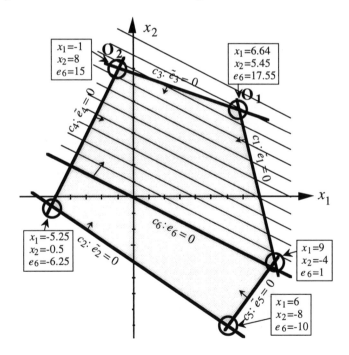

Figure X.2: Convex feasible space and iso-value lines.

The constraints define the **convex set** of **feasible solutions**, where a feasible solution is a pair (x_1, x_2) such that all of the five constraints are met; i.e., $\tilde{e}_i \geq 0$. The iso-value line for the aspiration level of maximum long-term profit (c_6), is indicated for increasing values (parallel lines) in the diagram in Figure X.2. In terms of the value functions discussed in Chapter V, we can say that a linear model assumes that the decision maker has a constant **marginal rate of substitution** ($MRS=0.5$) for the goal function (negative slope of the goal function). This means that there is a constant tradeoff between the two projects, with the value lines e_6=constant as indifference lines.

The graphical representation of our constraint-optimization decision problem, as shown in Figure X.2, allows one to find the optimal solution by graphical analysis. We have thus to move the e_6=0 line (by translation) as far as possible in the direction of maximal increase of effectiveness, but still within the space of feasible solutions. The direction of maximal increase is orthogonal to the goal line e_6=0. As it can be seen graphically in Figure X.2, the optimal solution (O_1) lies at the intersection of the lines $\tilde{e}_1$=0 and $\tilde{e}_3$=0. This is the point $x_1=73/11=6.64$, $x_2=60/11=5.45$. If we insert these values into e_6, we get $\max(e_6)=193/11=17.55$. If we insert the values $x_1=6.64$, and $x_2=5.45$ into our original tableau, we get the following tableau:

			$x_1=6.64$	$x_2=5.45$	1
$0 \leq$	$\tilde{e}_1 = 0$	$=$	-4	-1	32
$0 \leq$	$\tilde{e}_2 = 83.27$	$=$	4	6	24
$0 \leq$	$\tilde{e}_3 = 0$	$=$	-1	-3	23
$0 \leq$	$\tilde{e}_4 = 17.82$	$=$	2	-1	10
$0 \leq$	$\tilde{e}_5 = 37.82$	$=$	-4	3	48
goal function:	max	$e_6 = 17.55$	$=$	1	2	0

From this tableau we see in fact that the first and third criteria take on their extreme values (0) and thus determine the optimal solution. Criteria that determine the optimal solution are called **binding constraints**. The other criteria are not binding but simply in compliance with the constraints ($\tilde{e}_i \geq 0$). The maximal value for the long-term profit is the value of the goal function: $\tilde{e}_6=17.55$.

I we decrease the slope of e_6=0 (Figure X.2) so that it is parallel to $\tilde{e}_3$=0 ($e_6=x_1+3x_2$), we have an infinite number of optimal solutions. These are all the points on $\tilde{e}_3$=0 between O_1 (6.64,5.45) and O_2 (-1,8), where the latter point is the intersection of $\tilde{e}_3$=0 and $\tilde{e}_4$=0. If we decrease the slope of e_6=0 even more, than the optimal solution lies in O_2 (-1,8), and the maximal long-term profit is $\max(e_6)=15$.

In our example, the five constraints are all inequalities. However, constraint-optimization problems can also contain equalities. For example, let's assume the city government adds as an additional criterion the number of workers that must be employed by the two programs (c_7). We assume that both programs require the same departmental work-units per allocated resource unit, and that the city wants to employ ten workers.

Thus, we get an additional constraint which is an equality constraint:

$$0 = \tilde{e}_7 = -x_1 - x_2 + 10.$$

With this additional constraint, we get the tableau below, with the corresponding solution in O_3 (Figure X.3). The long-term profit (c_6) is now lower ($e_6=16.5$) than without this additional constraint ($e_6=17.55$).

			$x_1=3.5$	$x_2=6.5$	1
$0 \leq$	$\tilde{e}_1 = 11.5$	$=$	-4	-1	32
$0 \leq$	$\tilde{e}_2 = 77.0$	$=$	4	6	24
$0 \leq$	$\tilde{e}_3 = 0$	$=$	-1	-3	23
$0 \leq$	$\tilde{e}_4 = 10.5$	$=$	2	-1	10
$0 \leq$	$\tilde{e}_5 = 53.5$	$=$	-4	3	48
goal function: max	$e_6 = 16.5$	$=$	1	2	0
$0 =$	$\tilde{e}_7 = 0$	$=$	-1	-1	10

Let's now assume that we maximize safety (c_4) and omit c_6. Then, the optimal solution is in O_4 (Figure X.3) with: $x_1=22/3=7.33$, $x_2=8/3=2.67$, and the maximal safety $\tilde{e}_4=12$. If we now drop the equality constraint (c_7), the optimal solution lies in O_5. The values are: $x_1=9$, $x_2=-4$, and the maximal public safety is $\tilde{e}_4=22$.

It should be noted that if we wanted to minimize labor time (c_1), then we would have to change the coefficients in our tableau to 4 and 1.

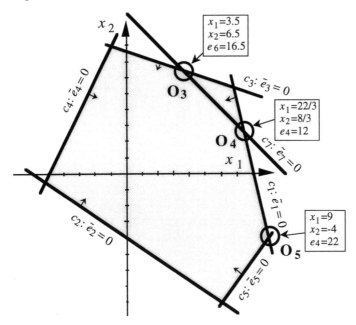

Figure X.3: Convex set of feasible solutions with equality constraint.

B) Multi-Criteria Optimization

To illustrate graphically in two dimensions the concept of linear constraint-optimization, we have used an example with only two decision variables. However, the concept holds for more than two decision variables. In our example, the aspiration level (max) is expressed with only one criterion, while all other criteria are used to define constraints ($\geq$, $>$, or $=$). In many policy analysis cases, however, multiple criteria are used to define the aspiration level. For example, we could introduce penalty weights for each unit-effectiveness per criterion that exceeds a certain aspiration value. Constraint-optimization with multiple goals is known in the literature as **goal programming** [Charnes and Cooper, 1977]. Even if there are multiple criteria contributing to the aspiration levels, these multiple objectives are always reflected in a single **goal function**.

To define a multi-criteria aspiration level (goal function), we could first take each criterion one-by-one as the sole aspiration level and compute the corresponding optimal policy. These values could then be used to define the joint multi-criteria aspiration level. However, we will not proceed in this way but assume that the city government has decided to use only c_4, c_6, and c_7 to define the aspiration level.

We saw that the optimal safety (c_4) has the value 22, the optimal long-term profit (c_6) the value 17.55, and that the required work-units (c_7) are 10. The city government would then define penalty weights for missing or exceeding these three **aspiration levels**, where the weights denote the proportional penalty. For example, if we have a penalty weight of 3, and a safety level of only 20, we get a total penalty of 6 units (2×3) for missing the optimal safety level of 22 units. Lets' assume the city government defines the following values with the corresponding penalty weights for exceeding (+) and missing (-) these aspiration levels:

	x_1	x_2	aspiration level	penalty weight
$e_4 =$	2	-1	≥ 22	3(-)
$e_6 =$	1	2	≥ 17.55	2(-)
$e_7 =$	-1	-1	$= 10$	4(+),5(-)

The overall goal is then to minimize the total weighted deviations from the aspiration levels. The aspiration levels are the values 22 (for e_4), 17.55 (for e_6), and 10 (for e_7). It should be noted that because the other constraints still hold, we can never get $e_4 > 22$ or $e_6 > 17.55$. Therefore, we could have defined our aspiration levels all in terms of equalities: $e_4 = 22$, $e_6 = 17.55$, and $e_7 = 10$. However, we will proceed with the inequalities to generalize the case.

Every aspiration level has two types of deviations, a positive ($y_i^+ \geq 0$) and a negative ($y_i^- \geq 0$) deviation. Therefore, for every aspiration level there are two positive **auxiliary decision variables**. For our example, we have: y_4^+, y_4^-, y_6^+, y_6^-, y_7^+, and y_7^-.

Our objective is to minimize those weighted deviations which do not meet the aspiration level. In other words, y_4^+ and y_6^+ can be as large as possible, because they always meet our goals.

So, the new goal function with these new auxiliary decision variables is:

$$\text{min: } e_8 = 3\,y_4^- + 2\,y_6^- + 4\,y_7^+ + 5\,y_7^-.$$

The additional constraints that we have to introduce are simply the definitions of the deviations. Thus:

$$c_4: 2x_1 - x_2 - (y_4^+ - y_4^-) = 22$$
$$c_6: x_1 + 2x_2 - (y_6^+ - y_6^-) = 17.55$$
$$c_7: x_1 + x_2 - (y_7^+ - y_7^-) = 10$$

The original constraints pertaining to the other criteria are simply added to the formal model. We have thus the following tabular notation (it should be noted that we did not add the constraints $y_i^+ \geq 0$ and $y_i^- \geq 0$ in order not to overload the table, although they also belong to the formal model):

		x_1	x_2	y_4^+	y_4^-	y_6^+	y_6^-	y_7^+	y_7^-	1
$0 \leq$	$\tilde{e}_1 =$	-4	-1	0	0	0	0	0	0	32
$0 \leq$	$\tilde{e}_2 =$	4	6	0	0	0	0	0	0	24
$0 \leq$	$\tilde{e}_3 =$	-1	-3	0	0	0	0	0	0	23
$0 =$	$\tilde{e}_4 =$	-2	1	1	-1	0	0	0	0	22
$0 \leq$	$\tilde{e}_5 =$	-4	3	0	0	0	0	0	0	48
$0 =$	$\tilde{e}_6 =$	-1	-2	0	0	1	-1	0	0	17.55
$0 =$	$\tilde{e}_7 =$	-1	-1	0	0	0	0	1	-1	10
goal fu.: min	$e_8 =$	0	0	0	3	0	2	4	5	0

The solution to this formal model cannot be determined graphically as we did before, because now we have eight decision variables. However, the same resolution model that will be introduced in Section 3 can be applied, because the general structure of this model is the same as for the two-dimensional case.

The solution is $x_1 = 7.33$, $x_2 = 2.67$, $y_4^+ = 0$, $y_4^- = 10$, $y_6^+ = 0$, $y_6^- = 4.83$, $y_7^+ = 0$, and $y_7^- = 0$. The optimal solution lies in O_4 (Figure X.3). The numerical solution is not graphically obvious and the results must be taken on faith at this point. Inserting these values into the first row of the tableau, and performing the corresponding computations, gives the following tableau.

		$x_1 =$ 7.33	$x_2 =$ 2.67	$y_4^+ =$ 0	$y_4^- =$ 10	$y_6^+ =$ 0	$y_6^- =$ 4.83	$y_7^+ =$ 0	$y_7^- =$ 0	1
$0 \leq$	$\tilde{e}_1 =$ 0 =	-4	-1	0	0	0	0	0	0	32
$0 \leq$	$\tilde{e}_2 =$ 69.33=	4	6	0	0	0	0	0	0	24
$0 \leq$	$\tilde{e}_3 =$ 7.67 =	-1	-3	0	0	0	0	0	0	23
$0 =$	$\tilde{e}_4 =$ 0 =	-2	1	1	-1	0	0	0	0	22
$0 \leq$	$\tilde{e}_5 =$ 26.67=	-4	3	0	0	0	0	0	0	48
$0 =$	$\tilde{e}_6 =$ 0 =	-1	-2	0	0	1	-1	0	0	17.55
$0 =$	$\tilde{e}_7 =$ 0 =	-1	-1	0	0	0	0	1	-1	10
min	$e_8 =$ 39.77=	0	0	0	3	0	2	4	5	0

2.2 Structural Goals: Actions and Content Goals

Structural goals define the structure of the decision problem. Thereby, two aspects are important: (1) the way of combining actions to make up potential multi-action solutions, and (2) the role that the content goals play in the decision problem.

A) Structural Goals referring to Actions

Throughout the previous chapters we have encountered several structural goals that referred to actions. For example, we introduced a structural goal saying that only one out of n actions should be taken. Let x_j be the binary decision variable for action a_j, $(j=1,...,n)$ where

$$x_j = \begin{cases} 1 \text{ if } a_j \text{ is chosen} \\ 0 \text{ otherwise} \end{cases}$$

Then, $\sum_{j=1}^{n} x_j = k$ says to choose exactly k out of the n actions. If we evaluate these actions with m criteria and then aggregate these evaluations across the criteria for each action to a utility value u_j, then

$$\max: \sum_{j=1}^{n} x_j u_j; \text{ s.t.: } \sum_{j=1}^{n} x_j = 1$$

says to maximize the utility (content goal), subject to (s.t.) only one action being taken (structural goal); that is, to select the action with greatest utility. Additional examples of structural goals for logistic statements were discussed in Chapter IV, Table IV.2.

Another situation where binary decision variables play an important role is in planning a transportation route. Let' s assume that a road network is given as a collection of intersections (nodes) and connections (links) between the nodes. An example of such a road network is given below. Because no orientations are shown, we can simply assume that a link between two intersections stands for a two-way road. Traveling along a link from intersection i to intersection j results in a certain effectiveness value, e_{ij}, for example travel time, cost, or risk.

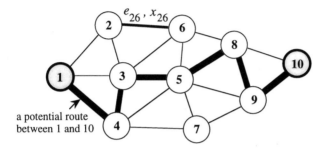

Figure X.4: Routing problem.

The content goal is to find a route from node 1 to node 10 which has the lowest overall effectiveness value (e.g., cost). To solve this problem we define the link between node i and node j as an elementary action, a_{ij}, to choose as part of the route. To each action, a_{ij}, we assign a binary **decision variable**:

$$x_{ij} = \begin{cases} 1 \text{ if link from node } i \text{ to node } j \text{ is part of the route} \\ 0 \text{ if link from node } i \text{ to node } j \text{ is not part of the route} \end{cases}$$

The content goal is obviously to minimize the total effectiveness value. This can be written as:

$$\text{min: } \sum_{j=1}^{10} \sum_{i=1}^{10} e_{ij} x_{ij}.$$

With only this content goal, we would choose all binary decision variables, x_{ij}, to be zero and would have as the solution that no route is the best route. Consequently, we must define what we mean by a route. This definition is part of the structural goal. A route is a connected sequence of links starting in node 1 and ending in node 10. For the example in Figure X.4, the route through the nodes, 1, 4, 3, 5, 8, 9, and 10 is a potential route from node 1 to node 10.

Node 1 is the only node along the route which has only one link leaving it, and node 10 is the only node along the route which has only one link entering it. All other nodes (intersections) along the route are left as often as they are entered. However, because we minimize the total effectiveness value of the route, we will certainly pass through an intersection at most once. Thus, the nodes 4, 3, 5, 8, and 9 are entered and left exactly once.

For a node k, the sum of the x_{ik}'s of the incoming links (i) minus the sum of the x_{kj}'s of the leaving links (j) equals -1 for node 1, +1 for node 10, and 0 for all other nodes on the network (not only the nodes 4, 3, 5, 8, and 9).

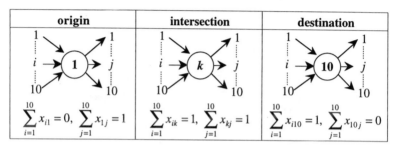

Table X.1: Logistic statements for route planning.

These structural goals for the feasibility of a route can be formalized as follows:

$$\sum_{i=1}^{10} x_{ik} - \sum_{j=1}^{10} x_{kj} = \begin{cases} -1 \text{ if } k \text{ is origin (1)} \\ 0 \text{ if } k \text{ is intersection (2,...,9)} \\ 1 \text{ if } k \text{ is destination (10)} \end{cases}$$

This structural constraint assures that a road consists of a connected sequence of links from the origin (node 1) to the destination (node 10). Because of this characteristic, this structural goal is called **flow conservation constraint**. It simply means that what flows into a node flows also out of it, except for the origin (only outflow) and the destination (only inflow).

B) Structural Goals referring to Content Goals

Structural goals can refer to constraints; for example, if we want to express that some of the content goals hold alternatively. Lets assume that we have m criteria (c_i), where m-1 criteria are used to define constraints with the following two types of inequalities (type 1: "$\leq$," and type 2: "$\geq$"):

$$\text{type 1: } e_{i1}x_1 + \ldots + e_{in}x_n \leq e_{CGi}$$
$$\text{type 2: } e_{i1}x_1 + \ldots + e_{in}x_n \geq e_{CGi}$$

The decision maker might require that only r ($\leq m\text{-}1$) of these m constraints must hold. The problem (given there is some goal function) is not only to find the optimal decision variables x_j ($j=1,...,n$) but also to decide which of the r constraints to consider.

To solve such constraint problems, we introduce for each constraint a binary decision variable, y_i ($i=1,...,m\text{-}1$) which takes on the value $y_i=1$, if the constraint is chosen, and $y_i=0$, if we decide to omit it. With these new structural decision variables, we would then reformalize our original constraints as follows:

$$\text{type 1: } e_{i1}x_1 + ... + e_{in}x_n \leq e_{CGi1} + M(1\text{-}y_i)$$
$$\text{type 2: } e_{i1}x_1 + ... + e_{in}x_n \geq e_{CGi1} + M(y_i\text{-}1)$$

where M is a very large number (e.g., 10^6). For example, type 1 inequality could be: $3x_1+2x_2+5x_3 \leq 10^6(1\text{-}y_i)$. We see then, if the decision variable is $y_i=1$ (i.e., the constraint is chosen), the additional terms, $M(1\text{-}y_i)$ and $M(y_i\text{-}1)$, are both equal to zero and the constraints hold just as defined for both type 1 and 2 inequalities. If $y_i=0$, then the additional term, $M(1\text{-}y_i)$, equals $+M$ for type 1 inequalities and $-M$ for type 2 inequalities. This means that the right-hand side of type 1 inequalities is a very large positive number. In this case, any value of the decision variable x_j meets this inequality; so, the constraint becomes obsolete. A very large negative right-hand-side for type 2 inequalities also means that for any value of the decision variable x_j the inequality holds and the constraint becomes obsolete. Thus, we need to add a new constraint, $\sum_{i=1}^{m-1} y_i = r$, saying that r out of $m\text{-}1$ constraints must be satisfied.

These modified constraints, together with this new constraint, are added to any other constraints that might hold for the decision problem. Together with the goal function, the problem can be solved for n original (x_i) and ($m\text{-}1$) structural (y_i) decision variables.

As a last consideration of structural goals referring to content goals, let's assume that the right hand side of an inequality has multiple alternative values, but only one can be chosen. For example, we might have the following constraint:

$$e_{i1}x_1 + ... + e_{in}x_n = e_{CGi1} \text{ or } ... \text{ or } e_{CGis},$$

where only one of the s right-hand side values, $e_{CGi1}, ... , e_{CGis}$ can hold. A numerical example is: $3x_1+2x_2+5x_3=10$ or 15 or 20. The selection of one (e.g., 15) and omission of the other right-hand side values (e.g., 10 and 20) is also a structural decision. We would therefore introduce, for each of the s values, a binary decision variable y_k, with $y_k=1$ when the value e_{CGik} (content goal value) is selected, and $y_k=0$ when the value e_{CGik} is not selected. This gives: $3x_1+2x_2+5x_3=10y_1+15y_2+20y_3$. Adding the additional structural constraint, $\sum_{k=1}^{s} y_k = 1$ (e.g., $y_1+y_2+y_3=1$), completes the formal model.

2.3 Additional Examples of Formal Models

A) Transportation Planning

A cement company has three production sites (S_1, S_2, and S_3), from which it dispatches cement trucks to four destinations (D_1, D_2, D_3, and D_4). The transportation distances between the production sites and the destinations are given in the figure below. Note that trucks can deliver cement only on the given network. The number of trucks (s_i) at each production site (S_i) and the demands of truck loads (d_j) at each destination (D_j) are given also. The transportation cost (c_{ij}) for one truck to go from production site S_i to destination D_j is also shown in the network below. For example, production site S_1 has $s_1=12$ trucks that can go either to delivery place D_1 (at a cost of $c_{11}=12$ units per truck) or to D_3 (at a cost of $c_{13}=14$ units per truck).

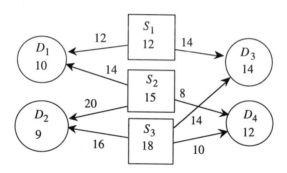

The problem is to decide for each production site, how many trucks should go to which destination, such that the total cost is minimized.

Let x_{ij} (the **decision variables**) be the number of trucks going from production site S_i to destination D_j. Then, the costs for all trucks going between S_i and D_j are: $c_{ij}x_{ij}$. The content goal is to minimize the sum of these costs. There are two structural goals. One is to send out all trucks and the other is to receive all trucks required. Obviously, there cannot be a solution if the sum of trucks to be dispatched (in this example 45) is not equal to the sum of trucks being requested (also 45). Thus, we have, with $x_{ij} \in \{0,1,2,3,...\}$:

- **content goal:** $\displaystyle \min: \sum_{j=1}^{4} \sum_{i=1}^{3} c_{ij}x_{ij};$

- **structural goals:** $\displaystyle \sum_{j=1}^{4} x_{ij} = s_i$, for all i; and $\displaystyle \sum_{i=1}^{3} x_{ij} = d_j$, for all j.

Using the values from the figure above, we get the following tableau:

		x_{11}	x_{13}	x_{21}	x_{22}	x_{24}	x_{32}	x_{33}	x_{34}	1
$0=$	$s_1=$	1	1	0	0	0	0	0	0	-12
$0=$	$s_2=$	0	0	1	1	1	0	0	0	-15
$0=$	$s_3=$	0	0	0	0	0	1	1	1	-18
$0=$	$d_1=$	1	0	1	0	0	0	0	0	-10
$0=$	$d_2=$	0	0	0	1	0	1	0	0	-9
$0=$	$d_3=$	0	1	0	0	0	0	1	0	-14
$0=$	$d_4=$	0	0	0	0	1	0	0	1	-12
goal func.	min	12	14	14	20	8	16	14	10	0

The solution to this transportation problem is: $x_{11}=7$, $x_{13}=5$, $x_{21}=3$, $x_{22}=0$, $x_{24}=12$, $x_{32}=9$, $x_{33}=9$, and $x_{34}=0$. The result can be verified by inserting these values into the above table. By adopting this policy, we get the minimum cost, which is 562 units.

B) Two-Actor Zero-Sum Conflicts

In Chapter IX, Section 3.1, we addressed two-actor zero-sum conflict situations and mentioned that a mixed strategy solution always exists. This type of conflict situation is characterized by the fact that the gain of one actor is the loss of the other actor. Given below is a **normal form representation** for a two-actor zero-sum conflict situation. The utilities u_{ij} are the wins for *Row* and the losses for *Col*, and p_i is the probability that *Row* chooses the pure strategy R_i, and q_j is the probability that *Col* chooses the pure strategy C_j. The problem is to find the **security levels** mixed strategies for *Row* and *Col*.

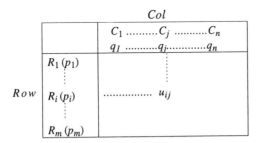

• _Row's point of view_:

Row can guarantee itself at least the security level u^* (> 0), with a mixed strategy, if there exists a probability distribution $(p_1,...,p_m)$, $p_i \geq 0$ and $\sum_{i=1}^{m} p_i = 1$, such that for whichever C_j _Col_ chooses, _Row's_ expected utility is at least u^*:

$$E[u_{R|C_j}] = \sum_{i=1}^{m} p_i u_{ij} \geq u^*, \text{ for } j=1,...,n.$$

We can set $p_i/u^* = x_i$. Thus, we have $\Sigma_i\, x_i = 1/u^*$. We can then write _Row's_ problem as (with $x_i \in \mathbf{R}^{\geq 0}$ being the **decision variables**, for $i=1,...,m$)

- **content goals**: $\min \sum_{i=1}^{m} x_i$, and $\sum_{i=1}^{m} x_i u_{ij} \geq 1$, for $j=1,...,n$.

- **structural goals**: none

• _Col's point of view_:

Let's remember that _Col's_ utility is the negative of _Row's_ utility. This means that _Col_ prefers low values of u_{ij} over high values in the conflict matrix. _Col_ can guarantee itself at least u^* (> 0), with a mixed strategy, if there exists a probability distribution $(q_1,...,q_n)$, $q_j \geq 0$ and $\sum_{j=1}^{n} q_j = 1$, such that for whichever R_i _Row_ chooses, _Col's_ expected utility is at least u^*:

$$E[u_{C|R_i}] = -\sum_{j=1}^{n} q_j u_{ij} \geq u^*, \text{ for } i=1,...,m.$$

We can set $q_j/u^* = y_j$. Thus, we have $\Sigma_j\, y_j = 1/u^*$. We then can write _Col's_ problem as (with $y_j \in \mathbf{R}^{\geq 0}$ being the decision variables, for $j=1,...,n$):

- **content goals**: $\max \sum_{j=1}^{n} y_j$, and $\sum_{j=1}^{n} y_j u_{ij} \leq 1$, for $i=1,...,m$.

- **structural goals**: none

• Example:

Let's apply this approach to the example discussed in Chapter IX, Section 3.1, C:

	C_1 q	C_2 $1-q$
$R_1: p$	4	17
$R_2: 1-p$	9	-6

For *Row's* problem we have the following formulation ($x_1, x_2 \geq 0$):

$$\text{min: } x_1 + x_2$$
$$\text{s.t.: } 4x_1 + 9x_2 \geq 1$$
$$17x_1 - 6x_2 \geq 1$$

The solution to this problem is $x_1=0.085$ and $x_2=0.073$. From, $x_1+x_2=1/u^*$, we compute $u^*=6.32$, and from $p=x_1u^*$, we get $p=0.538=15/28$.

For *Col's* problem we have the following formulation ($y_1, y_2 \geq 0$):

$$\text{max: } y_1 + y_2$$
$$\text{s.t.: } 4y_1 + 17y_2 \leq 1$$
$$9y_1 - 6y_2 \leq 1$$

The solution to this problem is $y_1=0.130$ and $y_2=0.028$. From, $y_1+y_2=1/u^*$, we compute $u^*=6.32$, and from $q=y_1u^*$, we get $q=0.82=23/28$. These are the same results which we obtained in Chapter IX, Section 3.1, C.

C) Nurse Scheduling,

A hospital manager is in charge of scheduling the work shifts for nurses. The six shifts and the minimum number of nurses required in each shift are given in the table below:

shift #	time	min. # of nurses
1	00:00 - 04:00	3
2	04:00 - 08:00	2
3	08:00 - 12:00	4
4	12:00 - 16:00	5
5	16:00 - 20:00	6
6	20:00 - 24:00	8

Each nurse works eight hours a day in two consecutive shifts. The problem is to find the smallest number of nurses needed for this plan.

The **decision variables** are the number of nurses in two consecutive shifts i and j: $x_{ij} \in \{0,1,2,...\}$. Thus, our aspiration is to minimize the total number of nurses:

$$\text{min: } x_{12} + x_{23} + x_{34} + x_{45} + x_{56} + x_{61}.$$

We must also ensure that the minimum number of nurses per shift is present. Because in each shift some nurses work their first shift while others work their second shift, the number of nurses, let's say in shift 3, must satisfy the inequality: $4 \leq x_{23}+x_{34}$.

These requirements for all six shifts are given below:

shift #	working # of nurses	min. # of nurses
1	$x_{61} + x_{12} \geq$	3
2	$x_{12} + x_{23} \geq$	2
3	$x_{23} + x_{34} \geq$	4
4	$x_{34} + x_{45} \geq$	5
5	$x_{45} + x_{56} \geq$	6
6	$x_{56} + x_{61} \geq$	8

An interesting aspect of this example is that we are dealing only with structural goals. There are no evaluations (expressed in cost or time) involved in this problem and therefore no content goals (such as minimize cost or time). Obviously, structural goals can also be expressed in terms of aspirations (to minimize the total number of nurses) and constraints (to guarantee that the minimum number of nurses is present in each shift).

As is the case with many problems that have integer decision variables, the solution to this problem is not unique. In the table below are eight optimal solutions to the nurse scheduling problem which all yield the same minimum total number of 15 nurses (each '+' indicates a surplus of one nurse per shift). It should be noted that all solutions have a total surplus of two nurses.

							nurses per shift needed/assigned					
solutions	x_{12}	x_{23}	x_{34}	x_{45}	x_{56}	x_{61}	1: 3	2: 2	3: 4	4: 5	5: 6	6: 8
1.	2	0	4	1	7	1	3	2	4	5	8^{++}	8
2.	0	2	4	1	5	3	3	2	6^{++}	5	6	8
3.	2	0	4	1	6	2	4^{+}	2	4	5	7^{+}	8
4.	0	2	2	3	3	5	5^{++}	2	4	5	6	8
5.	1	1	5	0	6	2	3	2	6^{++}	5	6	8
6.	1	1	3	2	4	4	5^{++}	2	4	5	6	8
7.	2	0	5	0	7	1	3	2	5^{+}	5	7^{+}	8
8.	2	0	5	0	6	2	4^{+}	2	5^{+}	5	6	8
9.	0	2	2	3	4	4	4^{+}	2	4	5	7^{+}	8
10.	1	1	3	2	6	2	3	2	4	5	8^{++}	8
11.	0	2	2	3	3	5	5^{++}	2	4	5	6	8

D) Mixing Substances

A chemical company is producing two types of substances (A and B) consisting of three types of raw materials (I, II, and III). The requirements on the compositions of the three substances and the profits are as follows:

Substance	compositions	profits [per kg]
A	• at most 20% of I • at most 10% of II • at least 20% of III	10
B	• at most 40% of I • at most 50% of III	8

Each material's available amount and treatment costs are as follows:

raw material	available amount [kg]	treatment costs [per kg]
I	400	4
II	500	5
III	300	6

The company's problem is to find out how much of each substance to produce at which composition, such that the profit is maximized.

Because we have two substances and three raw materials, we introduce 6 (=2×3) **decision variables** x_{ij}, i=A, B, and j=I, II, III. For example, x_{BIII} is the amount of raw material III in substance B. The goal function is thus to maximize the profits minus the treatment costs:

max: $10(x_{AI}+x_{AII}+x_{AIII}) + 8(x_{BI}+x_{BII}+x_{BIII}) - 4(x_{AI}+x_{BI}) - 5(x_{AII}+x_{BII}) - 6(x_{AIII}+x_{BIII})$.

Rewriting this in terms of the 6 decision variables gives:

max: $6x_{AI}+5x_{AII}+4x_{AIII}+4x_{BI}+3x_{BII}+2x_{BIII}$.

The material constraints are:

raw material	chosen amounts [kg]	available amounts [kg]
I	$x_{AI} + x_{BI} \leq$	400
II	$x_{AII} + x_{BII} \leq$	500
III	$x_{AIII} + x_{BIII} \leq$	300

The composition constraints are:

Substance	compositions	amounts	constraints
A	• at most 20% of I • at most 10% of II • at least 20% of III	$0.2(x_{AI} + x_{AII} + x_{AIII})$ $0.1(x_{AI} + x_{AII} + x_{AIII})$ $0.2(x_{AI} + x_{AII} + x_{AIII})$	$\geq x_{AI}$ $\geq x_{AII}$ $\leq x_{AIII}$
B	• at most 40% of I • at most 50% of III	$0.4(x_{BI} + x_{BII} + x_{BIII})$ $0.5(x_{BI} + x_{BII} + x_{BIII})$	$\geq x_{BI}$ $\geq x_{BIII}$

From the first inequality in the table above, we can derive the following constraint: $0 \leq -0.8x_{AI} + 0.2x_{AII} + 0.2x_{AIII}$. Our problem has therefore five constraints from the table above plus three material constraints, and one goal function. In addition, we require that the decision variables, x_{ij}, take on only integer values.

The optimal solution to this problem is: $x_{AI}=85$, $x_{AII}=42$, $x_{AIII}=299$, $x_{BI}=306$, $x_{BII}=458$, $x_{BIII}=1$, with a total gain of 4,516. This means that the company should produce 426 kg of substance A and 765 kg of substance B. Of the 400 kg of material I only 391 kg are use, while all of the 500 kg of material II and all of the 300 kg of material III are used.

3. Resolution Models

3.1 Real-Valued Decision Variables

A) Theoretical Considerations

If we have n decision variables, then the space of potential solutions is n-dimensional. A **convex set** of feasible solutions can be determined by $(n-1)$-dimensional **hyperplanes** in the n-dimensional space of potential solutions. Each of the hyperplanes splits the space into two parts - one with growing positive values of the slack variable and the other with growing negative values. A hyperplane in the n-dimensional space is defined as:

$$0 = \tilde{e}_i = e_{i1}x_1 + \ldots + e_{in}x_n - e_{CGi},$$

where values $(x_1, ..., x_n)$ resulting in a positive $\tilde{e}_i$ lie on one side of the hyperplane and values with a negative $\tilde{e}_i$ lie on the other side of the hyperplane. It should be noted that an equality constraint, $\tilde{e}_i = 0$, can be transferred into two inequality constraints, as follows: $0 \leq \tilde{e}_i$ and $0 \leq -\tilde{e}_i$.

In the 2-dimensional case (two actions) we have as the space of potential solutions the x_1-x_2 plane. The 1-dimensional hyperplanes are straight lines that split the x_1-x_2 plane into two parts. As we saw in Figures X.2 and X.3, if the hyperplanes (lines) define a convex set, then the extreme value (minimum or maximum) lies either on the intersection of at least two hyperplanes or on an interval of a hyperplane. For an n-dimensional space of potential solutions, we need at least $n+1$ hyperplanes to define a finite convex set. A point in the n-dimensional space is the intersection of at least n hyperplanes. A systematic way to solve such linear constraint-optimization problems would thus be to compute all corner points of the convex set and to take the one where the goal function is optimal (maximal or minimal).

However, computing all intersections and then searching for the one which optimizes the goal function is not an efficient procedure. Rather, we would like to progress in a promising direction and process some information at every computation step that helps us decide whether we have reached the optimum or whether we should continue with the search. This principle is exactly what optimization algorithms do. They start with a feasible solution, perform iterations followed by an optimality test, and end when the iteration no longer improves the goal function.

B) The Simplex Algorithm

Let us return to our 2-dimensional example at the beginning of Section 2.1, A, starting with the tableau (T_0). Then, we start with the policy $x_1=0$ and $x_2=0$, which, as we see from Figure X.5, is a **feasible solution**. If the origin is not a feasible solution we perform a coordinate transformation so that the origin falls somewhere within the space of feasible solutions.

T_0		$x_1 = 0$	$x_2 = 0$	1
$0 \leq$	$\tilde{e}_1 = 32 =$	-4	-1	32
$0 \leq$	$\tilde{e}_2 = 24 =$	4	6	24
$0 \leq$	$\tilde{e}_3 = 23 =$	-1	-3	23
$0 \leq$	$\tilde{e}_4 = 10 =$	2	-1	10
$0 \leq$	$\tilde{e}_5 = 48 =$	-4	3	48
goal function: max	$e_6 = 0 \quad =$	1	2	0

Departing from this initial tableau, we want to move to an adjacent corner point. Possible new corner points would therefore be (see Figure X.5):

- $x_1 = 0$ and $\tilde{e}_2 = 0$
- $x_1 = 0$ and $\tilde{e}_3 = 0$
- $x_2 = 0$ and $e_1 = 0$
- $x_2 = 0$ and $\tilde{e}_4 = 0$

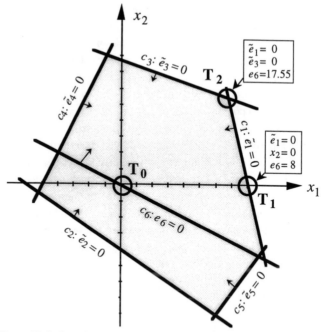

Figure X.5: Iterations in the simplex algorithm ($T_0 \rightarrow T_1 \rightarrow T_2$).

In graphical terms, using such a new corner point means that to 'walk' from the current point T_0 along the line $x_2=0$ to the adjacent point T_1. We could also go along $x_1=0$ to the point defined by $x_1=0$ and $\tilde{e}_3=0$. Of course, we want to go to the one that increases the goal function value e_6 the most.

To derive a general resolution model that tells us in which direction to 'walk,' we look at the general tableau:

T_0		$x_1 = 0$	$x_2 = 0$	1
$0 \leq$	$\tilde{e}_1 =$	e_{11}	e_{12}	b_1
$0 \leq$	$\tilde{e}_2 =$	e_{21}	e_{22}	b_2
$0 \leq$	$\tilde{e}_3 =$	e_{31}	e_{32}	b_3
$0 \leq$	$\tilde{e}_4 =$	e_{41}	e_{42}	b_4
$0 \leq$	$\tilde{e}_5 =$	e_{51}	e_{52}	b_5
goal function:	max $e_6 =$	e_{61}	e_{62}	b_6

The present intersection point is determined by the variables which stand at the top of the tableau. For tableau T_0 these are the decision variables x_1 and x_2. 'Walking' to an adjacent intersection point means to replace one of the two decision variables with one of the five slack variables. Let's say we want to go to the point $x_1=0$ and $\tilde{e}_3=0$. Then, we would have to exchange the current line (hyperplane), $x_2=0$, with the new line (hyperplane), $\tilde{e}_3 = 0$. To do this exchange, we have to solve the current equation $\tilde{e}_3 = f(x_1, x_2)$ for $x_2 = g(x_1, \tilde{e}_3)$; that is, we exchange the decision variable x_2 with the slack variable $\tilde{e}_3$. Thus, we get:

$$x_2 = -\frac{e_{31}}{e_{32}} x_1 + \frac{1}{e_{32}} \tilde{e}_3 - \frac{b_3}{e_{32}}$$

Inserting this expression for x_2 into the tableau T_0 gives us the new tableau T_1:

T_1		$x_1 = 0$	$\tilde{e}_3 = 0$	1
$0 \le$	$\tilde{e}_1 =$	$e_{11} - e_{12}\dfrac{e_{31}}{e_{32}}$	$\dfrac{e_{21}}{e_{32}}$	$b_1 - e_{12}\dfrac{b_3}{e_{32}}$
$0 \le$	$\tilde{e}_2 =$	$e_{21} - e_{22}\dfrac{e_{31}}{e_{32}}$	$\dfrac{e_{22}}{e_{32}}$	$b_2 - e_{22}\dfrac{b_3}{e_{32}}$
$free$	$x_2 =$	$-\dfrac{e_{31}}{e_{32}}$	$\dfrac{1}{e_{32}}$	$-\dfrac{b_3}{e_{32}}$
$0 \le$	$\tilde{e}_4 =$	$e_{41} - e_{42}\dfrac{e_{31}}{e_{32}}$	$\dfrac{e_{42}}{e_{32}}$	$b_4 - e_{42}\dfrac{b_3}{e_{32}}$
$0 \le$	$\tilde{e}_5 =$	$e_{51} - e_{52}\dfrac{e_{31}}{e_{32}}$	$\dfrac{e_{52}}{e_{32}}$	$b_5 - e_{52}\dfrac{b_3}{e_{32}}$
goal function	max $e_6 =$	$e_{61} - e_{62}\dfrac{e_{31}}{e_{32}}$	$\dfrac{e_{62}}{e_{32}}$	$b_6 - e_{62}\dfrac{b_3}{e_{32}}$

From this general formulation, let's derive some facts and rules that lead to a resolution model. We can summarize in words what we have to do when we exchange two lines (hyperplanes); that is, when we move to a new feasible intersection point (tableau):

- The element at the intersection of the gray shaded row and column in the tableau is called the **pivot element**. The pivot element assumes its reciprocal value ($e_{32} \rightarrow 1/e_{32}$).

- The new elements in the **pivot column** (e.g., e_{42}) are the original ones divided by the pivot element ($e_{42} \rightarrow e_{42}/e_{32}$).

- The new elements in the **pivot row** (e.g., e_{31}) are the original ones divided by the negative pivot element ($e_{31} \rightarrow -e_{31}/e_{32}$).

- The other new elements are the original ones (e.g., e_{41}) plus: the new element of the pivot row that is in their column ($-e_{31}/e_{32}$) multiplied by the old element in the pivot column that is in their row (e_{42}): $e_{41} \rightarrow e_{41} - e_{42}(e_{31}/e_{32})$.

We can now summarize the principles of the **simplex algorithm** and how to choose a pivot element that increases the goal function as much as possible:

- In the 2-dimensional case, we assume that we have a convex set which is defined by m-1 constraints. For the most general case, we also assume that all corner points of this set are determined as the intersection of only two lines (not three or more). In such a case, the simplex algorithm consists of exchanging basic (decision and slack) variables for non-basic slack variables. Because there are at most $m(m+1)/2$ corner points, the simplex algorithm stops after some iterations. A basic variable that has been exchanged for a non-basic variable never again becomes a basic variable.

- The entries in the column '1' (right-most column) must be positive for all slack variables (criteria) and for those decision variables x_j which are not free variables. A **free variable** can take on both positive and negative values. It should be noted that in many cases the decision variables are not free, but they must be positive ($x_j \geq 0$).

- If a decision variable is free, we could cancel its row in the new diagram because it no longer provides any information. If it is restricted to positive values, we simply keep it as if it were a constraint of a criterion. Thus, we must make a distinction between the exchange of a free decision variable and the exchange of a bounded decision variable ($x_j \geq 0$).

Case 1:

Exchanging a bounded basic variable, that is, $x_j \geq 0$ or $\tilde{e}_j \geq 0$ for a non-basic slack variable $\tilde{e}_i$:

The new element in the right-most entry of the pivot row must be positive (because the variables at the top of the tableau can be set to zero):

$$-\frac{b_i}{e_{ij}} \geq 0$$

However, because $b_i \geq 0$, we can conclude that $e_{ij} < 0$ must hold. Any other new element at the right-most position in row $k \neq i$ must also be positive:

$$b_k - e_{kj}\frac{b_i}{e_{ij}} \geq 0$$

If $e_{ij} \geq 0$, then this requirement is met because $b_i \geq 0$ and $e_{ij} < 0$. Moreover, the new right-most element in this row k is larger than the old one. This means that the new right-most element in the goal function row also grows. But if $e_{kj}<0$, then we have the requirement:

$$b_k \geq e_{kj} \frac{b_i}{e_{ij}} \text{ which gives: } \frac{b_k}{e_{kj}} \leq \frac{b_i}{e_{ij}}, k \neq i.$$

To have the goal function value grow (e_6 in our example) we must require: $e_{62}>0$. The quotient $\frac{b_k}{e_{kj}}$ is called **characteristic quotient**.

An exchange step performed according to these requirements is called a **simplex step**. In summary we can say for bounded basic variables:

- Choose as the pivot column any column that has a positive element in the goal function row ($e_{6j}>0$).

- Take all the <u>negative</u> elements in this pivot column ($e_{kj}<0$) and compute the characteristic quotients (b_k/e_{kj}). The <u>largest</u> of them determines the pivot column and thus the pivot element.

- Perform a simplex step on this pivot element.

Case 2:

Exchanging a free basic decision variable (i.e., x_j can be positive or negative) for a non-basic slack variable $\tilde{e}_i$:

If we exchange a free basic decision variable with a non-basic slack variable, we could omit the row in the new tableau because it does not provide any information (because a free variable can be both negative or positive). This would reduce the size of the tableau by one row. However, we want to keep the decision variable in the tableau because when the algorithm stops, we can read its final value right out of the tableau. It should also be noted that free basic variables can always be exchanged, regardless of the signs of the coefficients.

A simplex step for exchanging a free basic variable can be summarized as follows:

- The pivot column is determined by the position of the free variable.

  - If the goal-function element in this column is positive ($e_{6j}>0$), choose the pivot element as stated for bounded variables; that is, take all the <u>negative</u> elements in this pivot column ($e_{kj}<0$) and compute the characteristic quotients (b_k/e_{kj}). The <u>largest</u> of them determines the pivot column and thus the pivot element.

  - If the goal-function element in this column is negative ($e_{6j}>0$), take all the <u>positive</u> elements in this pivot column ($e_{kj}<0$) and compute the characteristic quotients (b_k/e_{kj}). The <u>smallest</u> of them determines the pivot column and thus the pivot element.

- After the simplex step, the row of the decision variable could be eliminated from the new tableau, because it does not constitute a constraint (the variable is 'free').

The principle of the **simplex algorithm** is summarized in Figure X.6, where the coefficients of the goal function (in our example e_{6j}) are referred to as z_j.

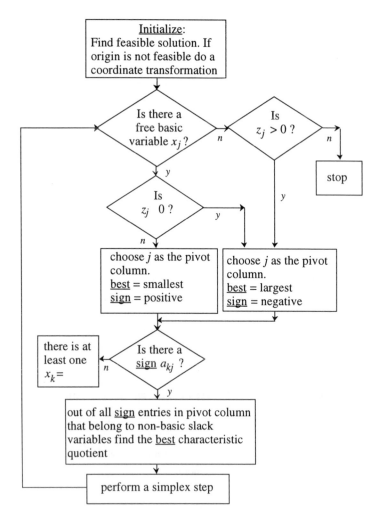

Figure X.6: Principle the of simplex algorithm.

C) Numerical Example

We return now to our original example and solve it with the simplex algorithm. The original tableau and the characteristic quotients for the first column are given below. Because both decision variables are free, we may choose either of the two for exchange, for example x_1. Once we decide to exchange x_1, we mark the pivot column with a vertical arrow ($\uparrow$). Then, we compute the characteristic quotients. Because the bottom entry in the pivot column (goal function) is positive, we compute only negative characteristic quotients and write them to the right of the tableau. The largest one determines the pivot row and thus the pivot element.

T_0		$x_1 = 0$	$x_2 = 0$	1	char. quotient
$0 \le$	$\tilde{e}_1 = 32 =$	-4	-1	32	-8: max
$0 \le$	$\tilde{e}_2 = 24 =$	4	6	24	
$0 \le$	$\tilde{e}_3 = 23 =$	-1	-3	23	-23
$0 \le$	$\tilde{e}_4 = 10 =$	2	-1	10	
$0 \le$	$\tilde{e}_5 = 48 =$	-4	3	48	-12
goal function: max	$e_6 = 0 \ \ =$	1	2	0	
		$\uparrow$	-1/4	8	

The new intersection is now T_1 (Figure X.5) which is determined by $x_2=0$ and $\tilde{e}_1=0$. The new tableau is given below. Now we want to exchange the decision variable x_2. We determine the pivot element as we did for the first tableau.

T_1		$\tilde{e}_1 = 0$	$x_2 = 0$	1	char. quot.
free	$x_1 \ = 8 =$	-1/4	-1/4	8	-32
$0 \le$	$\tilde{e}_2 = 56 =$	-1	5	56	
$0 \le$	$\tilde{e}_3 = 15 =$	1/4	-11/4	15	-60/11: max
$0 \le$	$\tilde{e}_4 = 26 =$	-1/2	-3/2	26	-52/3
$0 \le$	$\tilde{e}_5 = 16 =$	1	4	16	
goal function: max	$e_6 = 8 \ =$	-1/4	7/4	8	
		1/11	$\uparrow$	60/11	

The new intersection is thus T_2 (Figure X.5) which is determined by $\tilde{e}_1=0$ and $\tilde{e}_3=0$.

T_2		$\tilde{e}_1 = 0$	$\tilde{e}_3 = 0$	1
free	$x_1 =$		1/11	73/11
$0 \le$	$\tilde{e}_2 =$		-20/11	
free	$x_2 =$	1/11	-4	60/11
$0 \le$	$\tilde{e}_4 =$		12/22	
$0 \le$	$\tilde{e}_5 =$		-1/11	
goal function: max	$e_6 =$	-1/11	-7/11	193/11

Because both coefficients in the goal function row are negative, the simplex algorithm stops and we have to compute only the values for the decision variables and the goal function. The optimal solution is determined as: $x_1=73/11$, $x_2=60/11$, and $e_6=193/11$. This is the same result that we obtained with the graphical evaluation in Figure X.2, also with $\tilde{e}_1=0$ and $\tilde{e}_3=0$ being the binding constraints. If there was another positive entry in the bottom row of $\tilde{e}_1=0$ or $\tilde{e}_3=0$, we would perform another simplex step.

If we were interested in the minimum value of the goal function, rather than in the maximum, we would simply alter the goal function

$$\text{min: } e_i = \sum_{j=1}^{n} e_{ij} x_j$$

to the equivalent formulation:

$$\text{max: } (-e_i) = \sum_{j=1}^{n} (-e_{ij}) x_j,$$

and proceed with the simplex algorithm as we did before. The obtained maximum value e_i must then be multiplied by -1 to get the desired minimum value.

D) <u>Comments about the Simplex Algorithm</u>

Assuming the constraints define a **convex set**, then the simplex algorithm in two dimensions can be seen as a walk along straight lines ($x_j=0$ or $\tilde{e}_i=0$) from one corner to another. The corner points are the intersections of two straight lines (constraints). Because there is a finite number of decision variables and constraints, there also is a finite number of corner points. In the two-dimensional case we have two basic variables (x_1 and x_2) and m-1 constraints. Thus, there are at most

$$\binom{2+(m-1)}{2} = m(m+1)/2$$

corner points. This means that the simplex algorithm must stop after a finite number of iterations. It stops, when it fails the optimality test; that is, when the value of the goal function does not grow anymore. We already mentioned that the value of the goal function could be infinitely large if the feasible space is not bounded. This is the case if a pivot column can be identified but not a pivot element.

Another special case is if we have more than two lines that intersect in a corner point. Then, there are multiple equivalent tableaus and the value of the goal function may not grow after a simplex step. Such a corner point is called a degenerated corner point because the simplex algorithm stops there. In such cases, care must be taken not to fall into cycles, in which we return to a previous tableau after a few iterations.

In our numerical example, the origin was a feasible point. However, this is not always the case. We mentioned earlier that when the origin is not a feasible point, we have to do a coordinate transformation. This means that we must find a feasible point and translate the origin of our problem to this point. The determination of such a feasible point, especially in a higher dimensional problem, is done in a systematic manner. However, we will not elaborate on this issue, because advanced computer programs for constraint-optimization problems take care of this aspect automatically, even if we start from a non-feasible solution.

3.2 Duality and Sensitivity Analysis

A) General Concept

There are always two ways to look at a constraint-optimization problem. Nicholson [1995] uses the illustrative example of minimizing a rectangular area, A, subject to a given enclosure-length, U. The decision variables are the length (x_1) and the width (x_2) of the rectangular area. The **non-linear** constraint-optimization problem can be stated as:

$$\text{max: } A = x_1 x_2; \text{ s.t.: } 2x_1 + 2x_2 = U.$$

With the **Lagrange multiplier method** we compute the partial derivatives of

$$\mathcal{L} = x_1 x_2 + \lambda(U - 2x_1 - 2x_2),$$

set them equal to zero, and get the following solution $x_1 = x_2 = U/4$; that is, the field should be squared.

The fact that the field should be squared can also be obtained with a different formulation: minimize the enclosure-length U subject to a given area A:

$$\text{min: } U = 2x_1 + 2x_2; \text{ s.t.: } x_1 x_2 = A.$$

With the Lagrange multiplier method we compute the partial derivatives of

$$\mathcal{L}_{dual} = 2x_1 + 2x_2 + \lambda_{dual}(A - x_1 x_2),$$

set them equal to zero, and get the following solution $x_1 = x_2 = A^{.5}$; that is, the field should be squared, as we just found out before. This second definition of the decision problem is called the **dual problem**. While the original problem is defined as constraint-maximization, the dual problem is defined as constraint-minimization. They both lead to the same optimal solution.

This concept applies also to linear constraint-optimization problems. A typical example of duality is the conflict resolution example discussed in Section 2.3, B. There, *Col*'s problem (trying to minimize *Row*'s gain) is the dual of *Row*'s problem (trying to maximize his/her gain) and both have the same absolute expected utility.

For a linear constraint-optimization problem, the relation between original and dual formal model is the following:

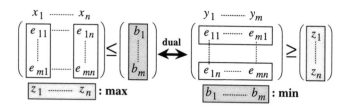

In matrix notation, we can write this as follows:

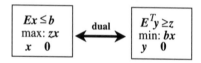

The components of the vector $y^{1 \times m}$ are called **shadow prices** because they indicate how much one could gain by relaxing the binding constraints in the original problem. The goal function (in our example e_6) in the final tableau is: $e_6 = y_1 \tilde{e}_1 + y_2 \tilde{e}_3 + z$. Because $\tilde{e}_1$ and $\tilde{e}_3$ are binding constraints ($\tilde{e}_1 = \tilde{e}_3 = 0$), we have $e_6 = z$. Now, the role of the shadow prices becomes clear. If we relax the constraint defined by c_1 by w, the optimal solution becomes: $e_6 = y_1 w + z$. Thus, it grows from the original value z by $y_1 w$. However, there is a limit to how much we can relax a constraint before it no longer is a binding constraint.

B) Numerical Example

For our numerical example introduced in Section 2.1, A, the initial tableaus and the solutions for the original and its dual problem are as follows:

$$\begin{bmatrix} 4 & 1 \\ -4 & -6 \\ 1 & 3 \\ -2 & 1 \\ 4 & -3 \end{bmatrix} \times \begin{bmatrix} x_1 \\ x_2 \end{bmatrix} \le \begin{bmatrix} 32 \\ 24 \\ 23 \\ 10 \\ 48 \end{bmatrix} \qquad \textbf{dual} \quad \longleftrightarrow \qquad \begin{bmatrix} 4 & -4 & 1 & -2 & 4 \\ 1 & -6 & 3 & 1 & -3 \end{bmatrix} \times \begin{bmatrix} y_1 \\ y_2 \\ y_3 \\ y_4 \\ y_5 \end{bmatrix} \ge \begin{bmatrix} 1 \\ 2 \end{bmatrix}$$

$$[1\ \ 2] \times \begin{bmatrix} x_1 \\ x_2 \end{bmatrix} : \max \qquad\qquad\qquad [32\ \ 24\ \ 23\ \ 10\ \ 48] \times \begin{bmatrix} y_1 \\ y_2 \\ y_3 \\ y_4 \\ y_5 \end{bmatrix} : \min$$

solution:　$x_1 = 73/11$, $x_2 = 60/11$　　　　solution: $y_1 = 1/11$, $y_3 = 7/11$, $y_2 = y_4 = y_5 = 0$

max: $e_6 = 17.55$　　　　　　　　　　　min: $e_b = 17.55$

shadow prices: $y_1 = 1/11$, $y_3 = 7/11$　　shadow prices: $x_1 = 73/11$, $x_2 = 60/11$

It should be noted that if we solve by hand or with a computer the problem as formulated in tableau T_0, with $b\text{-}Ex \ge 0$ instead of the original model $Ex \ge b$, the shadow prices (defined by the stop condition in the simplex algorithm to be negative) are $y_1 = -1/11$, $y_3 = -7/11$, which are the coefficients in the last tableau, T_2. Anyway, the interpretation of the results is the same.

Two types of sensitivity analyses can be performed, one for the constraints and one for the decision variables. The sensitivity analysis for the constraints includes the final values, the shadow prices, the constraints, and also the allowable increases and decreases of these constraints so that the optimal solution does not change. The sensitivity report corresponding to the solution of $b\text{-}Ex \ge 0$ is given in Table X.2.

constraints	final value	shadow price $y^{6\times 1}$	constraint $\ge$	allowable increase	allowable decrease
$\tilde{e}_1$	0	-1/11	0	28	27.73
$\tilde{e}_2$	83.27	0	0	83.27	∞
$\tilde{e}_3$	0	-7/11	0	26	32.67
$\tilde{e}_4$	17.82	0	0	17.82	∞
$\tilde{e}_5$	37.82	0	0	37.82	∞

Table X.2: Sensitivity report for the constraints.

The interpretation of the shadow prices and the allowable increases and decreases is illustrated in Figure X.7 for the binding constraint $\tilde{e}_1 = 0$. The allowable decrease means that the line $\tilde{e}_1 = 0$ can be translated to the intersection of the two lines $\tilde{e}_3 = 0$

and $\tilde{e}_5=0$ which is the point (14.2,2.93). For this point we have $\tilde{e}_1=-27.7$. If we keep moving even further, then c_3 and c_5 become the two binding constraints. The allowable increase is to move in the other direction to the point (-1,8) which is the intersection of $\tilde{e}_3=0$ and $\tilde{e}_4=0$. There we have $\tilde{e}_1=28$.

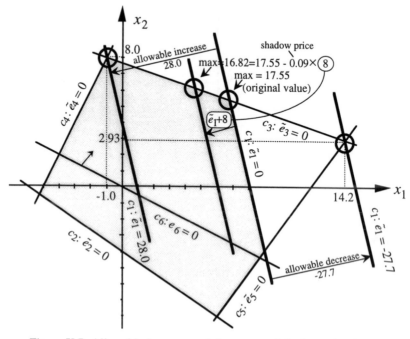

Figure X.7: Allowable increase and decrease and shadow price for c_1.

The shadow price for c_1 (-0.09) can be illustrated by moving $\tilde{e}_1=0$ such that we have $\tilde{e}_1=1$. The new maximum value is then 17.55-0.09=17.46. This can also be computed by replacing in the original tableau 32 with 31, and computing the optimal solution. The result is then 17.46. In Figure X.7, the constraint has been changed from $\tilde{e}_1=0$ to $\tilde{e}_1=8$. Thus, the new optimal value is 16.82=17.55-0.09×8. To compute this result, we replace in the original tableau 32 with 24.

The sensitivity report for the decision variables is given in Table X.3. The allowable increase of 7 for x_1 means that the original goal function max: $1x_1+2x_2$ can be changed to $7x_1+2x_2$ without affecting the solution.

decision variables	final value	allowable increase	allowable decrease
x_1	6.64	7	1/3
x_2	5.45	1	1.75

Table X.3: Sensitivity report for the decision variables.

3.3 Integer Decision Variables

If the decision variables are limited to integer values, and assuming that we have a limited space of feasible solutions, then the number of feasible solutions is finite, contrary to real-valued decision variables. One might assume that the optimal integer-valued solution is simply the rounded real-valued solution. Unfortunately, this is not necessarily the case, as can be seen from the example in Figure X.8. Due to the very 'flat' goal function and binding constraint, the optimal real-valued solution, max=9.5 at point (3,13/8), is different from the optimal integer-valued solution, max=8.0 at point (0,2). The closest neighboring integer point in the feasible space from the optimal rounded real-valued solution is the point (3,1) with max=7.0.

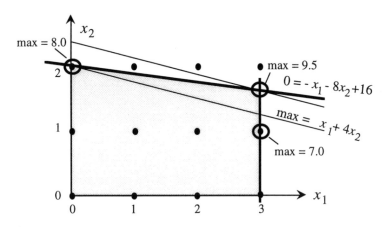

Figure X.8: Rounded real-valued solution being different from integer-valued

solution.

We must therefore conclude that rounding is not necessarily a feasible approach for integer-valued decision variables. Algorithms to solve linear constraint-optimization problems with integer-valued decision variables are based on the **branch-and-bound technique** [Hillier and Lieberman, 1986, Chapter 13.4]. The principle of this technique is to partition the set of all feasible solutions at each iteration step into several subsets. Then, the non-promising subsets are discarded. Among the remaining subsets, one is further partitioned. This process goes on until the optimal solution has been found.

Several heuristic algorithms have been developed which are quite efficient in finding a better solution than the rounded real-valued solution (the real valued solution is also called the **LP-relaxed solution**), although they might never find the true optimal solution. Commercial software packages for constraint-optimization problems use these heuristic algorithms to compute mixed-integer problems

---

(problems where some of the decision variables are integer-valued and some real-valued).

3.4 Binary Decision Variables

An example with only binary decision variables is finding an optimal (e.g., shortest) route from an origin to a destination on a network, as introduced in Section 2.2, A. Figure X.9 shows an example of a road network with the length of the links. The shortest route from the origin (A) to the destination (G) can be determined with an algorithm. The idea is to start at the origin and then to look for the nearest of all the adjacent nodes (B, C, or D); this is node C (1 distance unit). We can now 'close' node C because we have determined the shortest route from the origin to node C. Closed nodes are shaded in Figure X.8 and the shortest distances to the origin are also indicated.

The general principle of the algorithm is to find a neighbor of all the closed nodes which has the shortest distance to the origin. It should be noted that this is not necessarily the same as finding the nearest neighboring node. Figure X.9 (left) shows the shortest paths from the origin A of the two closed nodes C and F after two iterations. Figure X.9 (middle) shows the closed nodes and their shortest paths after four iterations. Ties are broken arbitrarily, which, for example, means that after closing nodes C and F, node B could be closed either from C or from F. In Figure X.9 we arbitrarily chose to close it from C.

Finally, Figure X.9 (right) shows the solution; that is, the shortest path from the origin (A) to the destination (G). The path goes along the nodes C and F, and it is 6 distance units long. For this example, the destination node was closed as the last of all the nodes (after six iterations). However, this is not generally the case. If the distance between C and G had been, for example, 1 instead of 6, node G would have been closed after two iterations (with the route going from A to C and to G).

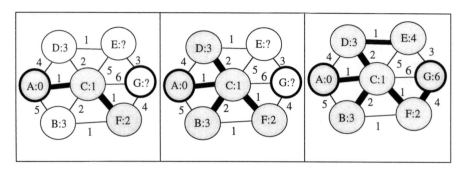

Figure X.9: Principle of shortest route algorithm.

The **shortest route algorithm** can be summarized as follows [Dantzig, 1955]. Let d_{ij} be the distance between two adjacent nodes i and j, r_{ik} the shortest distance between the nodes i and k, and S the set of closed nodes:

1. Initialize: Close the origin A ($r_{AA} = 0$); i.e., put it into S.

2. Find among all neighbors of the closed nodes $i \in S$ the node j which is nearest to the origin; i.e., $\min_{i \in S}(r_{Aj}) = (r_{Ai} + d_{ij})$. Put j into S.

3. If j = G, then stop, otherwise go to 2.

The number of comparisons needed to determine the next node to be closed at iteration step $k+1$ (i.e., k nodes are closed and thus in S) is k (comparing $k+1$ nodes from S). Thus, this type of algorithm is rather efficient because in an n-node network no more than $n(n-1)/2$ comparisons are needed; that is, the algorithm is polynomial.

3.5 Computer Implementation

There are many software packages available to solve constraint-optimization problems. For smaller problems, spreadsheet programs are sufficient. A brief discussion of how to solve the formalized problem in Section 3.1, C (i.e., b-$Ax \geq 0$) with the spreadsheet program EXCEL is as follows:

1. Enter the tableau T_0 as shown below but leave row A empty.

2. Define the formulas. In this case:
   - drag from A2 to A7, and then hit the '=' key
   - drag from B2 to B7, then hit the '*' key, then click on cell B1, then hit the '+' key
   - drag from C2 to C7, then hit the '*' key, then click on cell C1, then hit the '+' key
   - drag from D2 to D7, then hit the '*' key, then click on cell D1
   - if you work with Windows
        hit CONTROL+SHIFT+ENTER,
     else
   - if you work with Macintosh
        hit COMMAND+ENTER

   - now that the formula is defined, the column A gets filled in automatically and you see the tableau below. Should you make a mistake when defining the formulas, then repeat the whole process.

	A	B	C	D
1		0	0	1
2	32	-4	-1	32
3	24	4	6	24
4	23	-1	-3	23
5	10	2	-1	10
6	48	-4	3	48
7	0	1	2	0

3. Define the decision variables, goal function, and constraints

   3.1 from the menu, select Formula/Solver

   3.2 put the cursor in the space for 'Set Cell' and click on cell A7 (this defines that the optimal value is shown in cell A7. Note that in our example we had it in D7)

   3.3 define the goal function under 'Equal to:' as 'Max'

   3.4 put the cursor in the space for 'By Changing Cells:' and drag from B1 to C1 (this defines the decision variables)

   3.5 put the cursor in the space for 'Subject to the Constraints:' and click on 'Add'

   3.6 put the cursor in the space for 'Cell Reference:' drag from A2 to A6

   3.7 put the cursor in space for 'Constraint' and type '0'

   3.8 choose the relation '≥' (to define that the slack variables must be positive). If you need to define a binary variable, define it as an integer variable that must be ≥ 0 and ≤+1.

   3.9 if there are additional constraints (e.g., if the decision variables had to be positive) go back to 3.5 to add new constraints

4. Define the model type and run the model

   4.1 Click on 'Options...' and then click on 'Assume Linear Model'

   4.1 Click on 'Solve'

5. Results

When the 'Solver' is done it reports whether or not it has found a solution. You can then view different reports: 'Answer,' Sensitivity,' and 'Limits.' They are all shown in different windows. The solution tableau looks like this:

	A	B	C	D
1		6.636	5.455	1
2	0	-4	-1	32
3	83.27	4	6	24
4	0	-1	-3	23
5	17.82	2	-1	10
6	37.82	-4	3	48
7	17.55	1	2	0

The solution value for the goal function (i.e., the maximum) is shown in cell A7 (as was defined) and the values for the decision variables are shown in B1 and C2, respectively (also as defined).

If there are problems with solving a model, it is best to delete the formulas and then to redefine them. Sometimes a problem has no feasible solution. However, one has to be very careful when defining the formula, entering the coefficients, and specifying the goal and constraints.

4. Summary and Further Readings

Constraint-optimization policy analysis consists of optimizing (max. or min.) a goal function subject to a set of constraints which are defined in terms of criteria and actions (decision variables). A policy consists of multiple actions which are employed with a certain intensity. Each of the n actions provides a decision variable and the optimal policy is thus an n-dimensional vector stating the intensities with which each action should be employed. The literature on constraint-based optimization is very vast. Every book on operations research discusses linear and non-linear programming and network problems, e.g., [Ecker and Kupferschmid, 1988], [Hillier and Lieberman, 1986].

A linear constraint-optimization decision problem assumes that the overall effectiveness of an action is the product of its unit-effectiveness and the intensity with which it is employed. In addition, the aggregation of the effectiveness across the actions is assumed to be additive. This means that the subjective value function, as defined by the goal function, has a constant tradeoff (marginal rate of substitution) between the actions (decision variables).

The goal function can refer to only one criterion or to multiple criteria. For each criterion, a finite aspiration level can be defined, and the goal is to minimize the weighted deviations from these aspiration levels. For each criterion that is used to define the goal function in terms of deviations from aspiration levels, two auxiliary decision variables are introduced - one for positive deviations and the other for negative deviations.

One algorithm used to solve linear constraint-optimization problems is the simplex algorithm. It consists of a sequential exchange of basic decision or slack variables for non-basic slack variables. The algorithm assures that the value of the goal function keeps improving until the optimum is reached. The constraints that determine the optimum solution are called binding constraints. A decision variable which can take on both positive and negative values is called a free variable.

Sensitivity analysis refers to the allowable increases and decreases of the slack variables and coefficients of the goal function. Shadow prices tell how much could be gained by relaxing the binding constraints. Constraint-optimization problems have a dual problem that leads to the same results.

When the decision variables are restricted to integer or binary values, the simplex algorithm leads to the optimal solution only if the decision variables turn in fact out to be integer or binary. In general, the rounded optimal real-valued solution might not be identical to the optimal integer-valued solution. However, most computer programs, such as spreadsheet programs, can handle any type of decision variables.

5. Problems

1. a) List the assumptions made for linear constraint-optimization models and discuss their meaning in terms of the subjective value function introduced in Chapter V.

   b) If you look at Figure X.2, what are the investments in the two projects to reach the minimum long-term profit, and what is its value?

2. Draw the structural models and insert the formal models for the problems 2.1, A, B, 2.2, A, B, 2.3, A, B, C, D.

3. A company produces two articles A and B. Production cost for one item of A is $1/6 and for B $1, production time for one item of A is 1/2 minute and for B 1 minute, production capacity for one item of A is 1 storage and for B also 1 storage, and the benefit of one item of A is $1 and of B $3. The company has available at most $35, 50 minutes, and 80 storages. The company wants to maximize the benefit. Define the problem (i.e., write down the formal model) and find the solution graphically. Compute the shadow prices with the Lagrange multiplier method and give an interpretation.

4. Use a spreadsheet program to solve the problem shown in Figure X.2, both as $Ex \leq b$ and $b-Ex \geq 0$. Read from the sensitivity report the shadow prices. Then, insert any other points in the cells for the basic variables, recompute, and look at the value of the goal function. Also, change the binding constraints by one unit and notice that the value function changes as indicated by the shadow prices.

5. Add the following constraints to the route planning problem (Section 2.2, A):

   • if the route goes from node 6 to node 8 then it must also go from 8 to 10
   • the route cannot go both from 6 to 8 and from 8 to 10
   • the route must go either from 6 to 8 or 8 to 10

6. Formalize and solve the zero-sum conflict situation (see 2.3,B):

	C_1	C_2	C_3
R_1	7	2	1
R_2	5	8	6
R_3	2	3	7

7. Formalize and solve (e.g., with EXCEL) the routing problem of Figure X.9.

8. Revise the routing problem of Problem 7 such that it becomes the well known **traveling salesperson** problem. The traveling salesperson starts his/her route in one particular node (e.g., A), goes to all other nodes exactly once, and returns to the origin node. The goal is to minimize the total travel distance or costs.

9. In Chapter VII, Section 3, we introduced the *AlpTransit* project, where Switzerland plans to build some of the five transportation alternatives (L/S, GOT, YPS, SP1, and SP2). Let's assume that only four criteria are considered: gains, energy use, landloss, and risks. Each of the five alternatives gives a certain short-term gain if it is constructed. The data are the ones of scenario H (Chapter VII, Section 3.1).

	gains	energy use ≤ 300	landloss ≤ 1200	risks ≤ 15
L/S	50	103	125	7
GOT	60	145	249	3
YPS	45	170	540	4
SP1	18	135	415	10
SP2	12	145	434	9

Additional requirements are that if GOT is constructed than YPS cannot be chosen, and if SP1 is chosen than YPS must also be chosen. The question is which of the five alternatives to build, such that the short-term gain is maximized. This type of problem is referred to as **capital budgeting**.

10. A community must choose a certain number of locations to place police stations from a given set of three potential locations (A, B, and C), from where police units are sent to four sections of the city (I, II, III, and IV). The selection of a location involves a certain amount of fixed building costs. Each of the four sections in the city has a certain demand for police units. The objective is to minimize the total costs while meeting the demand of the four sections. The figure below shows the demands of the four sections and the travel costs from the three locations to the four sections. Write down the formal model.

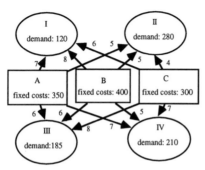

SOLUTIONS TO PROBLEMS

Chapter I

1. See Figure I.3. Important is that the goals become more concrete as we go down the hierarchy. 'Economically and environmentally sustainable agriculture' is very abstract, while 'Minimization of machinery area' is quite concrete (level 5 in the hierarchy). The lowest level in the goal hierarchy can be used to define potential and feasible alternatives. Therefore, goals are defined quite often like alternatives. For example, the reduction from three to two traffic lanes can be a goal or an alternative. It is a goal, if there are multiple ways to achieve this (e.g., by eliminating the right, middle, or left lane), otherwise it is an alternative (other alternatives would be no reduction or reduction to one lane). Quite often, decision makers perceive a problem (e.g., air pollution) and think of a possible alternative (reduce traffic volume). However, when they approach an analyst they tell him/her that the problem is how to go about to reduce the traffic volume (i.e., how to implement the alternative). If the analyst does not detect quickly what the real problem is, a lot of energy might get wasted working on the wrong problem. Also, decision makers refer often to scenarios (no control over the outcomes) but mean alternatives (control over the outcome); that is, they see alternatives *prior* to the choice process as events (things that can happen). For example, traffic reduction is an alternative, but the resulting long-term traffic volume is a scenario.

2. Explicit alternatives are defined 'one-by-one' in the problem description, as opposed to implicit alternatives (see Problem 3). Examples of explicit alternatives are several of landfill locations, different transportation modes, or alternative energy systems. Criteria can be costs, benefit, pollution, risks, etc. The goals can be defined in a hierarchy as in Problem 1.

3. Implicit alternatives are described by constraints, rather than 'one-by-one.' With the given constraints, potential or feasible explicit alternatives can be generated. Examples of binary implicit alternatives are: two out of five landfill locations, where there are ten explicit potential pairs of landfill locations to choose from; routes from city A to city B on a road network, where there are many potential routes between the two cities; presidential candidates, where for the U.S., the feasible explicit alternatives are all U.S. citizens who are older than 21 years and who were born in the U.S. Integer and real-valued alternatives are always described implicitly, such as the number of persons assigned to different projects (integer), or the hours assigned to different projects (real-valued).

4. An interval scale is characterized by the fact that it is invariant under positive linear transformations. This means that what we have measured in one unit does not loose

its meaning if we transform it to another unit by multiplying the original values with a positive constant and adding a constant. Thus, anything where the relative differences matter and the origin of the measurement is of no relevance refers to an interval scale. For example, the time to run a certain distance (e.g., 100 m) is measured on an interval scale. If one athlete is twice as fast as another measured in seconds on one day starting at 01:00 PM, then s/he is also twice as fast measured in minutes or hours on another day starting at 03:00 PM. The start time (zero point) does not matter and the proportions of time differences in seconds, minutes, or hours are the same.

5. a) a_5 (c_2: 3.2 > 3.0) and a_7 (c_5: 46 < 50) are not feasible.

   b) Of the six feasible alternatives, a_6 dominates all but a_8, and a_8 dominates all but a_4 and a_6. Thus the Pareto optimal set (set of non-dominated alternatives) is $\{a_6, a_8\}$.

   c) Both a_6 and a_8 are better than the others with respect to three criteria, making the choice quite subjective. However, we could use a model as discussed in Problem 6.

6. a) $n_{ij}=(e_{ij}-e_{i,\text{worst}})/(e_{i,\text{best}}-e_{i,\text{worst}})$. We could assign numerical values to the linguistic scale. For example 1 (worst), 2 (very low), 3 (low), 4 (med.), 5 (good), 6 (very good), 7 (best).

   b) With the transformation defined in 6.a) and added over all criteria, we get: a_6 (5.547)$\succ a_8$ (5.417)$\succ a_2$ (3.752)$\succ a_3$ (3.396)$\succ a_1$ (3.05)$\succ a_4$ (2.507)$\succ a_5$ (1.547)$\succ a_7$ (0.417).

   c) This linear model implies that relative changes for low evaluation values are perceived to be the same as relative changes of high evaluation values. In Chapters V and VII we will discuss normative models to construct other than just 'neutral' (linear) preference functions. The importance of criteria could be incorporated through the following formal relation: $n_j=\Sigma_i w_i n_{ij}$.

7. The graph looks like a spider net, where each alternative is visualized as a polygon. The larger the area covered by its polygon the more preferred the alternative.

8. None. If transitivity holds and the preferences of the alternatives are assessed in the sequence defined for $L=\{a_1\succ a_5,\ a_1\succ a_6,\ a_2\succ a_5,\ a_3\succ a_4,\ a_3\succ a_5,\ a_4\succ a_2,\ a_6\succ a_3\}$, then after assessing $a_4\succ a_2$, we can conclude $a_4\succ a_5$ and $a_3\succ a_2$. With the next assessment, $a_6\succ a_6$, all the remaining preference relations can be inferred transitively. The resulting preference order is: $a_1\succ a_6\succ a_3\succ a_4\succ a_2\succ a_5$.

9. Strong preference orders occur in situations where a decision maker is forced to prioritize alternatives, such as tasks that must be completed chronologically. Then, it is not possible to be indifferent between the priority of two tasks, one of the two must be done first. Weak preference orders occur when the decision maker is allowed to express indifferences. Indifference is different from indecision, where the decision maker cannot make up his/her mind. Pseudo orders occur when subjective preferences must be assessed for alternatives that are very similar to each other. Then, the indifference relation could be intransitive.

10. See Chapter III, Section 1.2.

Chapter II

1. a) See Section 4.1.

   b) The practical assessment of the model could be done by discussing it with colleagues, people with experience in organizing meetings, etc.

   c) The structural model of laws consists of the subjects, judges, courts, police, etc., and their relations. The formal model of laws is comprised of the law books, with statements as 'if ...., then ....' The resolution model of laws refers to the law enforcement with police and court activities.

2. a) The implicit alternatives (roads from city A to city H) are represented by the road network; for the explicit potential alternatives see b).

   b) There are 18 explicit roads from city A to city H.

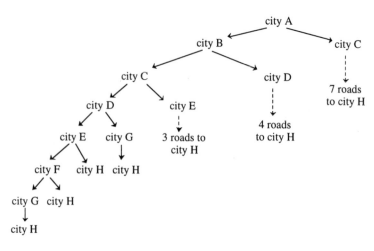

   c) The shortest route algorithm is discussed in Chapter X, Section 3.4.

3. See Chapter I, Figure I.2. The decomposition of criteria must be done from both a conceptual and an operational point of view. Conceptual decomposition means to create a very broad and deep hierarchy. However, too many evaluation criteria might not be feasible from an operational and practical point of view. For example splitting up safety to account for all possible types of accidents might not be practical because it would involve the collection of a large amount of data.

4. In I.a), the number of road accidents depends only on the traffic density, while in I.b) it also depends on the time of day. In other words, if in I.a) the traffic density is known, then the number of road accidents is also known, while in I.b) the number of road accidents is still uncertain because it depends on the time of day. In II.a), the benefit of project Type B depends on the chosen project and on the benefit of project Type A, while in II.b), it also depends on which project of Type A has been chosen.

5. -

6. -

7-10.

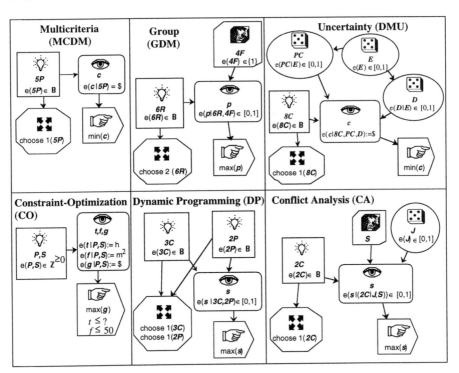

Multi-Criteria Decision Making (MCDM). The elements of decision making used to structure this problem are the costs (c, e.g., in $), the five projects (*5P*, whose decision variables take on binary values), the content goal (min(c)), and the structural goal (choose 1 out of the five projects). For each of the five projects one alternative element could have been defined. However, because they all would have had the same successor elements they can be grouped into one decision maker element. Multicriteria decision making will be addressed in Chapters III, IV, V, and VII.

Group Decision Making (GDM). The elements of decision making are the four friends (*4F*, whose logistic variables are of value 1, meaning that all friends take part in the assessments) grouped into one decision maker node, the six restaurants (*6R*) grouped into one alternative node, the preference (*p*) as the sole criterion which, for example, could take on values between 0 and 1, the content goal (max(*p*)), and the structural goal (choose 2 of out of the six restaurants). It should be noted that the *4F* element does not point into the two goal nodes. This means that the four friends do not have individual goals and that their assessments must first be aggregated to a group assessment with which the content goal is defined. The problem description does not specify whether the four friends must choose two different restaurants or whether they could go for lunch and dinner to the same place. The specification of this would be part of the structural goal. Formal and resolution models for group decision making will be addressed in Chapter IX.

Decision Making Under Uncertainty (DMU)
There are three scenario elements: project costs (*PC*), economy (*E*), and demand (*D*). In the formal model, the economy will be evaluated in terms of a marginal probability distribution, while project costs and demand will be evaluated as conditional probability distributions (conditioned on the economy). The criterion element refers to the costs (*c*) with which the eight cars (*8C*, binary decision variables) are evaluated for different outcomes of the production costs and demand. The content goal is to minimize the costs and the structural goal is to choose 1 car. Formal and resolution models for decision making under uncertainty will be addressed in Chapters VI, VII, and VIII.

Constraint-Optimization (CO)
There are two integer-valued decision variables, the number of pants (*P*) and the amount of shirts (*S*) that should be produced. Because both are evaluated for all three criteria, they can be grouped into one element. The three criteria are the time (*t*) (e.g., measured in hours), the amount of fabric (*f*, obviously measured in m^2), and the gain (*g*, e.g., measured in $). The content goals are: (1) to maximize the gain, (2) to use at most 50 m^2 fabric, and (3) to keep the total work time below a not yet specified value. It should be noted that there are no structural goals. A possible structural goal could be to produce a certain proportion of pants and shirts. Constraint-optimization problems will be addressed in Chapters V and X.

The decision elements are the three candidates ($3C$), the two projects ($2P$), the satisfaction (s), the content goal (max(c)), and the structural goal (choose 1 candidate and 1 project). For clarity reasons, the three candidates and the two projects are defined as separate alternative elements. However, because they both have the same successors, they could also be grouped into one element. The satisfaction could take on values between 0 and 1; the definition of the satisfaction function is part of the formal model. Formal and resolution models for dynamic decision problems will be addressed in Chapter VIII.

Conflict Analysis in a Game Setting (CA)

Because there are two independent (i.e., non-communicating) decision makers, Sue and Joe (S,J), we can model the problem as two independent decision problems, one for Sue and the other for Joe. For example, from Sue's point of view, Joe's choice for one of the two classes is uncertain. That is, Joe's decision options are scenarios for Sue's decision problem, where Sue has two decision variables (the two classes, $2C$). The structural goal is to choose one class, and the content goal is to maximize Sue's satisfaction. The formal and resolution models for these types of conflicts will be addressed in Chapter IX.

Chapter III

1. Use the following four approaches to determine the weights: (1) ordinal approach (Section 1.1); (2) scale from 0 to 100 (Section 1.1); (3) distributing a fixed value (Section 1.1); and (4) ratio scale approach according to the analytic hierarchy process (Section 1.2).

2. A is not reciprocal-symmetric, B is consistent, C is not ordinal transitive, and D is ordinal transitive. The weights, CR's, and λ's are: A: $[0.73, 0.17, 0.10]^T$, $CR_A=2.5\%$, $\lambda_A=3.03$; B: $[0.60, 0.30, 0.10]^T$, $CR_B=0\%$, $\lambda_B=3.00$; C: $[0.61, 0.24, 0.15]^T$, $CR_C=39\%$, $\lambda_C=3.45$; and D: $[0.65, 0.23, 0.12]^T$, $CR_D=0.3\%$, $\lambda_D=3.00$.

3. For each comparison of two elements a_i and a_j, we would not only ask which is more preferred but also the intensity of this preference. Thus, if the preference for a_1 is three times the one of a_2, then we get $\pi(a_1,a_2)=3$, and $\pi(a_2,a_1)=1/3$. If we also assess, for example, $\pi(a_2,a_3)=4$, then we can conclude that $\pi(a_1,a_3)=12$. In case of a contradiction of a transitively inferred preference, we would have to reverse all assessments as discussed for the ordinal approach.

4. With four elements, we have four triples to check ordinal transitivity:

$$1/2/3: \begin{bmatrix} 2 & 5 \\ - & 4 \end{bmatrix}, \ 1/2/4: \begin{bmatrix} 2 & 4 \\ - & 6 \end{bmatrix}, \ 1/3/4: \begin{bmatrix} 5 & 4 \\ - & 3 \end{bmatrix}, \text{ and } 2/3/4: \begin{bmatrix} 4 & 6 \\ - & 3 \end{bmatrix}.$$

Because 1/3/4 is not ordinal transitive we conclude that the matrix K is not ordinal transitive. If we change the value 4 in the upper right corner of K to a value $x \geq 6$ then the matrix is ordinal transitive.

5. In Section 1.3 we computed for the inconsistent matrix K in example of Figure III.1: $w_1=0.61$, $w_2=0.24$, and $w_3=0.15$, where w_1 is the sum of the elements in the first row of the matrix K^5 divided by the sum of all elements of K^5; w_2 is the sum of the elements in the second row of the matrix K^5 divided by the sum of all elements of K^5; and w_3 is the sum of the elements in the third row of the matrix K^5 divided by the sum of all elements of K^5.

The eigenvalue, λ_{max}, which belongs to the eigenvector $w=[w_1,w_2,w_3]^T = [0.61,0.24,0.15]^T$, can be computed using the relation: $Kw=\lambda_{max}w$. This means we have three equations which all give the same λ_{max}. From the first equation we get: $\lambda_{max}=(1\times0.61+5\times0.24+2\times0.15)/0.61=3.45$. This can be verified by computing λ_{max} with the second and third equation.

6. See Figure III.2 and corresponding text. When new criteria are added, determine which are evaluation criteria and which are not. Then assess the relative weights and compute the total weight as discussed in Section 2.2.

7. Because all alternatives lie on a straight line, only one dimension is needed to describe the differences among these alternatives. Thus, the result of a principal component analysis would be to use only one dimension (first principal component). The other two dimensions can be chosen arbitrarily but orthogonal. The total amount of sample variance explained by this new dimension is 100%.

8. The interpretation of the principal components is done by looking at the coefficients of the eigenvectors. The coefficients of the first eigenvector are almost the same which made the authors conclude that the first principal component affects all five stocks in about the same way. The first three coefficients of the second eigenvector are positive, while the fourth and fifth are negative. This made the authors conclude that the second component represents a contrast between the first three stocks (which are all from the chemical industry) and the last two stocks (which are both from the oil industry).

9. Use $w_1 = \hat{e}_1$ to compute $S_2 = S_1 - \lambda_1 \hat{e}_1 \hat{e}_1^T$. To compute numerically the largest eigenvalue and its corresponding eigenvector of S_2, we would compute $(S_2)^r$, where $r \rightarrow \infty$, and derive the components of the eigenvector and the corresponding eigenvalue as discussed in Problem 5.

10. See text in Section 3.1.

Chapter IV

1. Look at the example in Section 2.4, A (relative assessment), the example in Section 2.4, B (absolute assessment), and Section 3 for problems that could occur.

2. See Figure IV.2 and corresponding text. The problem with an additive dominance function is that we do assume that the decision maker's preference for one criterion is independent of the other criteria. If this were not the case, we could not aggregate the criterion-specific preferences with a linear model but we would also have to consider joint preferences of two or more criteria. The normative theories in Chapters V and VII treat this issue in more depth.

3. The value $k_{jk|i}$ is the dominance of alternative a_j over alternative a_k for criterion c_i. If $k_{jk|i}+k_{kj|i}=1$, then we have a linear and continuous dominance relation between the two alternatives between 0 and 1. This holds, for example, for the type C dominance function (Table IV.1). If $k_{jk|i}+k_{kj|i}=0$, then we have a linear function through the 0-point from $-f_i$ to $+f_i$. This means that if the dominance of alternative a_j over alternative a_k for criterion c_i is $k_{jk|i}$, then the dominance of alternative a_k over alternative a_j for criterion c_i is $k_{kj|i}=-k_{jk|i}$.

4. -

5. The resulting preference order is: $(a_1,a_6) \succ (a_4,a_5) \succ a_3 \succ a_2 \succ a_7$. Comparing this preference order to the preference order in Section 3.2, based on the k_j values, we see that a_1 improved a lot because it does not have as many tied values as other alternatives do. On the other hand, a_3 dropped in rank because its good values are tied many times by other alternatives.

6. Let $x_j \in \{0,1\}$ be the decision variable for a_j; then, $x_2+x_3-x_1 \leq 1$.

7. The formal model of type A gives for both the n and k values the following results: a_4 (1.000)$\succ a_5$ (0.075)$\succ a_1$ (-0.175)$\succ a_2$ (-0.425)$\succ a_3$ (-0.475). For type B model we get: a_4 (0.700) a_5 (0.125)$\succ a_2$ (-0.125)$\succ$ a_1(-0.225)$\succ a_3$ (-0.475). As we can see from Figure III.6, a_4 is the best alternative (even the dominating alternative), followed by a_5. It seems also that a_3 is the worst alternative. However, the preference order between a_1 or a_2 is not quite clear. While type A model favors a_2 over a_1, type B does the reversed.

8. -

9. For the first criterion with three alternatives we get the matrix below. Because the assessment is consistent, the priorities n_{ij} can be computed as the sum of the

---

elements in row j divided by the sum of all elements in the matrix. The same computation is done for the other three criteria and for the case with four alternatives.

c_1	a_1	a_2	a_3	n_{1j}
a_1	1	1/9	1/8	1/18
a_2	9	1	9/8	9/18
a_3	8	8/9	1	8/18

10. Descriptive assessments generally do not rely on axioms, mainly because they deal with dominances and not with preferences (such as the normative approaches in Chapters V and VII). An exception is the AHP which uses reciprocal-symmetry as an axiom (see Chapter III, Section 1.2).

The selection of a dominance function (such as those in Table IV.1 or the choice for a scale, ratio or interval) is not primarily based on the decision maker's preference structure but refers more to the characteristics of the evaluation values.

Preference aggregation is done with an additive weighted function. As discussed in Problem 2, the choice for an additive model is not based on the preference structure of the decision maker for the alternatives, and might thus cause questionable results.

Some approaches rank all alternatives from best to worst (PROMETHEE and AHP, see examples in Sections 2.2 and 2.4), while others allow for alternatives to be not comparable (the ELECTRE methods, see example Section 2.3).

Chapter V

1. First of all it should be noted that the subjective attitude of a decision maker refers to a range of outcomes per criterion and not to the criterion as such. For example, the subjective attitude towards a loss from $5.00 to $10.00 is different than from $0.00 to $10,000. For the interpretation of the value functions and the corresponding subjective attitudes see Figure V.9.

2. If one could conclude that an alternative is twice as good as another because its value is twice as large, we would deal with a ratio scale. However, value functions reflect primarily ordinal measures. The assessment of the values with the mid-value-splitting technique uses an interval approach. Thus, we might conclude that if the difference between two outcomes is twice as large as the difference between two other outcomes, then the relative value-differences are also of factor two. However, the most important aspect of a value function is that it preserves the preference order. Two value functions are called strategically equivalent if they can be transformed into each other by a strictly increasing transformation (see Section 2.1).

3. See explanation in Section 2.3.

4. See Figure V.4 and corresponding text.

5. Look at the example in Section 3.2. For six criteria we would need $2^6-2=62$ assessments. However, with the theorem mentioned, only 6-1=5 assessments are necessary. This means that we must show that $\{c_1,c_2\}$, $\{c_2,c_3\}$, $\{c_3,c_4\}$, $\{c_4,c_5\}$, and $\{c_5,c_6\}$ are preferentially independent of their complements. With the theorem mentioned we can then conclude that all criteria are mutually preferentially independent. It should be noted that to assess preferential independence of, let's say, $\{c_1,c_2\}$ from its complement $\{c_3,c_4,c_5,c_6\}$, we need to do several checks such as mentioned in Section 2.2, until we are convinced that preferential independence holds for the whole range of outcomes. A detailed discussion of the assessment of preferential independence (and of utility independence, see Chapter VII) can be found in [Keeney, 1977].

6. -

7. To show that c_1 is preferentially independent of c_2, we need to fix c_2 (for example at 6) and see that the preference order does not change if we fix c_2 at another level (for example at 7 or 8). Thus, because all values grow from top to bottom for all the three rows, we see that c_1 is preferentially independent of c_2. To show that c_2 is preferentially of c_1 we now fix c_1. For our numerical example, we can conclude this because all values in the rows grow from left to right. To see that the Thompsen condition does not hold, look at Figure V.6 and set $x=y=6$, $q=s=7$, and $r=t=8$.

8. -

9. -

10. The goal is to maximize the value function $v(x_1,x_2)=(x_1x_2)^{0.5}$. The constraint is that there is only \$200 to spend. Thus, the constraint function is: $100x_1+25x_2=200$. With the Lagrange multiplier method we get: $\mathcal{L}=(x_1x_2)^{0.5}+\lambda(200-100x_1-25x_2)$. Setting the partial derivatives equal to zero, we get:

- $\partial \mathcal{L}/\partial x_1= 0.5(x_2/x_1)^{0.5}-100\lambda=0$;
- $\partial \mathcal{L}/\partial x_2= 0.5(x_1/x_2)^{0.5}-25\lambda=0$;
- $\partial \mathcal{L}/\partial \lambda= 200-100x_1-25x_2=0$, which is the constraint equation.

Modifying and multiplying the first with the second equation gives $\lambda=0.01$; from this we get: $x_1=1$ and $x_2=4$, and $v(1,4)=2$. An interpretation of λ can be made if we derive λ analytically from these equations. We then get in general: $\lambda=(\partial v/\partial x_i)/k_i$ for all i, where $\mathcal{L}=v(x_1,x_2)+\lambda(K-k_1x_1-k_2x_2)$. Because λ is a constant, we can conclude that at the maximization point (1,4), no matter in which type of land (x_i) we invest one dollar, we always get the same value back. If this were not the case, one item would give more marginal value per dollar and the investment would not be optimally allocated. For our example, where $\lambda=0.01$, we see that an extra investment of \$100 yields an extra utility of 1.0. This means, for example, that if we invest an additional \$20, we would expect an increase of the value from 2.0 to 2.2. This is indeed the case. To see this, we repeat the computations above but replace the costs of \$200 by \$220, and get: $x_1^*=1.1$, $x_2^*=4.4$, and in fact $v(1.1,4.4)=2.2$.

Chapter VI

1. Wald's MaxMin rule refers to a conservative decision maker who wants to avoid low outcomes. Thus, his/her preference function is convex (see Chapter V, Figure V.9). Savage's rule is for a decision maker who wants to minimize possible regrets. Although it is not a MaxMax rule which would say to choose the alternative with the highest possible outcome, it applies to a progressive decision maker (i.e., concave preference function). Hurwicz's rule varies between Walds' rule and the MaxMax rule, depending on the factor α. Laplace's rule treats all outcomes as equally important which reflects the attitude of a neutral decision maker (linear preference function).

2. S/he could use a different rule (e.g., Savage, Hurwicz, or Laplace) to break ties between the two (or more) optimal alternatives.

3. Decreasing ordinal value functions are, for example, costs. Wald's MaxMin rule becomes then a MaxMin rule. The best alternative with Savage's rule is still the one with smallest regret but the regret is now the difference between outcome (e.g., costs of \$6) and missed outcome (e.g., costs of only \$2), which are costs of \$4. Hurwicz's rule says to minimize a weighted average between best and worst possible outcome, while Laplace's rule says to minimize the average outcome.

4. For example, we could choose the alternative which has the highest weighted sum between the MaxMin value and the max regret value:

$$\alpha(\max[\min(u_j)])+(1-\alpha)r_j\ ,\ \alpha\in[0,1].$$

If $\alpha=1$, we have Wald's rule and if $\alpha=0$ we have Savage's rule.

5. The ELECTRE I method (Chapter IV, Section 2.3) is based on a pairwise comparison approach of the alternatives for all criteria. The results of the comparisons are aggregated across all criteria after multiplying the criterion-specific values with the weights of the criteria. The same comparison of the alternative could be done for all scenarios. Then, we would aggregate across all scenarios by computing the expected value (i.e., instead of weights we have probabilities). This corresponds to Laplace's approach.

6. A technical component has a relatively high chance of failing at the beginning of its use because there might be something wrong with it. After the component has functioned properly for a while, the chance of failure decreases. As the component ages, the chance of failure increases again. Thus, the *pdf* of such a technical component starts at some high value, goes down monotonically to a low value, stays for a long time at this value, and then increases as it gets old. Because of its shape, such a *pdf* is also called a bathtub *pdf*.

7. If we discretize the function $E[x] = \int x\, dF(x)$ to $E[x] = \Sigma_i\, x_i\, (F(x_i) - F(x_{i-1}))$, we see that $x_i\, dF(x_i)$ is a rectangle under the survival function $S(x)$; see Figure VI.9. If we add up all these rectangles we get the total area under the survival function $S(x)$. This means that the expected value is the area under the survival function. This has some major consequences if we base our judgments on the expected value. For example, for the Dutch (Figure VI.9) and the Swiss (Figure VI.12) safety regulations, a technological system is considered safe if its survival function lies completely under the threshold line. However, it is possible to have a survival function complying with this requirement, while at the same time it has a larger expected value than the one which does not satisfy the safety goals. Or, even worse, we can make a technological system comply to the safety goals by lowering its function below the safety goal, and at the same time, increasing its expected damage.

8. The sum of all accidents is 100. Thus, the joint probabilities for accidents are the entries in the table divided by 100. The marginal probabilities are the sums in the rows and columns, respectively, divided by 100. The conditional probabilities $p(s|h)$ are computed as follows:

- $p(a|d)$ $= 10/(10+20) = 1/3$
- $p(m|d)$ $= 20/(10+20) = 2/3$
- $p(a|w)$ $= 15/(15+55) = 3/14$
- $p(m|w)$ $= 55/(15+55) = 11/14$

The relation $h \rightarrow s$, which reflects $p(s|h)$, is reversed to $s \rightarrow h$, which reflects $p(h|s)$, with Bayes' theorem as follows (for example for $p(d|a)$):

$$p(d|a) = \frac{p(a|d)p(d)}{p(a|d)p(d) + p(a|w)p(w)} .$$

- $p(d|a)$= (1/3)×(3/10)/[(1/3)×(3/10)+(3/14)×(7/10)]   = 2/5   = 10/(10+15)
- $p(w|a)$= (3/14)×(7/10)/[(3/14)×(7/10)+(1/3)×(3/10)]   = 3/5   = 15/(10+15)
- $p(d|m)$= (2/3)×(3/10)/[(2/3)×(3/10)+(11/14)×(7/10)]   = 4/15   = 20/(20+55)
- $p(w|m)$= (11/14)×(7/10)/[(11/14)×(7/10)+(2/3)×(3/10)]   = 11/15 = 55/(20+55)

The season and the humidity are not probabilistically independent because $p(s,h) \neq p(s) \times p(h)$.

9. For example, the extension of Yager's 'OR' operator is the one for n_{tot} used for the Swiss safety regulations (see at the end of Section 2.4).

10. Look at the Dutch safety regulation approach summarized in Figure VI.13, and replace N with n_{tot}.

Chapter VII

1. Take for example any lottery, $[x,p,y]$, and compare it to its monetary expected value, $[px+(1-p)y,1,-]$. If you prefer the lottery over the certainty equivalent then you are risk prone (i.e., you prefer the gamble), if you prefer the certainty equivalent over the lottery then you are risk averse, and if you are indifferent between the two, then you are risk neutral.

    For example, let's assume that one prefers [$100,0.5,$0] over [$50,1.0,-], then s/he is risk prone. But if s/he prefers [$50,1.0,-] over [$1000,0.5,-$900], then s/he is risk averse.

2. Both lotteries have the same expected monetary value but also the same expected utility value. Thus, if you are rational (i.e., you comply with the axioms of expected utility theory), then you should be indifferent between the two lotteries.

3. The decision maker is irrational; s/he should by definition be indifferent between the two lotteries because they have the same expected utility.

4. By computing the expected utility values we get from the first comparison $u(\$60)=0.8$, from the second comparison $u(\$30)=0.3$, and from the third comparison $u(\$50)=3/5 \times 0.8 + 2/5 \times 0.3 = 3/5$. The decision maker is risk averse.

5. See also Figure VII.6 for the interpretation. Case 1: risk averse; case 2: risk prone; and case 3: risk averse for low values and risk prone for high values.

6. Because mutual utility independence holds, we know that the multiplicative model is appropriate. If, in addition, the component scaling constants add up to 1, then the additive model is appropriate (first decision maker). To see this, we insert the scaling constants of the first decision maker ($k_1=0.4$, $k_2=0.35$, $k_3=0.25$) into the equation $1+k=(1+kk_1)(1+kk_2)(1+kk_3)$, (see 2.6, B):

$$1+k=(1+k0.4)(1+k0.35)(1+k0.25).$$ If we solve for k we get $k=0$.

If the three component scaling constants do not add up to 1.0, then the multiplicative model is not appropriate and k must be computed from a polynomial of third order (see 2.6, B): $1+k=(1+kk_1)(1+kk_2)(1+kk_3)$. For the second decision maker ($k_1=0.3$, $k_2=0.25$, $k_3=0.20$; $\Sigma k_i=0.75$), we get $k=1.23$, and for the third decision maker ($k_1=0.6$, $k_2=0.5$, $k_3=0.4$; $\Sigma k_i=1.5$), $k=-0.772$. Verify these results by inserting them into the equation for k.

7. The idea is to alter the conditional utility functions shown in Section 3.4 such that they reflect the risk attitudes (risk averse, prone, or neutral) for the different actors; assume the additive model holds and see the scaling constants as a sort of weights.

8. Value theory is a special case of utility theory if we see an outcome under certainty as a degenerated lottery.

9. -

10. If the host does not open any door, your chance to pick the right door is 1/3. Some people argue that after the host opens a door (which of course is not the door with the car), your chance to win the car would be 50-50 between the two remaining doors.

   However, the game should be seen as having two strategies: 'stay' (to stay with your initial choice) or 'switch' (to take the door which is <u>not</u> opened by the host). Let's assume that for both strategies you don't pay any attention to the host opening the doors (assuming s/he does not open the door with the car).

   The 'stay' strategy has a 1/3 chance to win the car. With the 'switch' strategy, you win the car if in your initial choice you missed it (which has a chance of 2/3); therefore, the strategy to adopt is the 'switch' strategy which has a 2/3 chance to win the car.

Chapter VIII

1. Traditional influence diagrams are also called conceptual models. They visualize the essential elements and their mutual influences. There are no real rules how to build them. A '+' or a '-' sign next to an oriented arrow could indicate positive or negative influences. Causal models are also referred to as covariance structure models or structural equation models, and visualized in form of path diagrams. They depict causal relations which are often formalized with linear relations. However, care must be taken because correlations do not necessarily imply causal relations (see discussion in Section 1.2). Probabilistic influence diagrams depict probabilistic dependencies among chances and decision nodes. A probabilistic influence diagram with no decision nodes is also called a relevance diagram.

2. Take the simple example of a probabilistic influence diagram with a node A (having two states: a_1 and a_2) and a node B (also with two states: b_1 and b_2). Assume there is an arrow from A to B. Then, we would evaluate the states of node A with a marginal probability distribution; e.g., $p(a_1)=0.2$ and $p(a_2)=0.8$. The evaluation of node B would be done with a conditional probability distribution; e.g., $p(b_1|a_1)=0.3$; $p(b_2|a_1)=0.7$; and $p(b_1|a_2)=0.6$; $p(b_2|a_2)=0.4$. With these probabilities we can compute the joint probability distribution of the two partitions A and B as: $p(a_i,b_j)=p(b_j|a_i)p(a_j)$. With the joint probability distribution we can compute the conditional probability distributions $p(a_i|b_j)$ and the marginal probability distribution $p(a_i)$. To see how this is done, see Problem 5. A cycle in the diagram would occur if we would define an arrow from B to A. However, we cannot do that because this part of the model is already formulated and thus also defined.

3. Diagram A: regular; diagram B: by reversing arc between nodes 4 and 1, node 4 becomes barren and can be omitted, node 3 does not contribute to the decision problem; diagram C: regular; diagram D: decision node 3 is barren and can be omitted, optimal decision is conditioned on node 4; diagram E: cycle defined by nodes 1, 3, and 4; diagram F: all decisions are optimal.

4. The number of evaluations are for diagram A: 8 (for the marginal values and the decision node)+2^4(for the utilities)=24; diagram B: $2+2+2+4+2^2=14$; diagram C: $2+2+2+4+2^3=18$; diagram D: $2+2+2+2+2^3=16$; diagram F: $2+2+4+4+2^1=14$.

5. With two partitions, each with two states, we have four scenarios. If the four probabilities in the table add up to 1, then they stand for the scenario probabilities $p(a_i,b_j)$, see table I (below). From a joint distribution we can compute all the marginal probabilities $p(a_i)$ and $p(b_j)$, and the conditional probabilities $p(a_i|b_j)$ and $p(b_j|a_i)$. With the values in table I we get: $p(a_1)=(0.1+0.3)=0.4$, $p(a_2)=0.6$,

$p(b_1)$=0.5, $p(b_2)$=0.5, $p(a_1|b_1)$=0.1/(0.1+0.4)=0.2, $p(a_2|b_1)$=0.8, $p(a_1|b_2)$=0.6, $p(a_2|b_2)$=0.4, $p(b_1|a_1)$=0.25, $p(b_2|a_1)$=0.75, $p(b_1|a_2)$=2/3, and $p(b_2|a_2)$=1/3. As we see from these probabilities, the states of the two partitions are not probabilistically independent. If they were, then we would have $p(a_i,b_j)=p(a_i|b_j)p(b_j)$, which is not the case (for $i=j$=1: 0.1≠0.4×0.5=0.2).

I	a_1	a_2
b_1	0.1	0.4
b_2	0.3	0.2

II	a_1	a_2
b_1	0.2	0.8
b_2	0.3	0.7

III	a_1	a_2
b_1	0.4	0.1
b_2	0.6	0.9

If the elements in each of the two rows add up to 1 (table II), then we have the conditional probabilities $p(a_i|b_j)$. In this case we can choose the marginal probabilities $p(b_j)$. With these marginal and conditional probabilities we can compute the joint probabilities $p(a_i,b_j)=p(a_i|b_j)p(b_j)$. From the joint distribution we can compute any other probabilities as we did above. If the elements of each column add up to 1 (table III), then we have the conditional probabilities $p(b_j|a_i)$. Now we can choose the marginal probabilities $p(a_i)$. From these we can compute the joint probabilities and then any other probabilities.

6. The following abbreviations are used: EUP (European traffic policy), NP (National policy), AP (Air pollution), and SL (Speed limit). An example of how to read the table below is as follows: If EU traffic policy is *relaxed* then choose National policy *engine*; with National policy being *engine* and Air pollution being *high*, then choose Speed limit *110*.

• Decision Tree:
The optimal policy is the one given in table in the text of Section 1.2. It is described in terms of six decisions, indicated by D1,..., D6 (see table and decision tree below). D1 says to choose SL=110, given AP=high and NP=LPG, because u(SL=110)=20 > u(SL=90)=10.

The same reasoning leads to D2, D3, and D4. D5 says to choose NP=LPG, given EUP=stringent, because E[NP=LPG]=0.2×20+0.8×50=44 > E[NP=engine]= 0.1×80+0.9×30=35. The same reasoning leads to D6. The expected utility when adopting this policy is: E[u]=0.3×44+0.7×47.5=46.45.

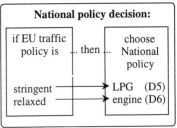

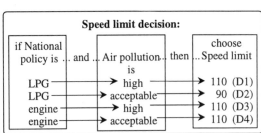

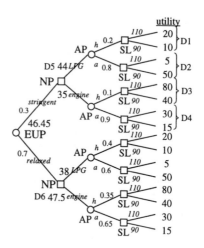

• Node elimination algorithm:

Compare the node elimination steps with the computations for the decision tree.

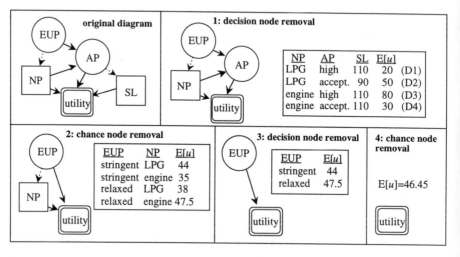

7. Probabilistic sensitivity can be assessed for marginal chance nodes. These are for diagram A: 1, 3, 4 (value of information is gain); B: 1 (after arc reversal and removal of barren node 4, value of information is gain); C: 3, 4 (value of information is gain); D: 1 (value of information is gain), 4 (value of information is loss).

8. -

9. Admissible transformations to a probabilistic influence diagram do not affect its characteristics; thus, after the transformations we can still call the diagram a probabilistic influence diagram. However, the causal meaning might get lost. The

formal model gets more compact as nodes are eliminated. The solution of the transformed probabilistic diagram is still the same as the one of the original diagram.

10. A possible influence diagram is given below.

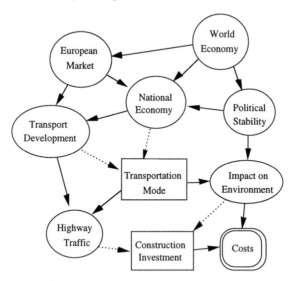

Assuming each node has two outcomes (states or decision options), the numbers of assessments are (the names are abbreviated by their initials): WE (2), EM (4), PS(4), NE (16), TD (8), TM (2), IE (8), HT (8), CI (2), C (4). The chance node WE is formalized with a marginal probability distribution; all other chance nodes with conditional probability distributions. The costs could be transferred into utility values.

There are two decision nodes TM and CI, and we get the following types of solutions, assuming that NE is either strong or weak, TD either high or low, TM either road or rail, IE either heavy or light, and CI either high or low:

if	NE	...and...	TD	...then...	TM
	strong		high		road
	strong		low		road
	weak		high		rail
	weak		low		road

if	IE	...then...	CI
	heavy		high
	light		low

A sequence for solving the diagram is the following: 1: elimination of HT; 2: elimination of CI; 3: elimination of IE; 4: arc reversal between PS and NE; 5: elimination of PS; 6: arc reversal between WE and EM; 7: arc reversal between WE and NE; 8: elimination of WE; 9: arc reversal between EM and NE; 10: arc reversal between EM and TD; 11: elimination of EM; 12: elimination of TM; 13: elimination of TD; and 14: elimination of NE; 15. Sensitivity analysis could be done by computing the probabilistic sensitivity for WE.

Chapter IX

1. Clinton and Dole were very close in first place, while Perrot was quite far behind in the polls. Of the five voters, three favor Clinton over Dole over Perrot, and two favor Dole over Clinton over Perrot. The three favoring Clinton rank Clinton first, Dole second and Perrot third. Those favoring Dole know from the polls that their candidate is slightly behind and thus could pretend to favor Perrot over Clinton (although in fact they might not). This gives the following rankings (table on the left), where Dole wins over Clinton. If the votes for the third candidate (Perrot) are canceled, and the rankings readjusted for the first two candidates, we see that Clinton wins over Dole. Thus, a fairer procedure would be to have the five voters repeat their vote for the best two candidates which are Clinton and Dole.

	V_1	V_2	V_3	V_4	V_5	tot
C	1	1	1	3	3	9
D	2	2	2	1	1	8
P	3	3	3	2	2	13

	V_1	V_2	V_3	V_4	V_5	tot
C	1	1	1	2	2	7
D	2	2	2	1	1	8

2. a) The techniques we discussed to aggregate preferences across criteria are the descriptive approaches based on paired comparisons and the normative approaches based on preference functions. The paired comparison technique as employed in the analytic hierarchy process (Chapters III and IV) was also used to aggregate preferences across decision makers (Chapter IX, Section 2.2, C). The normative techniques to aggregate preferences across criteria as discussed in Chapters V and VII were also used to aggregate preferences across decision makers (Chapter IX, Section 2.2, B).

  Borda's count, the majority rule (Section 2.1, A), and Kendall's rank correlation method (Section 2.1, B) could also be used to aggregate preferences across criteria. The fuzzy aggregation principle (Section 2.3) was used for aggregating preferences both across criteria and decision makers.

  b) The concepts of concordance and discordance are based on the idea that each element has both strength (concordance) and weakness (discordance) when compared to other elements. The assessment of the strength and weakness is done with paired comparisons using an ordinal and an interval scale (outranking methods), or a ratio scale analytic hierarchy process (Chapter IV). However, any other set of preference orders, for example determined with the utility theory approach (Chapter VII), could be transferred into an ordinal scale and analyzed for rank correlation with Kendall's approach.

3. The probability distribution of S can be determined by computing S for all possible combinations of two rankings. For decision maker DM_1 we choose an arbitrary rank, while for decision maker DM_2 we determine all 24 different ranks.

	decision maker 1				decision maker 2				$S = P - Q$		
	a_1	a_2	a_3	a_4	a_1	a_2	a_3	a_4	S	P	Q
1	1	2	3	4	1	2	3	4	6	6	0
2					1	2	4	3	4	5	1
3					1	3	2	4	4	5	1
4					1	4	2	3	2	4	2
5					1	3	4	2	2	4	2
6					1	4	3	2	0	3	3
7					2	1	3	4	4	5	1
8					2	1	4	3	2	4	2
9					3	1	2	4	2	4	2
10					4	1	2	3	0	3	3
11					3	1	4	2	0	3	3
12					4	1	3	2	-2	2	4
13					2	3	1	4	2	4	2
14					2	4	1	3	0	3	3
15					3	2	1	4	0	3	3
16					4	2	1	3	-2	2	4
17					3	4	1	2	-2	2	4
18					4	3	1	2	-4	1	5
19					2	3	4	1	0	3	3
20					2	4	3	1	-2	2	4
21					3	2	4	1	-2	4	2
22					4	2	3	1	-4	1	5
23					3	4	2	1	-4	1	5
24					4	3	2	1	-6	0	6

It should be noted that the extreme values in the table above are $\pm n(n-1)/2$, which for $n=4$ are ±6. From this table we can derive the histogram for S.

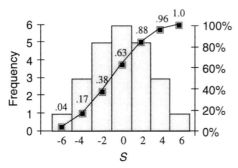

We see that the values for S cluster around 0. Thus, if the ranks of the two decision makers are independent (i.e., random), we would expect a small $|S|$. If $|S|$ is large (i.e., if S lies in the tails), we would reject the hypothesis of independent ranks, meaning that the ranks are either almost the same or almost reversed.

Let's assume that the two decision makers assessed the third set of ranks in the table above, which gives $S=4$. Then, from the histogram we can determine the probability for a more extreme S value: $p(|S|{\geq}4)=8/24=1/3$. Usually, for S to be extreme, the probability of getting such a value should be at most 0.05. However, with $p=1/3$, we would not reject the hypothesis of independent ranks. If the decision makers had assessed the first set of ranks, where $S=6$, we would get $p(|S|{\geq}6)=1/12=0.08$, and we would still not reject the hypothesis of independent ranks.

If a positive correlation between the two ranks can be expected (i.e., we know that the ranks will not be reversed, such that $S{\geq}0$), we would compute the probability for the one-sided interval; that is: $p(S{\geq}4)=4/24=1/6=0.17$, which is still rather large. For $S=6$, we would get $p(S{\geq}6)=1/24=0.04$, which would lead us to reject the hypothesis of independent (i.e., random) assessments, and we would conclude that the two decision makers agree on the preference order.

4. For the first kind of comparison we get: $a_1 \succ a_2$ and then $a_3 \succ a_1$, which means that a_3 is the best alternative. For the second kind of comparison we get: $a_2 \succ a_3$ and then $a_1 \succ a_2$, which means that a_1 is the best alternative. Arrow [1951] addresses this problem and finds that the majority rule works also sequentially if the decision makers' preferences are single-peaked. From the figure below we see that this is not the case for our example.

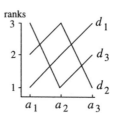

5. We know that all two-actor zero-sum conflict situations have at least a mixed strategy equilibrium. A pure strategy equilibrium does not exist if the MinMax equilibrium test fails. That is, there is no element which is the minimum value in its row and the maximum value in its column. We can easily construct such a situation from the conflict situation given in Section 3.1, B, by changing the value of the equilibrium pair R_2/C_2 from 4 to, say, 8. In Problem 6 (Chapter X) you will be asked to compute the mixed-strategy solution of this conflict situation.

	C_1	C_2	C_3
R_1	7	2	1
R_2	5	4	6
R_3	2	3	7

$\rightarrow$

	C_1	C_2	C_3
R_1	7	2	1
R_2	5	8	6
R_3	2	3	7

6. The expected utilities for the two non-cooperative decision makers as a function of p (the probability for Row to choose R_1) and q (the probability for Col to choose R_1) are:

- $E[u_{Row}] = pqu_{R_{11}} + (1-p)qu_{R_{21}} + p(1-q)u_{R_{12}} + (1-p)(1-q)u_{R_{22}}$
- $E[u_{Col}] = pqu_{C_{11}} + (1-p)qu_{C_{21}} + p(1-q)u_{C_{12}} + (1-p)(1-q)u_{C_{22}}$

A **dynamic plot** consists of the lines ranging from $p=0|q=k\times0.05$ to $p=1|q=k\times0.05$, for $k=0,1,...,20$; and the lines from $q=0|p=k\times0.05$ to $q=1|p=k\times0.05$, for $k=0,1,...,20$. The figure below shows the dynamic plot for the prisoner dilemma conflict situation discussed in Section 3.2, A.

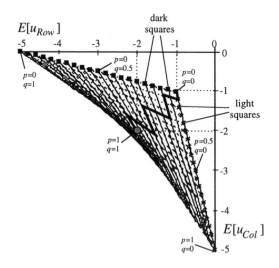

The dark squares indicate points for which $p=0$, at an increment of $q=0.05$, while the light squares indicate points for which $q=0$, at an increment of $p=0.05$. The non-Pareto Nash equilibrium is at (-2,-2) and the best compromise for the two decision makers is at (-1,-1). This dynamic plot representation could be used for **virtual negotiations** because it shows for any point how *Row* (by changing p) and *Col* (by changing q) can influence the utilities for the two non-cooperative decision makers.

Starting in (-2,-2), where $p=q=1$, *Row* could change p to 0.8, indicating that s/he is willing to give up some utility (from -2 to -2.6). In response, *Col* could accept the virtual deal and also give up some utility.

The polygon in the figure above shows a possible virtual negotiation path that leads to the point (-1,-1).

A meaningful interpretation of the probabilities p and q is percentages for the two options; for example, if the two options are two different roads A and B, then p could stand for the percentage of vehicles assigned to road A by one company (*Row*) and 1-p to B, and q the percentage of vehicles assigned to road A by another company (*Col*) 1-q to B. For more on this issue see [Beroggi, 1997 b].

7. If we compute the MaxMin solutions for both actors we get:

- *Row*:   MaxMin$[2p-(1-p),-p+(1-p)] \to p=2/5$, $E[u_R]=1/5$.
- *Col*:   MaxMin$[q-(1-q),-q+2(1-q)] \to q=3/5$, $E[u_C]=1/5$.

The probabilities $p=2/5$ and $q=3/5$ form a non-Pareto Nash equilibrium. This can be seen from the figure below, which shows that the lines of move through the point $(0.2,0.2)$ are for both decision makers orthogonal to their utility axis; that is, neither of the two can improve his/her position unilaterally.

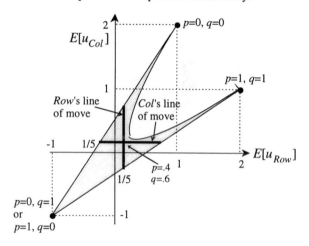

8. a) R_2 is a dominating strategy for *Row*. Because *Col* knows that *Row* chooses R_2, *Col* will choose C_1 to minimize the loss.

   b) There is no dominating pure strategy. The Nash equilibrium is $p=3/4$, $q=3/8$, $E[u_R]=-E[u_C]=4.25$. If *Row* has reasons to assume that $q>3/8$ then s/he chooses R_2, and if s/he can assume that $q<3/8$ then s/he chooses R_1. If *Col* has reasons to assume that $p>3/4$ then s/he chooses C_1, and if s/he can assume that $p<3/4$ then s/he chooses C_2.

9. a) The Nash point lies on the intersection of the 45^0 slope and the efficient frontier. From the graph we see that $E[u_R]=E[u_C]=3/2$.

   b) The security level has been computed as $E[u_R]=E[u_C]=1/5$ (Problem 7). Thus, we are looking for: max $(u_R^*-1/5)(u_C^*-1/5)$, and $u_C^*=-u_R^*+3$ (negotiation set), which gives: max $(u_R^*-1/5)(-u_R^*+14/5)=-u_R^{*2}+3u_R^*-14/25$. Setting the derivative equal to zero gives: $3=2u_R^*$, and thus $u_R^*=3/2$. From $u_C^*=-u_R^*+3$, we get $u_C^*=3/2$.

10. a) All sets of axioms express more or less the same idea of a 'fair' aggregation method for multiple decision makers. The most relevant axioms discussed in these sections are: transitivity, Pareto optimality (the aggregated assessment of

unanimous votes must be this vote), binary relevance (to avoid rank reversal), non dictatorship, universal domain (all preferences are possible), and recognition (all actors are considered).

b) In an ad-hoc crisis management center for an off-shore oil spill, where several decision makers are involved (e.g., coast guard, port authority, cargo owner, local fishermen, ecologists), different preference aggregation axioms would be employed. For example, the non-dictatorship axiom might be relaxed because a crisis management organization is based on hierarchical structures where one decision can override another. For the same reasons, the recognition axiom might also not be considered. Transitivity might not hold because irrational aspects are crucial in a dynamic decision situation under stress.

Chapter X

1. a) The assumptions for linear constraint-optimization problems are proportionality within actions and additivity across actions. Proportionality means that if the effectiveness of $x_j=1$ (employing alternative a_j with intensity 1) with respect to criterion c_i is e_{ij} (unit effectiveness), then employing a_j with intensity $x_j \in \mathbf{R}$ is $x_j e_{ij}$. Additivity means that the total effectiveness of employing a_j with intensity $x_j \in \mathbf{R}$ and a_k with intensity $x_k \in \mathbf{R}$ is $x_j e_{ij} + x_k e_{ij}$. The goal function is what we called in Chapter V the value function. Linearity means that we have a linear value function and thus a constant marginal rate of substitution.

b) To determine the minimum value of the goal function we move the goal line in the negative direction to the extreme point which lies at the intersection of the two criteria c_2 and c_5 ($x_1=6$, $x_2=-8$); the long-term profit is -10. This means that 6 units are invested in the social program and 8 units are taken away from the transportation program which results in a maximum long-term loss of 10 units.

2. Below is the structural model of the problem in 2.1, B with the auxiliary decision variables. It should be noted that there is no structural goal.

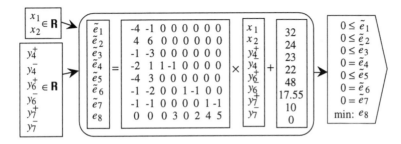

The routing model, defined in 2.2, A, is shown below.

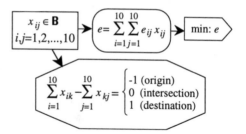

The structural model of the problem in 2.2, B is given below for the problem in 2.1,A (Figure X.2), assuming that only 4 out of the 5 constraints must hold.

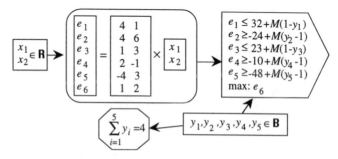

You should verify (e.g., compute with a spreadsheet program) that the optimal solution is $x_1=3.67$, $x_2=17.33$, $y_1=1$, $y_2=0$, $y_3=0$, $y_4=1$, $y_5=0$, $e_6=38.33$; the binding constraints are therefore c_1 and c_4, and the optimal solution lies at the intersection of $\tilde{e}_1=0$ and $\tilde{e}_4=0$ (see Figure X.2, Chapter X).

The structural models of the problems in 2.3,A and 2.3,D can be derived straight forward from the tableaus in the text.

The structural model of the non-cooperative conflict situation in 2.3,B is given in Figure IX.2. Note, however, that the formal model is different (see text in 2.3,B and in Chapter IX, Section 1.3).

The structural model of the nurse scheduling problem (2.3,C) is given below. It consists of only an alternative and a structural goal node; there are no criteria at all.

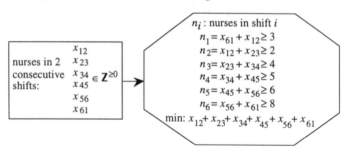

3. The two decision variables are the number of articles to produce of type A (x_1) and B (x_2). There are four criteria: production time, production capacity, cost, and benefit. The formalization of the problem is:

• time:	$50 \geq$	$1/2x_1 + x_2$
• capacity:	$80 \geq$	$x_1 + x_2$
• cost:	$35 \geq$	$1/6x_1 + x_2$
• benefit:	max:	$x_1 + 3x_2$

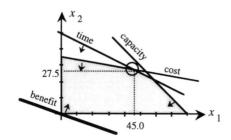

The optimal solution lies at the intersection of the cost and time constraints which is the point (45,27.5). That is, we have as solution $x_1=45$ and $x_2=27.5$ (assuming that half an article can be produced and sold), which gives a benefit of 127.5 (if the decision variables are restricted to integer values, the optimal solution is $x_1=46$ and $x_2=27$ (which is not the rounded solution!) which gives a benefit of 127; shadow prices for integer decision variables are not meaningful. With the Lagrange multiplier method we compute the shadow prices:

$$\mathcal{L}=x_1+3x_2 + \lambda_1[50-1/2x_1-x_2]+\lambda_2[35-1/6x_1-x_2].$$

We compute the partial derivatives and set them equal to zero:

- $\partial\!/\, x_1=1-1/2\lambda_1-1/6\lambda_2=0$
- $\partial\!/\, x_2=3-\lambda_1-\lambda_2=0$
- $\partial\!/\, \lambda_1=50-1/2x_1-x_2=0$
- $\partial\!/\, \lambda_2=35-1/6x_1-x_2=0$

From the third and fourth equations we confirm the result from above ($x_1=45$, $x_2=27.5$). From the first two equations we can compute $\lambda_1=\lambda_2=1.5$. This means that if we relax the time or the cost constraint by 1 unit then we gain 1.5 units on the benefit. For example, if we set the cost constraint to 51 instead of 50, the benefit will be 129 instead of 127.5. The corresponding solution would then be $x_1=48$ and $x_2=27$.

4. -

5. The formalization of these logic statements is discussed in Table IV.2, Chapter IV. From there we get, with $x_{i,j} \in \{0,1\}$:

- if 6 to 8 then 8 to 10: $x_{6,8} - x_{8,10} \leq 0$
- at most one of the two: $x_{6,8} + x_{8,10} \leq 1$
- either 6 to 8 or 8 to 10: $x_{6,8} + x_{8,10} = 1$

6. Let (p_1, p_2, p_3) be the probability distribution for *Row* (i.e., *Row* chooses R_1 with probability p_1, R_2 with probability p_2, and R_3 with probability p_3), and (q_1, q_2, q_3) the probability distribution for *Col* (for C_1, C_2, and C_3). Then, *Row* gets the security level u^* (>0) independently of *Col*'s choice (where $p_i/u^* = x_i$ and thus $\Sigma_i x_i = 1/u^*$):

- $E[u_{R|C_1}] = 7p_1 + 5p_2 + 2p_3 \geq u^* \quad \rightarrow = 7x_1 + 5x_2 + 2x_3 \geq 1$
- $E[u_{R|C_2}] = 2p_1 + 8p_2 + 3p_3 \geq u^* \quad \rightarrow = 2x_1 + 8x_2 + 3x_3 \geq 1$
- $E[u_{R|C_3}] = 1p_1 + 6p_2 + 7p_3 \geq u^* \quad \rightarrow = 1x_1 + 6x_2 + 7x_3 \geq 1$
- the goal function is: min: $x_1 + x_2 + x_3$

Col wants to minimize the security level u^* (>0) which s/he gets independently of *Row*'s choice (where $q_i/u^* = y_i$ and thus $\Sigma_i y_i = 1/u^*$.):

- $E[u_{C|R_1}] = 7q_1 + 2q_2 + 1q_3 \leq u^* \quad \rightarrow = 7y_1 + 2y_2 + 1y_3 \leq 1$
- $E[u_{C|R_2}] = 5q_1 + 8q_2 + 6q_3 \leq u^* \quad \rightarrow = 5y_1 + 8y_2 + 6y_3 \leq 1$
- $E[u_{C|R_3}] = 2q_1 + 3q_2 + 7q_3 \leq u^* \quad \rightarrow = 2y_1 + 3y_2 + 7y_3 \leq 1$
- the goal function is: max: $y_1 + y_2 + y_3$

If we solve these formal models with a spreadsheet program, we get for the security level $u^* = 5.29$, $p_1 = 0.14$, $p_2 = 0.86$, $p_3 = 0$, and $q_1 = 0.71$, $q_2 = 0$, $q_3 = 0.29$, $E[u_R] = E[u_C] = 5.29$.

7. The decision variables are the links connecting two nodes; for example, $x_{AD} = 1$ if the road goes through nodes A and D. The links in the network of Figure X.9 do not have an orientation. This means that we would have to define twice as many decision variables as if there were only single oriented links. To reduce the number of decision variables we make some meaningful assumptions about the orientations of the links. For example, we assume that only the link between nodes C and D can be traveled in both directions.

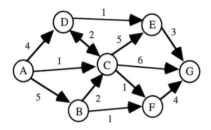

The formal model is discussed in Section 2.2, A. With d_{ij} being the distance between the nodes i and j, we have the following formalization in tabular form:

	x_{AB}	x_{AC}	x_{AD}	x_{BC}	x_{BF}	x_{CD}	x_{CE}	x_{CF}	x_{CG}	x_{DC}	x_{DE}	x_{EG}	x_{FG}
A: -1=	-1	-1	-1	0	0	0	0	0	0	0	0	0	0
B: 0=	1	0	0	-1	-1	0	0	0	0	0	0	0	0
C: 0=	0	1	0	1	0	-1	-1	-1	-1	1	0	0	0
D: 0=	0	0	1	0	0	1	0	0	0	-1	-1	0	0
E: 0=	0	0	0	0	0	0	1	0	0	0	1	-1	0
F: 0=	0	0	0	0	1	0	0	1	0	0	0	0	-1
G: 1=	0	0	0	0	0	0	0	0	1	0	0	1	1
min:	5	1	4	2	1	2	5	1	6	2	1	3	4

The decision variables must be defined as binary variables. Binary decision variables x_j can be defined in terms of integer variables ($\mathbf{Z}$) as follows (e.g., EXCEL does not know binary variables): $x_j \in \mathbf{Z}$, $x_j \geq 0$, $x_j \leq 1$. The solution is in fact what we determined for Figure X.9:

d_{tot}	x_{AB}	x_{AC}	x_{AD}	x_{BC}	x_{BF}	x_{CD}	x_{CE}	x_{CF}	x_{CG}	x_{DC}	x_{DE}	x_{EG}	x_{FG}
6	0	1	0	0	0	0	0	1	0	0	0	0	1

8. Let d_{ij} be the distance between node i and node j and x_{ij} the binary decision variable for which $x_{ij}=1$ if the link between i and j is part of the route and $x_{ij}=0$ otherwise. The goal function remains the same, that is, to minimize the total distance of the route which connects all n nodes:

$$\min: \sum_{j=1}^{n}\sum_{i=1}^{n} d_{ij}x_{ij}.$$

The flow conservation constraint must now say that all nodes are entered and left exactly once. This can be done as follows:

$$\sum_{i=1}^{n} x_{ij} = 1 \text{ (enter all nodes once)}, \quad \sum_{j=1}^{n} x_{ij} = 1 \text{ (leave all nodes once)}, \quad i,j=1,...,n.$$

These constraints, however, could lead to several subtours; all we require is that each node is entered and left exactly once which could lead to the following solution:

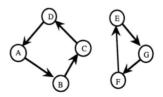

To assure that there will be no subtours, we have to introduce additional constraints which, however, are quite technical (for more on this issue see [Wagner, 1969, Chapter 13].

9. Let x_j be the binary decision variable for which $x_j=1$ if alternative a_j is chosen and $x_j=0$ otherwise. Then, the formalization and the resulting solution (i.e., to build GOT and L/S at a gain of 110) are given in the table below.

	L/S $x_1=1$	GOT $x_2=1$	YPS $x_3=0$	SP1 $x_4=0$	SP2 $x_5=0$	resulting impacts
energy use: 300 $\geq$	103	145	170	135	145	248
landloss: 1200 $\geq$	125	249	540	415	434	374
risks: 15 $\geq$	7	3	4	10	9	10
at most one of GOT or YPS: 1 $\geq$	0	1	1	0	0	
if SP1 then also YPS: 0 $\geq$	0	0	-1	1	0	
gain: max	50	60	45	18	12	**110**

10. Let x_j be the binary decision variable for which $x_j=1$ if location a_j is chosen and $x_j=0$ otherwise. The number of police units dispatched from location a_j to section s_i is y_{ji} (≥ 0, integer) and the costs to ship one unit from location a_j to section s_i is c_{ji}. The fixed costs for building the police station in location a_j is K_j.

The goal is to minimize the total travel costs plus the fixed costs of the chosen locations:

$$\min: \sum_{j=1}^{3}\sum_{i=1}^{4} c_{ji} y_{ji} + \sum_{j=1}^{3} K_j x_j .$$

The constraints are that the demands at section s_i, d_i, be met:

$$\sum_{j=1}^{3} y_{ji} = d_i, \; i=1,2,3,4.$$

Finally, we must assure that police units are dispatched only from chosen locations; that is from locations for which $x_j=1$. This can be done by requiring that the total of what gets dispatched, $\Sigma_i \, y_{ji}$, is smaller than $x_j M$, where M is a very large number. If $x_j=1$, the constraint is satisfied for any amount shipped from location a_j, and if $x_j=0$, then nothing can be shipped. However, the maximum that could be dispatched from any location is $\Sigma_i \, d_i$, which is the case if only one location is chosen. Thus, we can write the additional constraint, which assures that units are dispatched only from chosen locations, as:

$$\sum_{i=1}^{4} y_{ji} \le x_j \sum_{i=1}^{4} d_i, \; j=1,2,3.$$

The numerical example with two sets of decision variables x_j and y_{ji} gives then the following formulation and solution in tabular form:

	A	B	C	AI	AII	AIII	AIV	BI	BII	BIII	BIV	CI	CII	CIII	CIV
	x_A =0	x_B =1	x_C =1	x_{AI} =0	x_{AII} =0	x_{AIII} =0	x_{AIV} =0	x_{AI} =0	x_{AII} =0	x_{BIII} =185	x_{BIV} =210	x_{CI} =120	x_{CII} =280	x_{CIII} =0	x_{CIV} =0
120=	0	0	0	1	0	0	0	1	0	0	0	1	0	0	0
280=	0	0	0	0	1	0	0	0	1	0	0	0	1	0	0
185=	0	0	0	0	0	1	0	0	0	1	0	0	0	1	0
210=	0	0	0	0	0	0	1	0	0	0	1	0	0	0	1
$0\le$	795	0	0	-1	-1	-1	-1	0	0	0	0	0	0	0	0
$0\le$	0	795	0	0	0	0	0	-1	-1	-1	-1	0	0	0	0
$0\le$	0	0	795	0	0	0	0	0	0	0	0	-1	-1	-1	-1
min:	350	400	300	7	5	6	7	8	5	6	5	6	4	8	7

The best decision is to locate police stations in B and C. From B we would then dispatch units only to the districts III (185) and IV (210), while from C we would dispatch units only to the districts I (120) and II (280). The corresponding minimum costs are 4,700.

It should be noted that instead of requesting that units be dispatched only from chosen locations, we could have required that units be received only from chosen locations. This could have been formalized with the constraint $\Sigma_j \, x_j y_{ji} = d_i$ ($i=1,2,3,4$). However, this is a non-linear constraint and we therefore better stick with the linear constraint which assures that units be dispatched only from chosen locations.

REFERENCES

Allais M., 1953. "Le comportement de l'homme rationel devant le risque: critique des postulats et axioms de l' écôle Américane." *Econometrica*, 21, 509-546.

Arrow K.J., 1951. *Social Choice and Individual Values*. First Edition, Second Edition 1963. Wiley, New York.

Avison D. E., Golder P. A. and Shah H. U., 1992. "Towards an SSM Toolkit: Rich Picture Diagramming." *European Journal of Information Systems*, 1/6, 397-407.

Axelrod, 1984. *The Evolution of Cooperation*. Basic Books, New York.

Bagozzi R.B., 1980. *Causal Models in Marketing*. Wiley, New York.

Bazerman M., 1998. *Judgment in Managerial Decision Making*. Wiley, New York.

Beach L.R., 1997. *The Psychology of Decision Making: People in Organizations*. Foundation of Organization Science Series, Sage Publications, London.

Belton V. and Gear T., 1983. "On a Shortcoming of Saaty's Method of Analytic Hierarchies." *Omega*, 11/3, 228-230.

Beroggi G.E.G., 1997 a. "Visual Interactive Strong Preference Ordering Approach for Rapid Policy Decision Analysis." Technical Report, School of Systems Engineering, Policy Analysis, and Management, Delft University of Technology, Netherlands.

Beroggi G.E.G., 1997 b. "Dynamic Plots in Virtual Negotiations." Technical Report, School of Systems Engineering, Policy Analysis, and Management, Delft University of Technology, Netherlands.

Beroggi G.E.G. and Aebi M., 1996. "Model Formulation Support in Risk Management." *Safety Science*, 24/2, 121-142.

Bodily S.E., 1985. *Modern Decision Making: A Guide to Modeling with Decision Support Systems*. McGraw-Hill, New York.

Brans J.P. and Mareschal B., 1991. The PROMCALC and GAIA Decision Support System for Multicriteria Decision Aid. Free University Brussels, CSOOTW/254.

Brans J.P. and Vincke Ph., 1985. "A Preference Ranking Organization Method." *Management Science*, 31/6, 647-656.

Brodley R.F., 1982. "The Combination of Forecasts: A Bayesian Approach." *Journal of the Operational Research Society*, 33, 171-174.

Bronstein I.N. and Semendjajew K.A., 1981. *Taschenbuch der Mathematik*. Verlag Harri Deutsch, Frankfurt/Main.

BUWAL, 1991. Handbuch I zur Störfallverordnung. Bundesamt für Umwelt, Wald und Landschaft, Eidg. Drucksachen- und Materialzentrale, 3000 Bern, Switzerland.

Charnes A. and Cooper W.W., 1977. "Goal Programming and Multiobjective Optimizations." *European Journal of Operational Research*, 1, 39-45.

Chatfield C., 1995. *Problem Solving: A Statistician's Guide*. Chapman & Hall, New York.

Checkland P.B., 1988. "Soft Systems Methodology: An Overview." *Journal of Applied Systems Analysis*, 15.

Clemen R.T., 1996. *Making Hard Decisions: An Introduction to Decision Analysis*. Second Edition, Duxbury Press, Wadsworth Publishing Company, Belmont, CA.

Dalkey N.C., 1970. The Delphi Method: An Experimental Study of Group Opinion. Technical Report, RM-5888-PR, Rand Corporation, Santa Monica, CA.

Dantzig G.B., 1955. "On the Shortest Route Through a Network." The Mathematical Association of America, Studies in Graph Theory, Part I, 89-93.

Dombi J., 1982. "A General Class of Fuzzy Operations, the DeMorgan Class of Fuzzy Operators, and Fuzzy Measures Induced by Fuzzy Operators." *Fuzzy Sets and Systems*, 8, 149-163.

Dubois D. and Prade H., 1980. *Fuzzy Sets and Systems: Theory and Applications*, Academic Press, New York.

Dyer J.S., 1990. "Remarks on the Analytic Hierarchy Process." *Management Science*, 36/3, 249-258.

Dyer J.S. and Wendell R.E., 1985. "A Critique of the Analytic Hierarchy Process." Working Paper 84/85-4-24, Department of Management, The University of Texas at Austin.

Ecker J.G. and Kupferschmid M., 1988. *Introduction to Operations Research*. Wiley, New York.

Eden, C., 1988. "Cognitive Mapping." *European Journal of Operational Research*, 36, 1-13.

Edwards W., 1992. *Utility Theory: Measurements and Applications*, Kluwer Academic Publishers, Boston.

Evans M., Hastings N., and Peacock B., 1993. *Statistical Distributions*. Wiley, New York.

Farquhar P.H., 1984. "Utility Assessment Methods." *Management Science*, 30/11, 1283-1300.

Fischhoff B, Lichtenstein S., Slovic P., Derby L., and Keeney R.L., 1981. *Acceptable Risk*. Cambridge University Press.

Frank M.J., 1979. "On the Simultaneous Associativity of $F(x,y)$ and $x+y-F(x,y)$." *Aequat. Mathematical*, 19, 194-226.

French S., 1988. *Decision Theory: An Introduction to the Mathematics of Rationality*. Ellis Horwood Series in Mathematics and its Applications. Ellis Horwood, New York.

Glantz S.A. and Slinker B.K., 1990. *Primer of Applied Regression and Analysis of Variance*. McGraw-Hill, New York.

Golden B.L., Wasil E.A., and Harker P.T., 1989. *The Analytic Hierarchy Process: Applications and Studies*. Springer Verlag, New York.

Gorman W,M., 1968. "The Structure of Utility Functions." *Review of Economic Studies*, 35, 367-390.

Gower J.C. and Hand D.J., 1996. *Biplots*. Monographs on Statistics and Applied Probability, 54, Chapman & Hall, London.

Gustafson D.H., Shukla R.K., Delbecq A., and Walster G.W., 1973. "A Comparative Study of Differences in Subjective Likelihood Estimates Made by Individuals, Interacting Groups, Delphi Groups, and Nominal Groups." *Organizational Behavior and Human Performance*, 9, 200-291.

Hamacher H., 1978. "Über logische Verknüpfungen unscharfer Aussagen und deren zugehörige Bewertungsfunktionen." In Trappl R., Klir G.J., and Ricciardi L. (eds.), *Progress in Cybernetics and Systems Research*, Vol. 3, Hemishpere, Washington, D.C.

Harker P.T., 1989. "The Art and Science of Decision Making: The Analytic Hierarchy Process." In Golden B.L., Wasil E.A., and Harker P.T. (eds.), *The Analytic Hierarchy Process: Applications and Studies*. Springer Verlag, New York.

Harker P.T. and Vargas L.G., 1987. "The Theory of Ratio Scale Estimation: Saaty's Analytic Hierarchy Process." *Management Science*, 33/11, 1383-1403.

Harsanyi J.C., 1955. "Cardinal Welfare, Individualistic Ethics, and Interpersonal Comparisons of Utility." *Journal of Political Economy*, 63, 309-321.

Heineman R.A., Bluhm W.T., Peterson S.A., and Kearny E.N., 1997. *The World of the Policy Analyst: Rationality, Values, and Politics*. Second Edition, Chatham House Publishers, Chatham, New Jersey.

Hillier F.S. and Lieberman G.J., 1986. *Introduction to Operations Research*. Holden-Day, Oakland, CA.

Howard R.A., 1992. "In Praise of the Old Time Religion." In Edwards W. (ed.), *Utility Theory: Measurements and Applications*, Kluwer Academic Publishers, Boston.

Howard R. A., 1989. "Knowledge Maps." *Management Science*, 35/8, 903-922.

Hwang C.-L. and Lin M.-J., 1987. *Group Decision Making Under Multiple Criteria.* Lecture Notes in Economics and Mathematical Systems. Springer-Verlag, New York.

Johnson R.A. and Wichern D.W., 1998. *Applied Multivariate Statistical Analysis.* Fourth Edition, Prentice Hall, Englewood Cliffs, New Jersey.

Kahan J.P. and Rapoport A., 1984. *Theories of Coalition Formation.* Lawrence Erlbaum Associates, Publishers, Hillsdale, New Jersey.

Kahneman D., Slovic P., and Tversky A., 1982. *Decision Making Under Uncertainty: Models and Choices.* Prentice-Hall, Inc., Englewood Cliffs, New Jersey.

Kahneman D. and Tversky A., 1979. "Prospect Theory: An Analysis of Decision Under Risk." *Econometrica*, 47, 263-291.

Keeney R.L. and Raiffa H., 1993. *Decisions with Multiple Objectives: Preferences and Value Tradeoffs.* Cambridge University Press, Cambridge, UK.

Keeney R.L., 1977. "The Art of Assessing Multiattribute Utility Functions." *Organizational Behavior and Human Performance*, 19, 267-310.

Kendall M. and Gibbons J.D., 1990. *Rank Correlation Methods.* Oxford University Press, New York.

Kirchgraber U., Marti J., and Hoidn H.P., 1981. *Lineare Algebra.* Verlag der Fachvereine, Zürich, Switzerland.

Kirkwood C.W. and Sarin R.K., 1980. "Preference Conditions for Multiattribute Value Functions." *Operations Research*, 28, 225-232.

Lootsma F.A., 1992. "The REMBRANDT System for Multi-Criteria Decision Analysis via Pairwise Comparisons or Direct Rating." Report 92-05, Faculty of Technical Mathematics and Informatics, Delft University of Technology, Delft, Netherlands.

Luce R.D. and H. Raiffa, 1985. *Games and Decisions: Introduction and Critical Survey.* Dover Publications, Inc., New York.

MCord M.R. and de Neufville R., 1983. "Empirical Demonstration that Expected Utility Decision Analysis is Not Operational." In Stigum P.B. and Wenstøp F. (eds.), *Foundations of Utility and Risk Theory with Applications.* Reidel, Dordrecht (NL), 181-199.

Miller G.A., 1956. "The Magical Number Seven Plus or Minus Two: Some Limits on Our Capability for Processing Information." *The Psychological Review*, 63, 81-97.

Morgan M.G. and Henrion M., 1990. *Uncertainty - A Guide to Dealing with Uncertainty in Quantitative Risk and Policy Analysis*. Cambridge University Press, Cambridge.

Morris P.A., 1983. "An Axiomatic Approach to Expert Resolution." *Management Science*, 29/1, 24-32.

Morrison D., 1976. *Multivariate Statistical Methods*. McGraw-Hill, New York.

Mumpower J.L., 1991. "The Judgment Policies of Negotiators and the Structure of Negotiation Problems." *Management Science*, 37/10, 1304-1324.

Nicholson W., 1995. *Microeconomic Theory: Basic Principles and Extensions*. The Dryden Press, London.

Oliver R.M. and Smith J.Q., 1990. *Influence Diagrams, Belief Nets, and Decision Analysis*. Wiley, New York.

Olson D.L., Fliedner G., and Currie K., 1995. "Comparison of the REMBRANDT System with the Analytic Hierarchy Process." *European Journal of Operational Research*, 82, 522-539.

Patton C.V. and Sawicki D.S., 1993. *Basic Methods of Policy Analysis and Planning*. Second Edition, Prentice-Hall, Englewood Cliffs, N.J.

Payne J.W., Bettman J.R., and Johnson E.J., 1993. *The Adaptive Decision Maker*. Cambridge University Press, New York.

Raiffa H., 1968. *Decision Analysis: Introductory Lectures on Choices Under Uncertainty*. McGraw-Hill, Inc., New York.

Ramanathan R, and Ganesh L.S., 1994. "Group Preference Aggregation Methods Employed in AHP: An Evaluation and an Intrinsic Process for Deriving Members' Weightages." *European Journal of Operational Research*, 79, 249-265.

Resnik M.D., 1987. *Choices: An Introduction to Decision Theory*. University of Minnesota Press, Minneapolis.

Rietveld P. and Ouwersloot H., 1992. "Ordinal Data in Multicriteria Decision Making, A Stochastic Dominance Approach to Siting Nuclear Power Plants." *European Journal of Operational Research*, 56, 249-262.

Roubens M., 1982. "Preference Relations on Actions and Criteria in Multicriteria Decision Making." *European Journal of Operational Research*, 10, 51-55.

Roy B. and Vincke Ph., 1981. "Multicriteria Analysis: Survey and New Directions." *European Journal of Operational Research*, 8, 207-218.

Roy B., 1978. "Electre III: Algorithme de classement basé sur une représentation floue des préférences en présence de critières multiples." *Cahiers du CERO*, 20/1, 3-24.

Roy B., 1974. "Critières multiples et modélisation des préférences: l'apport des relations de surclassement." *Revue d'Economie Politique*, 1.

Roy B. and Bertier P., 1973. "La méthod Electre II, une application au média-planning. In M. Ross (ed.), *OR 72*, North-Holland, 291-302.

Saaty T.L., 1989. "Group Decision Making and the AHP." In Golden B.L., Wasil E.A., and Harker P.T. (eds.), *The Analytic Hierarchy Process: Applications and Studies*, Springer-Verlag, New York, 59-67.

Saaty T.L., 1987. "Concepts, Theory, and Techniques: Rank Generation, Preservation, and Reversal in the Analytic Hierarchy Decision Process." *Decision Sciences*, 18, 157-177.

Saaty T.L., 1996. *The Analytic Network Process*. RWS Publications, Pittsburgh.

Saaty T.L., 1980. *The Analytic Hierarchy Process*. McGraw-Hill, New York.

Saaty T.L. and Vargas L.G., 1984. "Comparison of Eigenvalue, Logarithmic Least Square and Least Square Methods in Estimating Ratios." *Journal of Mathematical Modeling*, 5, 309-324.

Schachter, R. D., 1986. Evaluating Influence Diagrams. *Operations Research*, 34/6, 871-882.

Schweizer B. and Sklar A., 1961. "Associative Functions and Statistical Triangular Inequalities." *Publication Mathematical Debrecen*, 8, 169-186.

Senge P.M., 1990. *The Fifth Discipline: The Art and Practice of the Learning Organization*. Doubleday/Currency, New York.

Simon H.A., 1972. "Theories of Bounded Rationality." In McGuire C.B. and Radner R. (eds.), *Decision and Organization*. University of Minnesota Press, Minneapolis, 161-176.

Smidts A., 1997. "The Relationship Between Risk Attitudes and Strength of Preference: A Test of Intrinsic Risk Attitude." *Management Science, 43/3*, 357-370.

Sugden R. and Williams A., 1978. *The Principles of Practical Cost-Benefit Analysis*. Oxford University Press, New York.

Torgerson W.S., 1958. *The Theory and Measurement of Scaling*. Wiley, New York.

Van den Honert R.C. and Lootsma F.A., 1996. "Group Preference Aggregation in the Multiplicative AHP: The Model of the Group Decision Process and Pareto Optimality." *European Journal of Operational Research*, 96, 363-370.

van Huylenbroek G., 1995. "The Conflict Analysis Method: Bridging the Gap Between ELECTRE, PROMETHEE, and ORESTE." *European Journal of Operational Research*, 82, 490-502.

Vincke Ph., 1992. *Multicriteria Decision-Aid*. Wiley, New York.

Von Neumann J. and Morgenstern O., 1947. *Theory of Games and Economic Behavior*. Princeton University Press, Princeton (second edition, first edition 1944).

Von Nitzsch R. and Weber M., 1993. "The Effect of Attribute Ranges on Weights in Multiattribute Utility Measurements." *Management Science*, 39/8, 937-943.

Vos Savant M., 1996. *The Power of Logical Thinking*. St. Martins Press, New York.

VROM, 1989. Premises for Risk Management. Directorate General for Environmental Protection at the Ministry of Housing, Physical Planning and Environment, The Hague, Netherlands.

Walker W.E., 1988. "Generating and Screening Alternatives." Chapter 6 in Miser H.J. and Quade E.S. (eds.), *Handbook of Systems Analysis: Craft Issues and Procedural Choices*. North-Holland, New York.

Willemain T.R., 1995. "Model Formulation: What Experts Think About and When." *Operations Research*, 43/6, 916-932.

Wagner H.M., 1969. *Principles of Operations Research*. Prentice-Hall, Englewood Cliffs, NJ.

Weber E.U. and Miliman R.A., 1997. "Perceived Risk Attitudes: Relating Risk Perception to Risky Choice." *Management Science*, 43/2, 123-144.

Wolters W.T.M. and Mareschal B., 1995. "Novel Types of Sensitivity Analysis for Additive MCDM Methods." *European Journal of Operational Research*, 81, 281-290.

Yager R. and Filev D.P., 1994. *Essentials of Fuzzy Modeling and Control*. Wiley, New York.

Yager R.R., 1993. "Non-Numeric Multi-Criteria Multi-Person Decision Making." *Group Decision and Negotiation*, 2, 81-93.

Yager R.R., 1980. "On a General Class of Fuzzy Connectives." *Fuzzy Sets and Systems*, 4, 235-242.

Zadeh L.A., 1965. "Fuzzy Sets." *Information and Control*, 8, 338-353.

Symbol Index

Symbol	Definition
a_j	j-th alternative, j=1,...,n (action, decision option, policy, strategy, tactic) Note that in Chapters VI and VIII a_j stands for the j-th partition of set A
x_j	decision variable for a_j
c_i	i-th criterion (attribute) , $i = 1,...,m$. Note that in Chapter VI c_i stands for the i-th partition of set C and in Chapter VIII for the i-th decision of set C
s_k	k-th scenario, $k = 1,...,v$
d_l	l-th decision maker, $l = 1,...,w$
$e_i^{kl}(a_j) \equiv e_{ij}^{kl}$	evaluation or effectiveness value of a_j for c_i, s_k, and d_l.
$e_{ij} \equiv e_{j|i}$	evaluation value of a_j for c_i
$k_{jk|i}$	preference (dominance) function of (e_{ij}, e_{ik})
k_i	preference intensity of c_i
$k_{ij} = w_i/w_j$	relative preference intensity of c_i over c_j
$u(a)$	utility function over all criteria (attributes)
$u(a_j) \equiv u_j$	utility of a_j
$u_{j|ik}$	utility of a_j for c_i and c_k
$u_i(e_i)$	component utility function for c_i
$u_i(a_j) \equiv u_{ij}$	utility of a_j for c_i
$u(a)$	value function over all criteria (attributes)
$v(a_j) \equiv v_j$	value of a_j
$v_{j|ik}$	value of a_j for c_i and c_k
$v_i(e_i)$	component value function for c_i
$v_i(a_j) \equiv v_{ij}$	value of a_j for c_i
w_i	weight of c_i
$A^{1 \times n}$	set of decision options: $[a_1,...,a_j,...,a_n]^T$
$C^{1 \times m}$	set of criteria (attributes): $[c_1,...,c_i,...,c_m]^T$
$E^{m \times n}$	evaluation matrix or score table for m criteria (attributes) and n alternatives
$N^{m \times n}$	normalized evaluation matrix or preference score card
$U^{m \times n}$	utility matrix or preference score card under uncertainty
$V^{m \times n}$	value matrix or preference score card under certainty
$K_i^{n \times n}$	paired preference (dominance) matrix for c_i
$K^{m \times m}$	paired preference (dominance) matrix across all criteria (attributes)
$\Re$	general preference relation $(\succ, \succsim, \sim)$
$\succsim$	weak preference

$\succ$	strong preference
$\sim$	indifference
R	set of real numbers: $\{...,-1/3,...0,...,2/7,...\}$
Z	set of integer numbers: $\{...,-2,-1,0,1,2,...\}$
B	set of binary numbers: $\{0,1\}$
L_j	j-th lottery

Subject Index

abstraction, 36
action, 14,39
actor, 4,232
aggregation of criteria, 9,76
ALARA, 165
algorithm, 53
   node elimination, 213,221
   shortest-route, 299
   simplex, 288,290
allowable changes, 32
alternative, 15
   explicit, 15
   implicit, 15
   inferiority,
analytic hierarchy process, 67
   absolute, 100
   relative, 100
analytic network process, 85
anchoring and adjustment, 197
arc
   conditional, 203,204
   informational, 42,202,203,204
   no-forgetting, 204
   reversal, 158
arithmetic average, 241
Arrow's impossibility theorem, 236
aspiration level, 11,12,272
at least as great as, 25
attribute, 8,10
availability, 197
averaging out, 207,219
axiom, 23,49
   AHP, 68,75
   Kolmogorov, 152
   social choice, 236
   utility theory, 174
   value theory, 116

bargaining set, 258
Bayes, 151
   theorem, 154
behavior
   sub-optimal, 12
bi-plot, 80
Borda count, 235
branch-and-bound technique, 297

capital budgeting, 303
cardinality, 248
certainty, 16
   effect, 197
   equivalent, 178
   outcome, 175
characteristic quotient, 289
coalition theory, 259
collectively exhaustive, 16,144,154
collinearity, 78
complement, 120
computationally complex, 55
conceptualization phase, 38
concordance, 88
   relative measure of, 237
conditional expectation, 219
Condorcet's method, 235
conflict ($\rightarrow$ game)
   analysis method, 89
   resolution
      communication, 256
      dominance/inferiority, 250
      MaxMin theorem, 252
      MinMax equilibrium test, 251
      tit-for-tat strategy, 256
   situation, 7
      cooperative, 7,234
      dynamic, 233
      non-cooperative, 234
      perfect cooperation, 257
      strictly competitive, 7,234,249
      zero-sum, 7,234
   solution
      equilibrium pair, 251
      Nash equilibrium, 253,258
      Pareto optimality, 250
      security level, 146
conjunction fallacy, 197
consistency, 66,68,69
   index, 72
   ratio, 72,244
constraint, 11,12
   binding, 270
   flow conservation, 276
convex set, 136,270,284
convexification, 258
correlation, 77,120
cost-benefit analysis, 125

criteria, 8,10,39
   dominance, 71
   dominant, 26
   evaluation, 9,44
   root, 9,44
   weight, 48,65,71

decision, 38
   analysis, 2
   maker, 7,39,232,233
      conflict, 45,232
      supra, 242,244,248
      group, 7,45,232
      under risk, 144
   one-at-the-time, 6
   optimal, 12
   problem, 1
      identification, 6,38
      definition, 17,38
      dual, 293
      dynamic, 42
      resolution, 26
      solving process, 3
   rule
      majority rule, 89,235
      MaxMin, 146
      MinMax Regret, 147
      optimism-pessimism, 148
      principle of insuff. reasoning, 148
   support system, 56
   tree, 207
   variable, 2,14,41,52,254,264,268, 278,280,282
      auxiliary, 272
      binary, 14
      integer, 14
      real, 14
decomposition, 8
Delphi method, 241
democratic theory, 233
diagram
   F/N, 165
   probabilistic influence, 203,211
      regular, 205
   relevance, 159
dimension reduction, 82
direct assessment approach, 183
discordance, 88

discount
   factor, 125
   rate, 125
dominance, 48,90
dynamic plot, 262,325

effectiveness
   additivity of, 52,265
   proportionality of, 52,264
   unit-, 52,264
efficient
   alternative
      criteria, 27
      stochastic, 28,146
   frontier, 137,162,257,258
eigenvalue, 71,80
eigenvector, 71,80,244
elements, 39
empirical, 151
evaluation, 39
   matrix, 8,18,264
   measure, 10,17
   scenarios, 150
   table, 8

feedback, 75
folding back, 207,220
formalization, 37
fuzzy set theory, 166,246

game, ($\rightarrow$ conflict)
   setting, 7
   theory, 233
goal, 10
   content, 11,12,39,52,268
   function, 267,272
   hard, 12
   soft, 12
   structural, 11,12,39
gradient, 126
graph, 39

human, 5
hyperplane, 284

ill-defined, 55
ill-structured, 55
implementation, 6
impossibility theorem, 7

optimism, 148,197
outranking, 89,92
   flow, 89

paired comparison, 2,22,45
paradox of
   Allais, 197
   Condorcet, 235
Pareto optimal set, 27,162
   non- optimal, 254
partition, 16,44,144
path, 70
perception
   human, 5
   sensors, 5
pivot
   column, 287
   element, 287
   row, 287
play, 249
policy, 15
   analysis, 3
pre-planning, 6
preference, 48
   aggregation, 26
   criteria
      across, 49
      within, 49
   decision makers (across)
      Borda count, 235
      descriptive, 243
      eigenvector meth., 244
      geometric mean, 243
      linguistic, 248
      majority rule, 89
      normative, 112
      ordered weighted average,
         248
      process, 240
      utility approach, 112
      value approach, 112
      weighted average mean, 243,
         244
   scenarios (across)
      exp. monetary value,
      expected value, 159
      expected utility, 115
   dependence, 120
   disjoint, 58

elicitation, 22
   descriptive, 22,48
   normative, 23,49
   function, 23,24,50
   graph, 39
   oriented, 28
   weighted, 29
   intensity, 65
   order, 29,88
   complete
      strong, 25
      weak, 25,112
   incomplete, 99
   partial
      strong, 25
      weak, 25
   pseudo, 26
   ordinal, 236
   strong, 24
   valued, 92
   weak, 25,112
principal components technique, 80
principle of insuff. reasoning, 148
problem
   identification, 38,39
   definition, 38
probability
   comparison method, 182
   conditional, 152,204,205
   joint, 152,157,205
   marginal, 152,204,205
   objective, 150
   subjective, 150
programming
   binary integer, 265
   goal, 272
   linear, 265
   mixed integer, 265
projection, 81
proportionality, 52
prospect theory, 197

rank-reversal, 32,104,105,155,235
rational, 23,50,112,115,179
rationality bounded, 12
relation, 39
   binary, 24
   causal, 205
   incomparability, 24

marginal, 181
multilinear, 189
multiplicative, 189
graph
  concave, 185
  convex, 185
  linear, 185
theory,
  expected, 115,176
  mutliattribute, 115

value
  marginal, 127
  expected monetary, 178
  function, 112
    additive, 121
    component, 113
    explicit, 127
    exponential, 134
    implicit, 127
    linear, 123
    multiattribute, 115
    ordinal, 112
    perfect complement, 123
    perfect substitute, 124
  of information, 208,225
  theory, 115
  graph
    concave, 128
    convex, 128
    linear, 129
  tradeoff, 127
variance, 79
  -covariance matrix, 79
variable
  basic, 269
  → decision
  free, 288
  linguistic, 166
  non-basic, 269
  slack, 268
virtual negotiation, 262,325
voting procedure, 234

walk, 70
  intensity, 70